A FIRST COURSE IN
Geometry

Edward T. Walsh

DOVER PUBLICATIONS, INC.
Mineola, New York

Bibliographical Note

This Dover edition, first published in 2014, is an unabridged republication of the work originally published in 1974 by Rinehart Press, San Francisco. The author has provided a new Preface and a list of Errata for this edition.

Library of Congress Cataloging-in-Publication Data

Walsh, Edward T., 1934–
    A first course in geometry / Edward T. Walsh. — Dover edition.
        pages cm
    Originally published: San Francisco : Rinehart Press, 1974.
    Includes bibliographical references and index.
    ISBN-13: 978-0-486-78020-7
    ISBN-10: 0-486-78020-1
     1. Geometry. I. Title.
QA453.W22 2014
516.2—dc23

                                  2014011022

Manufactured in the United States by Courier Corporation
78020101   2014
www.doverpublications.com

To Mark and Patrick
who were present at the creation

CONTENTS

v

Mighty is geometry, joined with art, resistless.
EURIPIDES (C. 480 – 406 BCE)

PREFACE TO THE DOVER EDITION

Demonstrative proof and the logical development of mathematical ideas goes back at least as far as Thales of Miletus (c. 624 – 547 BCE), who is believed to have proven *The Isosceles Triangle Theorem* which appears in Chapter 4 of this text. Three hundred years later, Euclid of Alexandria (c. 325 – 265 BCE), in a logically coherent manner, gathered together mathematics that originated with him and earlier mathematicians in a multi-volume work we call the *Elements.* It is the oldest existing axiomatic deductive treatment of mathematics.

In the late 12th century, the works of Euclid, which had been lost in the West but survived in the Muslim world, were translated from Arabic into Latin. The first English version was published in 1570. For centuries afterward knowledge of at least part of the *Elements* was expected of all university students. In the early twentieth century, the thirteen books of Euclid's *Elements* appeared in an English translation by Sir Thomas Heath (still available in a 1956 Dover Paperback edition) and was widely used as a textbook for decades. Indeed, the present author's father first studied geometry using Heath's translation.

Most of the ideas presented in Chapters 1 – 8 of this volume can be found (in a much different form) in the first six books of Euclid's *Elements*. Euclid's Book XI contains the Chapter 9 material.

Nearly two millennia after Euclid's *Elements* was written, *Discourse on Method* (1637) was penned by René Descartes. Algebraic analysis had begun to flower and Descartes proposed to "borrow all that is best in Geometrical Analysis and Algebra, and correct the errors of the one by the other." We now call this method, *analytic geometry.* Chapter 10 of *A First Course in Geometry* rests on this method. With his creation of analytic geometry and its "transform – invert – solve" technique, Descartes is credited with being the inventor of modern mathematics

In the forty years since *A First Course in Geometry* was originally published there have been many changes in the introductory geometry curriculum. Transformation geometry, the most significant of these changes, has now claimed its rightful place as a unifying tool relating the intuitive ideas of congruence, similarity and symmetry.

The last thing one knows when writing a book is what to put first.

BLAISE PASCAL (1623–1662)

PREFACE

This book is an outgrowth of the author's experience with college students who enroll in a first course in geometry. Traditionally, such a course has been little more than a swifter version of a high school geometry course and the textbooks used have usually missed the opportunity to exploit the fuller measure of maturity and sophistication possessed by college students. Frequently accompanying this maturity is a history of frustration with mathematics. For this reason, the opening chapters move slowly and very carefully. This early tender loving care helps the student build the understanding and self-confidence needed to proceed with assurance through the more rapid development that follows.

The book begins with a chapter on the algebra of logic and sets. The ideas and vocabulary introduced in this chapter are used both explicitly and implicitly throughout the subsequent chapters. For example, in Chapter 2 the student is eased into the process of constructing formal proofs through the device of "three-column form." The third column gives the basic argument forms involved, forms that were cultivated in Chapter 1.

An instructor who does not wish to include the sections of Chapter 1 that deal with logic should omit the discussion of "three-column form" and ignore the third column used in some of the proofs of Chapter 2. He should also provide an alternative introduction to indirect proof (Section 4.5) since the one given employs the symbols of logic.

Classes that have had a previous introduction to sets and set operations can omit Sections 1.6 and 1.7.

The use of metric postulates makes constructions unnecessary. Nevertheless, constructions are included in Chapter 8 because of their aesthetic appeal and their obvious applicability to engineering and applied science. Recognizing that many instructors will want to omit the material on constructions, the Chapter 8 Review Exercises and Chapters 9 and 10 have been written with this in mind.

Although the text contains enough material for a five-hour semester course, it is also well suited for shorter courses. For a three-hour semester or quarter course, a number of options are possible.

Option 1:
Chapters 1 through 8, Section 7.
Option 2 (For classes that omit Chapter 1):
Chapters 2 through 9, or Chapters 2 through 10, omitting Chapter 9.
Option 3 (For classes familiar with sets):
Chapter 1, Sections 1 through 5, and Chapters 2 through 8.

The author thinks of mathematics as an activity: something that people do. Therefore, the problem sets are constructed in such a way that the student has an opportunity to *do* something. He may make a discovery (e.g., Pascal's triangle in problem 4 on pp. 30–31), or he may develop the proof of a theorem to be found in a subsequent section (e.g., the Exterior Angle Theorem in problem 12 on p. 120).

Excluding the chapter introductions, each section of the text is followed by a problem set, in which the problems are arranged in the order of increasing difficulty, the earlier ones being more routine or copiously supplied with hints, the latter ones requiring more thought. In addition, every chapter has a set of review exercises. The author believes that the student should be provided hints or outlines for most of the proofs as well as answers to all the other problems contained in the Problem Sets and Review Exercises.

The student is given some problems that he can answer intuitively but is unable to formally verify because they depend upon a postulate not yet introduced. For example, several problems require Euclid's Fifth Postulate for their complete solutions. The frustrations the student faces in such situations lead to a clearer understanding of the lack of arbitrariness in our choice of assumptions.

Most chapters are followed by an annotated set of references for related outside study. Footnotes, unless otherwise indicated, refer to entries in these sets of references.

Finally, some attention has been given to the historical context in which certain ideas have surfaced. Mathematics does not grow in a vacuum, and, like any other human endeavor, is inextricably related to the rest of the world around it. The evolution of geometry aptly illustrates this fact.

ERRATA

358   In the answer to Problem 13 of Problem Set 5.3, change T42 to T45.

359   Change the answer of Problem 11(a) of Problem Set 7.2 to: $15 + 5\sqrt{3}$.

360   Change the answer of Problem 15 of Problem Set 7.4 from 5 to 60.
       The answer to Problem 2(i) of Chapter 7 Review Exercises is 264.

361   The answer to Problem 1(e) of Problem Set 8.6 should be $\dfrac{75\sqrt{3}}{2}$.

The essence of mathematics is its freedom.

G. CANTOR (1845–1918)

CHAPTER ONE

THE LANGUAGE
OF MATHEMATICS

1.1 INTRODUCTION

If your intuition about physical space is fairly well developed, many of the ideas developed in this book will seem quite familiar to you. For although the geometry presented here is not the only way to think about space (more will be said about this in Chapter 5), it coincides with the notions most of us come to take for granted as we are growing up. On the other hand, if your geometric intuition is not very good, your formal study of geometry here should improve the situation.

In order that we may proceed with some confidence, our approach will be systematic. Specifically, we will employ what is often called "the axiomatic method." This method was apparently invented by the Greeks of the fourth, fifth, and sixth centuries B.C. to securely systematize the geometry they were developing, in the face of some paradoxes they were unable to resolve.[1]

[1] See R. L. Wilder, *Evolution of Mathematical Concepts* (New York: John Wiley & Sons, 1968), pp. 103–110.

The elements of the axiomatic method are few. We begin with some *primitive* or *undefined terms*, some *definitions*, and a set of non-contradictory assumptions called *postulates* or *axioms*.

We then endeavor to make plausible conjectures about relationships among objects of our study, employing the defined and undefined terms. Our attempts to establish the validity of these conjectures are referred to as *proofs*, and the valid conjectures are called *theorems*.

Since we want to discuss these conjectures with some measure of clarity and precision, it is obvious why we should first agree on the meanings of technical terms involved (definitions), as well as how these terms may be related to each other (axioms). But why must we also begin with some undefined terms? It would certainly be convenient to have a definition for every mathematical term that we encounter. But is this possible? Consider the following definitions from *Webster's Collegiate Dictionary*

happening, *n.* Occurrence.

occurrence, *n.* Appearance or happening.

appearance, *n.* The action or an instance of appearing; an occurrence.

Note that these definitions haven't really told us anything since each of the words is defined in terms of the others. Such sets of definitions are referred to as circular definitions, and although our trio is far less poetic, it is about as logically meaningful as Gertrude Stein's "A rose is a rose is a rose . . ."

This problem of circularity arises whenever we attempt to define every term we use. To avoid this difficulty we will agree that certain terms will remain undefined, and, in particular, that definitions and other general statements (axioms, theorems, etc.) can be expressed using these primitive undefined terms. The meanings of the undefined terms will, however, become clearer as the subject unfolds.

In this book the undefined terms adopted are point, line, plane, and set. Our axioms, definitions, and theorems will employ these terms in a manner that is probably consistent with your view of the real world, for, after all, geometry has been studied and expanded primarily in order to provide men with a clearer vision of physical space. We will use the axiomatic method merely as a tool to guide us to this clearer vision. It will provide us with some measure of certainty that the conclusions we are coming to are not contradictory, and that they do follow logically from the geometric facts previously established (logic will be discussed at some length in the following pages). Some care must be taken not to let the formalism of the axiomatic method obscure the beauty of the geometrical structure.

1.2   LOGIC: PROPOSITIONS AND TRUTH TABLES

In the fourth century B.C., the Greek philosopher Aristotle systematized deductive logic, sometimes referred to as the art and science of reasoning. The models of argumentation shown to us by Aristotle in his development of *formal logic* are still acceptable to most literate Westerners. Moreover, despite periodic attacks on Aristotelian logic by writers of the past few centuries, to study logic was to study the formal logic of Aristotle until, in 1847, George Boole published a thin volume, entitled *The Mathematical Analysis of Logic*. In his book Boole, who was an English elementary school teacher at the time, attempted to give a mathematical formulation to Aristotelian logic. This mathematical mode of systematizing the "laws of thought" reached maturity in the years 1910–1913, just after the publication of the great *Principia Mathematica*, by Alfred North Whitehead and Bertrand Russell.

A brief study of some of the more elementary and fundamental aspects of mathematical logic should lead you to understand how to use some of the important rules of logic employed so often in mathematical arguments. To simplify our work and to free ourselves somewhat from the ambiguities of our language, we will use a symbolic language akin to the one you learned while studying the algebra of the real number system. We will call this algebra the *algebra of propositions*. A proposition is merely a declarative sentence that is unambiguous; for example:

> 2 is an even integer.
>
> Some Californians are not wealthy.
>
> Our president is a man of peace.

In your study of the algebra of the real number system you used variables to represent real numbers. In this second algebra, the variables will represent propositions. We will agree that these variables must take one (and only one!) of the two values *true* (T) and *false* (F). That is, a proposition must be either true or false, but not both. We will consider some special operations defined on propositions, for just as we combine numerical expressions algebraically through the operations of addition and multiplication, we combine sentences with operations known as *conjunction, disjunction, implication,* and *equivalence.*

In a *conjunction* we combine two given propositions by placing the word "and" between them. Consider the propositions

> Geometry is interesting.
>
> Logic is a game.

Their conjunction is as follows:

> Geometry is interesting and logic is a game.

If we denote the first two propositions by p and q, then the proposition which is their conjunction is denoted $p \wedge q$, read "p and q." We will agree that a conjunction is true whenever both the sentences that make it up are true; otherwise, the conjunction is false. Let us gather these ideas into a single definition.

DEFINITION 1.1   If p and q are propositions, the sentence "p and q," denoted $p \wedge q$, is called the *conjunction* of p and q, and its truth values are given by the following table:

p	q	$p \wedge q$
T	T	T
T	F	F
F	T	F
F	F	F

In other words, the conjunction $p \wedge q$ is true only in the case that both p and q are true; otherwise, it is false.

NOTATION RULE 1.1   If we want a variable to represent a specific proposition, we will use the following notation:

> p: Some men are mathematicians.

Example 1   Consider the sentences

$$r: 2 + 3 = 5$$
$$s: 5 - 1 = 3$$

Obviously, r is true and s is false. Therefore, the second row of the table defining the conjunction tells us that the following sentence is false:

   $r \wedge s$

From a different row of the table, you should be able to determine whether the following sentence is true or false by replacing s and r by p and q, respectively:

   $s \wedge r$

Suppose we wish to write the denial of a proposition such as

   Surgery is recommended.

We write

   Surgery is not recommended.

This negation of the original sentence was easily obtained. However, the denial or negation of a proposition is not always so transparent. Consider the proposition

   All ducks waddle.

Which of the following sentences would you choose as the negation of this sentence?

   p_1: All ducks do not waddle.

   p_2: Some ducks waddle.

   p_3: Some ducks do not waddle.

   p_4: Not all ducks waddle.

   p_5: It is not the case that all ducks waddle.

If you chose p_3, p_4, *and* p_5, you are thinking clearly. In any case, the point is that some care must be taken when constructing the denial of a proposition. With these ideas in mind we can frame the following definition.

DEFINITION 1.2 If p is a proposition, the sentence "not p," denoted $\sim p$, is called the *negation* of p, and its truth values are given by the following table:

p	$\sim p$
T	F
F	T

This table elucidates the nearly obvious observation that if a proposition is true, its denial is false, and vice versa. Tables such as this one and the one given in Definition 1.1 are called *truth tables*.

Example 2 Consider the sentences

p: Some oysters are silent.

$\sim p$: No oysters are silent.

Here the temptation is to write

$\sim p$: Some oysters are not silent.

We would be mistaken, however, since it would then be possible for both p and $\sim p$ to be true.

Forming the negation of a proposition is sometimes a tricky task. Truth tables can be used to determine under what conditions the new sentence is true, and then can be used as an aid in giving alternative English-language formulations of such a sentence. Before considering an example, let us clarify some of the notation involved.

NOTATION RULE 1.2 By the notation

$(\sim p) \wedge q$

we wish to denote only the negation of p in the sentence $p \wedge q$. If we wish to denote the negation of $p \wedge q$ (or the negation of any other sentence involving more than one variable), we will use parentheses as follows:

$$\sim(p \wedge q)$$

Example 3 This notational device exists to remove the possible ambiguity from such sentences as $\sim p \wedge q$. Consider

> p: Mark is intelligent.

> q: Mark is wealthy.

Then we have:

> $(\sim p) \wedge q$: Mark is not intelligent and Mark is wealthy.

and

> $\sim(p \wedge q)$: It is not the case that Mark is intelligent and wealthy.

It should be clear that these two sentences do not have the same meaning. The conditions under which $\sim(p \wedge q)$ is true can be inferred from the following truth table:

p	q	$p \wedge q$	$\sim(p \wedge q)$
T	T	T	F
T	F	F	T
F	T	F	T
F	F	F	T

Thus unless Mark is wealthy *and* intelligent, the sentence $\sim(p \wedge q)$ is true. This fact should be a clue to the reader who desires to translate sentences such as $\sim(p \wedge q)$ into English sentences less contrived than those beginning "It is not the case that . . ."

As we shall soon see, such truth tables are enormously effective in the determination of the validity of certain forms or patterns of argumentation.

PROBLEM SET 1.2

In each of the following problems (1–6), you are given two propositions. The truth values of these propositions should be obvious to you. Use the truth table defining the conjunction to decide whether the conjunction of each pair is true or false. Indicate which row of the table helped you decide.

1. Some Americans are Republicans. Lansing is the capital of Michigan.

2. A triangle has three sides. Abraham Lincoln was the thirty-third president of the United States.

3. Some men are not handsome. Rudyard Kipling is the author of *Gulliver's Travels*.

4. $15 - 13 = 6; 21 + 5 = 26$

5. College students do not study geometry. Getting an education is a waste of time.

6. Liberia is on the western coast of Africa. The United Nations headquarters is in New York City.

Try to write a sentence that is the negation of each of the following sentences (7–12) without using such crutches as "it is not the case that."

7. Mathematics is interesting.

8. There is no further need to worry.

9. Some Marxists live in California.

10. Every line is a set of points.

11. Nuts can be found in grocery stores and on nut trees.

12. A buffalo can always toss one over a gate.

13. Complete the following truth tables:

(a)

p	q	$\sim p$	$(\sim p) \wedge q$
T	T		
T	F		
F	T		
F	F		

(b)

p	q	$\sim q$	$p \wedge (\sim q)$
T	T		
T	F		
F	T		
F	F		

14. In the manner of Example 3, and problem 13, construct the truth table for the sentence $(\sim p) \wedge (\sim q)$.

1.3 PROPOSITIONS AND TRUTH TABLES CONTINUED: DISJUNCTION, IMPLICATION, AND EQUIVALENCE

In the English language, the word "or" is used in a manner that is often somewhat ambiguous, as in the following sentence:

> Bertrand Russell or Alfred North Whitehead wrote the *Principia Mathematica*.

Here it is possible to interpret the "or" in the "either . . . or . . ." sense as well as the "and/or" sense. We will always assume that "or" is used in this second inclusive sense. A sentence employing "or" in this manner is called a *disjunction*.

DEFINITION 1.3 If p and q are propositions, the sentence "p or q," denoted $p \vee q$, is called the *disjunction* of p and q, and its truth values are given by the following table

p	q	$p \vee q$
T	T	T
T	F	T
F	T	T
F	F	F

As the table indicates, an "or" sentence is false whenever both the sentences making it up are false; otherwise, the disjunction is true.

Example 1 Consider the sentences

$$r: 2 + 3 = 5$$
$$s: 5 - 1 = 3$$

The second row of the table defining the disjunction tells us that the following sentence is true:

$$r \vee s$$

Example 2 It should be clear that the sentence

$$p \vee (\sim p)$$

is always true, since the definition of negation assures us that one of the component parts of this sentence is true. Thus if you ask the mathematician the question: Do you want to stay home or go out this evening? his answer may well be a truthful but un-communicative "Yes!"

When one idea is the consequence of another, we often say that the first idea implies the second. Such a sentence is called an *implication* or a *conditional sentence*. The sentence "A falling barometer implies that a storm is on the way" is an implication.

DEFINITION 1.4 If p and q are propositions, the sentence "p implies q," or alternatively "if p then q," denoted $p \rightarrow q$, is called an *implication*, and its truth values are given by the following table:

p	q	$p \rightarrow q$
T	T	T
T	F	F
F	T	T
F	F	T

p is called the *hypothesis* or *assumption* of the sentence $p \rightarrow q$, and q is called its *conclusion*.

Example 3 Consider the sentences

p: It is snowing.

q: It is cold.

The symbolic sentence $p \rightarrow q$ then represents

If it is snowing, then it is cold.

The implication often appears in forms other than "if ... then." Some examples follow:

> All men are mortal.
>
> Whenever it rains, it pours.
>
> Every even integer is divisible by two.

Be alert for such disguises.

The implication will be the most important form of logical proposition we will examine. Study the table carefully. You will notice that if the *hypothesis* of an implication is true and its *conclusion* is false, then the entire proposition is false. Otherwise, it is true. The third and fourth rows may seem mildly surprising to you. How can we, as the third row suggests, assign the value "true" to a sentence with a true conclusion coming from a false hypothesis?

Example 4

$$p:\ 3 = 5 \text{ and} \atop 5 = 3 \quad \Big\} \quad \text{False}$$

$$q:\ 8 = 8\} \qquad \text{True}$$

$$p \to q:\ \text{If } 3 = 5 \text{ and } 5 = 3, \text{ then } 8 = 8 \quad \text{True}$$

Starting with p, and using the addition law (see Appendix A if the terminology is unfamiliar), we have reached the true conclusion, q.

Example 5   Using the same algebraic reasoning as in Example 4, we can derive a false conclusion from a false assumption:

$$p:\ 3 = 5 \text{ and} \atop 3 = 5 \quad \Big\} \quad \text{False}$$

$$q:\ 6 = 10\} \qquad \text{False}$$

$$p \to q:\ \text{If } 3 = 5 \text{ and } 3 = 5, \text{ then } 6 = 10 \quad \text{True}$$

Examples 4 and 5 serve merely to suggest the reasonableness of the third and fourth rows of the table. They do not *prove* anything.

The last operation we will define in this algebra we have been introducing is the operation of *equivalence*.

DEFINITION 1.5 If p and q are propositions, the sentence "p is equivalent to q," denoted $p \leftrightarrow q$, is called the *equivalence* of p and q, and its truth values are given by the following table:

p	q	$p \leftrightarrow q$
T	T	T
T	F	F
F	T	F
F	F	T

In other words, the proposition $p \leftrightarrow q$ is true whenever both p and q have the same truth value; otherwise, it is false. The sentence $p \leftrightarrow q$ is also denoted by p iff q, which is read "p if and only if q."

Example 6 "Mark is happy if and only if Mark is eating" is an example of a sentence in the form of an equivalence. This sentence is false if Mark is ever unhappy while eating, or happy while not eating. Otherwise it is true.

Example 7 Consider the sentences

p: 6 is an even integer.

q: 6 is divisible by 2.

The first row of the table defining "$\leftrightarrow$" assures us that $p \leftrightarrow q$ is true.

The symbol "$\leftrightarrow$" suggests that the equivalence somehow connects the sentences "$p \rightarrow q$" and "$q \rightarrow p$." A glance at the following table confirms this suspicion:

p	q	$p \rightarrow q$	$q \rightarrow p$	$(p \rightarrow q) \wedge (q \rightarrow p)$
T	T	T	T	T
T	F	F	T	F
F	T	T	F	F
F	F	T	T	T

A comparison with the table for $p \leftrightarrow q$ clearly indicates that the two statements are equivalent, and we may write

$$(p \leftrightarrow q) \leftrightarrow [(p \rightarrow q) \wedge (q \rightarrow p)]$$

Specifically, the sentence "Mark is happy if and only if Mark is eating" is equivalent to the sentence "If Mark is happy, then Mark is eating, and if Mark is eating, then Mark is happy." Although the second form is more consistent with common usage, the first form is somewhat less cumbersome, especially in its symbolic representation.

Example 8 Every definition is an equivalence. Consider the following definition of "even integer":

An integer is *even* iff it is divisible by 2.

Although definitions are not often written in such a form, it will serve you well to keep this idea of equivalence in mind.

Many times when we are attempting to decide if a given proposition is true, an equivalent proposition will be easier to deal with. The truth value of the two sentences is the same; thus the examination of the second tells us what we wish to know about the first.

PROBLEM SET 1.3

1. Suppose we consider the sentences

p: 2 is an even integer.

q: James Joyce wrote "Death of a Salesman."

r: $3(5 + 2) = 21$

Write English sentences for each of the following and decide whether they are true or false. Use p, q, and r as just defined.

(a) $p \rightarrow q$ (b) $p \vee r$ (c) $q \vee (\sim r)$

(d) $q \rightarrow p$ (e) $p \leftrightarrow r$ (f) $(\sim p) \vee q$

(g) $(p \vee r) \leftrightarrow [q \vee (\sim r)]$ [*Hint:* Use (b) and (c).]

(h) $(\sim q) \rightarrow (\sim p)$

2. Construct truth tables for each of the following statements:

(a) $p \rightarrow (\sim q)$ (b) $q \rightarrow (\sim p)$ (c) $(\sim q) \vee (\sim p)$

(d) $\sim(p \wedge q)$ (e) $(\sim p) \leftrightarrow q$

(f) $[(\sim p) \rightarrow q] \wedge [q \rightarrow (\sim p)]$ (g) $p \rightarrow (q \rightarrow p)$

(h) $(p \vee q) \rightarrow p$

3. Which of the statements in problem 2 are equivalent?

4. The second of the three disguised implications listed on page 11 may be rewritten as follows:

If it rains, then it pours.

Write the other two sentences of this list in if . . . then form.

1.4   ARGUMENTS AND THEIR VALIDITY

The method of constructing a truth table to determine under what conditions a proposition is true,[2] can be extended to help us evaluate arguments. An argument is merely a list of assumptions followed by a conclusion.

Example 1

> a1. All students read books.
>
> a2. Fred is a student.
> _____
> c.   Fred reads books.

The horizontal line stands for the word "therefore." Symbolically, we may write this argument as follows:

> a1. $p \rightarrow q$
>
> a2. p
> _____
> c. q

In the above example, the first assumption is a disguised implication. In if . . . then form, it becomes

> a1. If a person is a student, then that person reads books.

---

[2] This device originated with the American logician and philosopher C. S. Peirce (1839–1914).

The second assumption can be written

   a2. A person (Fred) is a student.

The conclusion then becomes

   c. That person (Fred) reads books.

To rewrite an argument in such a way does have the advantage of making it more rigidly adhere to the argument form; however, it also removes it from the language of ordinary discourse. With a little practice, you should be able to learn to recognize the context in which statements in arguments are made and then derive conclusions without making them strictly adhere to the form.

Most people agree that it is desirable for our reasoning to follow paths that will not lead us from true assumptions to conclusions that are false. Consequently, we will agree that the form of an argument is *valid* if and only if a false conclusion never follows from assumptions that are all true. Otherwise, the argument form is *invalid*. We now construct a truth table for the argument of Example 1:

		a1	*a2*	*c*
p	*q*	$p \to q$	*p*	*q*
T	T	T	T	T
T	F	F	T	F
F	T	T	F	T
F	F	T	F	F

You should notice that in the cases where the conclusion is false (rows 2 and 4), at least one of the assumptions is false, so the argument (form) is valid. This form is so fundamental and pervasive in argumentation that it has been given the name *modus ponens* (m.p.)

Example 2   Of course, not all argument forms are valid. Consider

   a1. $p \lor \sim q$
   a2. p
   _____
   c. q

The truth table for this argument is as follows:

			$a1$	$a2$	c
p	q	$\sim q$	$p \vee \sim q$	p	q
T	T	F	T	T	T
T	F	T	T	T	F
F	T	F	F	F	T
F	F	T	T	F	F

In the second row, a false conclusion follows from assumptions that are all true; hence the argument is invalid.

As Examples 1 and 2 indicate, the truth of the conclusion of an argument does not depend upon the validity of the argument. Using valid argument forms assures us only that if we begin our arguments with true assumptions, the consequent conclusions will also be true.

A three-part argument such as modus ponens is called a *syllogism*. Two other syllogisms used very often in the following chapters are *modus tollens* (m.t.) and the *hypothetical syllogism* (h.s.). These forms are as follows:

Modus tollens

a1. $p \to q$
a2. $\sim q$
c. $\sim p$

Hypothetical syllogism

a1. $p \to q$
a2. $q \to r$
c. $p \to r$

You are given the opportunity to determine the validity of these forms in the next problem set.

An example of an m.t. argument is the following:

All geometry teachers are logical.

Patrick is not logical.

Therefore, Patrick is not a geometry teacher.

Notice that the second assumption is a denial of the conclusion of the implication that makes up the first assumption, and that this leads us to a denial of the hypothesis of the first assumption.

Now let us consider an example of an h.s. argument:

> If all war ends, then unemployment will increase.
>
> If unemployment increases, then wages will be lowered.
>
> Therefore, if all war ends, then wages will be lowered.

Here the pattern should be very clear. Each statement is an implication. The conclusion of the argument has as *its* hypothesis the hypothesis of one of the assumptions, and as its conclusion, the conclusion of the other assumption. Another mark of this argument form is that the hypothesis of one assumption is the conclusion of the other.

PROBLEM SET 1.4

1. Use truth tables to determine the validity of each of the following argument forms:

(a) a1. $p \wedge q$
 a2. $\sim q$
 c. p

(b) a1. $p \leftrightarrow q$
 a2. q
 c. p

(c) a1. $p \rightarrow q$
 a2. $\sim p$
 c. $\sim q$

(d) a1. $p \rightarrow q$ (m.t.)
 a2. $\sim q$
 c. $\sim p$

(e) a1. $p \rightarrow q$
 a2. q
 c. p

(f) a1. $p \rightarrow q$ (h.s.)
 a2. $q \rightarrow r$
 c. $p \rightarrow r$

The truth table for (f) has three variables and should be constructed as shown at the top of page 18.

2. Symbolize the following argument and then use a truth table to decide if it is valid or invalid:

> If taxes decrease, then there is inflation.
>
> If there is inflation, the cost of living will increase.
>
> Taxes will decrease.
>
> Therefore, the cost of living will increase.

[*Hint:* The truth table begins as in part (f) of problem 1.]

			$a1$	$a2$	c
p	q	r	$p \rightarrow q$	$q \rightarrow r$	$p \rightarrow r$
T	T	T			
T	T	F			
T	F	T			
T	F	F			
F	T	T			
F	T	F			
F	F	T			
F	F	F			

3. Give two examples of arguments in each of the following forms: (a) modus ponens; (b) modus tollens; and (c) hypothetical syllogism. Use English sentences, *not* the symbols of logic.

4. Using one of the basic argument forms, m.p., m.t., or h.s., supply, if possible, a valid conclusion for each of the following sets of assumptions. If no conclusion follows from the use of one of these forms, so indicate.

   (a) Our President is a man of peace. Omar is not a man of peace.
   (b) All ducks waddle. Donald waddles.
   (c) If $x + 3 = 7$, then $x = 4$; if $x = 4$, then $x + 1 = 5$
   (d) Attractive women do not smoke fat cigarettes. Wilma is un-attractive.
   (e) If the Prince is an idiot, then the Queen is a fool. If the Queen is a fool, the King is blameless.
   (f) Parallel lines do not intersect. Lines l and m do not intersect.
   (g) If an integer is not even, it is odd. The integer 3 is not odd. (Recall the distinction between validity and truth value!)
   (h) If Mark studies, he is happy. If Mark studies, then he grows old.
   (i) If Patrick works, then he earns money. Patrick works.

5. Consider the following implication:

   If Patrick dances, then he sings.

   Which of the following sentences do you think is equivalent to this implication?

   (a) If Patrick sings, then he dances.
   (b) If Patrick does not dance, then he does not sing.
   (c) If Patrick does not sing, then he does not dance.

1.5 DERIVED IMPLICATIONS: CONVERSE, INVERSE, AND CONTRAPOSITIVE

Given the implication $p \to q$, there are several other implications which can be derived. These new implications may or may not be true when the original implication is true. The most significant of these derived implications are the *converse*, the *inverse*, and the *contrapositive*:

Converse: $\qquad q \to p$

Inverse: $\qquad (\sim p) \to (\sim q)$

Contrapositive: $\quad (\sim q) \to (\sim p)$

Comparing the truth tables of $p \to q$ and its converse

p	q	$p \to q$	$q \to p$
T	T	T	T
T	F	F	T
F	T	T	F
F	F	T	T

we see that a statement and its converse can have different truth values. We see this in Example 1.

Example 1

If a bird is a sparrow, it flies.

If a bird flies, it is a sparrow.

The second statement is patently false. Such is not the case for the first statement.

Sometimes a statement and its converse have the same truth values.

Example 2

If an integer is even, it is divisible by 2.

If an integer is divisible by 2, then it is even.

Both statements are true.

Since, as we have seen, a statement and its converse sometimes have different truth values, it follows that they are not, in general, equivalent. This could have been inferred from problem 1(e) of Problem Set 1.4.

A comparison of the truth tables for $p \rightarrow q$ and $(\sim p) \rightarrow (\sim q)$ leads us to a similar conclusion about a sentence and its inverse:

p	q	$p \rightarrow q$	$(\sim p)$	$(\sim q)$	$(\sim p) \rightarrow (\sim q)$
T	T	T	F	F	T
T	F	F	F	T	T
F	T	T	T	F	F
F	F	T	T	T	T

From our point of view, the most valuable derived implication will be the contrapositive. A glance at the following truth table establishes this claim:

p	q	$p \rightarrow q$	$(\sim q)$	$(\sim p)$	$(\sim q) \rightarrow (\sim p)$
T	T	T	F	F	T
T	F	F	T	F	F
F	T	T	F	T	T
F	F	T	T	T	T

As the table demonstrates, a proposition and its contrapositive are equivalent.

Example 3 Consider the sentence

Parallel lines do not intersect.

The contrapositive of this sentence is (in a similar disguise)

Intersecting lines are not parallel.

It happens that both these sentences are true. But we need only verify one of them, inasmuch as they are contrapositives of one another, and thereby equivalent.

One of the meanings the uncritical reader often takes from the sentence "Parallel lines do not intersect" is that nonintersecting lines are parallel. Such an assertion is not only false, as we shall see, but it is an attempt to identify "Parallel lines do not intersect" with its converse. *Remember:* A proposition and its converse are *not* equivalent. It is a mistake to substitute one for the other.

PROBLEM SET 1.5

1. Write the converse, inverse, and contrapositive of the following sentences:

    (a) If it rains, then it is cold.
    (b) All men are mortal.
    (c) If a student does not study, then he will not pass this course.

2. Write the contrapositive of "If a^2 is even, then a is even."

3. "If $x = 2$, then $x^2 = 4$" is a true implication.

    (a) Is its converse true?
    (b) Is its inverse true?
    (c) Is its contrapositive true?

4. Give an example of a true implication that has a false converse. Is its inverse also false? Can you think of a true implication that has a false converse and a true inverse?

5. Compare the truth table for the inverse of $p \rightarrow q$ with the truth table for its converse. What do you conclude?

6. The words *necessary* and *sufficient* are used throughout mathematics. Consider the sentence "If $x = 3$, then $x^2 = 9$." Which of the following do you think is equivalent to the given implication?

    (a) $x = 3$ is necessary for $x^2 = 9$
    (b) $x = 3$ is sufficient for $x^2 = 9$

7. (a) Which of the following conditions would you regard as *necessary* for a student to receive a passing grade in this course?

    (i) To be interested in the subject.
    (ii) To score passing grades on most of the exams.
    (iii) To score better than everyone else on every exam.
    (iv) To be liked by the instructor.
    (v) To be able to find the classroom.

(b) Which of the conditions in (a) do you regard as *sufficient*?

(c) Write an implication suggested by each of the choices you made in (a) and (b). For example, if you concluded that (v) was *sufficient*, write

> If a student is able to find the classroom, he (or she) will receive a passing grade in this course.

If you concluded that (i) was *necessary*, write

> If a student receives a passing grade in this course, then he (she) is interested in the subject.

8. The assertion that a statement and its contrapositive are equivalent is actually an outgrowth of one of the three valid syllogisms of Section 1.4. Which one?

1.6   SETS

Closely related to the algebra of propositions is what is called the *algebra of sets*. As mentioned earlier, the term "set" is left undefined by mathematicians. It will be convenient to think of a set as a collection of objects, such as

> All the geese in New York City.
>
> The set of counting numbers: 1, 2, 3, ...

NOTATION RULE 1.3   We will use the symbol ". . ." to mean "and so forth" whenever we have displayed enough of a pattern for the reader to recognize the sequel.

Other examples of sets are the following:

> The set of whole numbers: 0, 1, 2, 3, ...
>
> The members of the U.S. Senate.

If your high school mathematics background is quite recent, you probably already know something about sets. If you are somewhat older and have your own children—or soon will—your study here can help narrow "the generation gap," for the algebra of sets is one of the mainstays of "the new math" currently in vogue in school mathematics.[3] Notions about sets serve our purposes here as convenient characterizations of the objects of geometry: points, lines, planes, triangles, and so on.

NOTATION RULE 1.4 It is convenient to use capital letters to name sets. For example, we might wish to let A represent the set of the first three counting numbers. We will employ braces in order to name sets. There are three methods of writing A with braces:

1. *Roster method:* $\{1, 2, 3\}$.

2. *Set description method:* {The first 3 counting numbers}.

3. *Rule method:* $\{x : x$ is a counting number less than 4$\}$.

The notation "$x : x$ is a . . ." is read "all x such that x is a . . . ," and is sometimes called *set-builder notation*.

Example 1 Using these three methods, the set of even whole numbers may be written as

$$\{0, 2, 4, 6, 8, \ldots\}$$

{The even whole numbers}

or $\{x : x$ is a whole number and x is even$\}$

The objects contained in a given set are called its *elements* or *members*.

NOTATION RULE 1.5 If x is an element of the set A, we write $x \in A$. Thus, for example, $2 \in \{1, 2, 3\}$. To deny this, we write, for example, $4 \notin \{1, 2, 3\}$.

Of our original four examples of sets in this section, the first and second differed in a fundamental way: A whole number names the number of elements of the first set, but such is not the case for the set of counting numbers.

---

[3] For a brief introduction to the so-called "new math," see Carl B. Allendoerfer's slender volume, *Mathematics for Parents* (New York: Macmillan Co., 1965).

DEFINITION 1.6 A set is said to be *finite* if the number of its elements can be named by a whole number. Otherwise, it is said to be *infinite*.

Consequently, we can say that the set $\{\triangle, +, *, \square\}$ is finite, whereas the set $\{0, 2, 4, 6, \ldots\}$ is infinite.

In the following pages we shall often refer to what are called *correspondences* between sets. Consider the sets

$$C = \{4, 5, 6\}$$

$$D = \{x, y, z\}^4$$

Members of C can be associated with elements of D in many ways. For example,

NOTATION RULE 1.6 Given any two sets A and B such that $a \in A$ and $b \in B$, if we wish to indicate that a is associated with b, we will connect a and b with the symbol $\leftrightarrow$, and will read this as "a corresponds to b," or "a is matched with b."

Other examples of matchings between the elements of C and D are

and

Although the scheme on the left is more "natural," both illustrate a concept that we now define.

DEFINITION 1.7 Let X and Y be sets. If there exists a matching that associates each element of X with a unique element of Y and vice versa, then that matching is called a *one-to-one correspondence*, and such sets are called *matching sets*.

Thus C and D are matching sets.

---

[4] In the algebra of real numbers, one writes $x = y$ whenever x and y are names for the same number. Here we use "$=$" in a looser and more general sense. We will agree that $A = B$ simply means that A and B are names for the same object.

Such a simple concept can lead to quite unanticipated conclusions. In fact, later we shall encounter some matching sets that are guaranteed to surprise or delight or provoke you.

You should notice that two finite sets cannot be put into a one-to-one correspondence with one another unless they have the same number of elements. Also, it should be fairly clear to you that no infinite set matches any finite set. This fact can be used to build a perfectly suitable, but not preferable, definition of "infinite set," to wit: A set is *infinite* if there is no finite set that matches it.

When a set A is in a one-to-one correspondence with a set B, it is conventional usage to say that "A has as many elements as B." We can illustrate this concept with a rather surprising example.

Example 2 Let $W = \{0, 1, 2, \ldots\}$ and let $N = \{1, 2, 3, \ldots\}$. Suppose for every $x \in W$ we write $x \leftrightarrow (x + 1)$, $(x + 1) \in N$. This defines a one-to-one correspondence between W and N. Thus N has as many elements as W.

Both N and W are infinite sets. It is natural to wonder if all infinite sets are matching sets. For a readable discussion of this question, consult George Gamow's *One Two Three . . . Infinity*.[5]

PROBLEM SET 1.6

1. Name each of the following sets using the roster, rule, and set description methods:

   (a) The set of whole numbers.
   (b) The set of letters of our alphabet which may be used as vowels.
   (c) The set of even counting numbers.

2. Let N represent the set of counting numbers, and let W represent the set of whole numbers. Name each of the following sets by the roster method:

   (a) $\{2n : n \in W\}$ (b) $\{2n - 1 : n \in N\}$
   (c) $\{n \in W : n^2 \text{ is less than } 4\}$
   (d) $\{n \in N : n^2 + 1 = 2n\}$

3. Which of the following are finite sets?

   (a) $\{1, 2, 3, \ldots, 10\}$ (b) $\{1, 2, 3, \ldots\}$ (c) $\{3n : n \in W\}$
   (d) The set of all words ever written.
   (e) $\{n : (n \in N) \wedge n \text{ is less than } 100\}$

[5] (New York: The New American Library, Mentor Books, 1947), pp. 25–34.

4. Can the set of counting numbers be written as follows?

$$N = \{x : x \text{ is a positive number}\}$$

Explain.

5. Give an original example of each of the following:

(a) A pair of matching finite sets. Indicate the correspondence, using "$\leftrightarrow$."
(b) A pair of nonmatching finite sets.
(c) A pair of matching infinite sets. Indicate the matching in a manner similar to that of Example 2.

6. How many different one-to-one correspondences are there between each of the following pairs of sets?

(a) $\{1\}$, $\{2\}$ (b) $\{1, 2\}$, $\{3, 4\}$ (c) $\{1, 2, 3\}$, $\{4, 5, 6\}$
(d) $\{a_1, a_2, a_3, \ldots, a_n\}$, $\{b_1, b_2, b_3, \ldots, b_n\}$ Enough of a pattern should have emerged from (a), (b), and (c) for you to make a plausible guess. (Every good mathematician is, first of all, a good guesser!)

7. How could a person who had not learned to count determine whether or not the number of red cards in a deck of cards is equal to the number of black cards?

1.7 OPERATIONS WITH SETS

The fact that we concern ourselves with counting the number of elements that a given set contains would seem to indicate that there is a kind of arithmetic that we can do with sets. As a matter of fact, there is an algebra of sets that is identical in structure to the algebra of logic. Akin to the operations of "or" and "and" are the set-theoretic operations of *union* and *intersection*. The definitions of these operations follow.

DEFINITION 1.8 Let A and B be sets. The *union* of A and B, denoted $A \cup B$, is the set of all elements x satisfying at least one of the conditions below:

1. $x \in A$

2. $x \in B$

Example 1 If $A = \{1, 2, 3\}$ and $B = \{3, 4, 5\}$, then $A \cup B = \{1, 2, 3, 4, 5\}$.

You should notice that although 3 is an element of both sets, it is listed only once in their union. The reason for this is that an element is either in a set or it is not, and to list it more than once would not shed any more light on the situation.

DEFINITION 1.9 Let A and B be sets. The intersection of A and B, denoted $A \cap B$, is the set of all elements x satisfying both the conditions below:

1. $x \in A$
2. $x \in B$

Example 2 Using A and B as in Example 1, we have $A \cap B = \{3\}$.

When reading these two new symbols we usually read $\cup$ as "union" or "cup," and read $\cap$ as "intersection" or "cap."

The manner in which Definition 1.9 is written tells us that something such as $\{1\} \cap \{6\}$ is a set; yet it does not have any elements. We have a special name and notation for this set.

DEFINITION 1.10 The set which contains no elements is called the *empty set* and is denoted by $\varnothing$.

Statements involving the operations of union and intersection can be expressed pictorially by what are called *Venn diagrams*.

In Figs. 1.1, 1.2, and 1.3, sets A, B, and C are represented by points inside the circles so labeled. The region inside the rectangle represents whatever large collection of objects we are allowing to serve as the source of all the sets we are possibly going to consider. This large collection, U, we shall call the *universal set*.

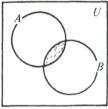

FIGURE 1.1

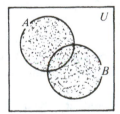

FIGURE 1.2

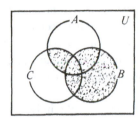

FIGURE 1.3

The shaded region in Fig. 1.1 obviously represents $A \cap B$. What do you think Fig. 1.2 represents? Figure 1.3 requires some analysis. The shaded region covers the points inside circle C and those inside the intersection of A and B, so we would refer to the set represented by the region as $(A \cap B) \cup C$. The parentheses are necessary, for, as a quick sketch will tell you, $(A \cap B) \cup C \neq A \cap (B \cup C)$, and hence $A \cap B \cup C$ would be ambiguous.

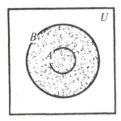
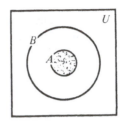

FIGURE 1.4   FIGURE 1.5

Consider Figs. 1.4 and 1.5. Figure 1.4 represents $A \cup B$, and Fig. 1.5 represents $A \cap B$. These diagrams differ from the previous ones, in that all the points inside A are inside B as well. A set that consists entirely of elements of some other set is called a *subset* of that set. Introducing a new symbol, we have the following definition of subset.

DEFINITION 1.11 A is a *subset* of B, written $A \subseteq B$, iff every element of A is also an element of B.

To deny this, we write $A \nsubseteq B$.

Notice that we can now state that every set is a subset of itself, and also that $(A \cap B) \subseteq (A \cup B)$ for all sets A and B.

Let $B = \{\ldots, -4, -2, 0, 2, 4, \ldots\}$. Let $A = \{2, 4, 6\}$. Then $A \subseteq B$. Also, $B \subseteq B$. Is $\varnothing$ a subset of B? Is there any element of $\varnothing$ that is not an element of B? Since $\varnothing$ has no elements, it has none that are not in B. Therefore, $\varnothing \subseteq B$. For the same reason $\varnothing \subseteq A$. Indeed, $\varnothing$ is a subset of every set.

Suppose we let the universal set $U = \{1, 2, 3, 4, 5\}$ and let $A = \{1, 2, 3\}$. We can define a new set consisting of all the members of the universe (or universal set) that are not in A. We call this set the *complement* of A denoted $\sim A$. In our example, $\sim A = \{4, 5\}$. We can put all these facts together and formulate the following definition.

DEFINITION 1.12 Let U be the universal set and let A be a subset of U. Then the *complement* of A, denoted $\sim A$, is the set of all elements of U that are not elements of A. Symbolically, with a slight change in the "rule method,"

$$\sim A = \{x \in U : x \notin A\}$$

NOTATION RULE 1.7   The following table summarizes the notation we will use when referring to some specific sets of numbers:

NOTATION	SET
N	The set of counting numbers: $\{1, 2, 3, \ldots\}$.
W	The set of whole numbers: $\{0, 1, 2, 3, \ldots\}$.
J	The set of integers: $\{\ldots, -2, -1, 0, 1, 2, \ldots\}$.
Q	The set of rational numbers:
	$\{x : x = p/q, p \in J, q \in J, q \neq 0\}$.
H	The set of irrational numbers. Irrational numbers are numbers that cannot be written in the form p/q, with $p \in J$ and $q \in J$. Numbers such as $\sqrt{2}$, $\sqrt{3}$, π, are irrational. No irrational number is rational.
R	The set of real numbers: $R = Q \cup H$.

Since no irrational number is rational, we may write $H \cap Q = \varnothing$. If, as in this case, the intersection of two sets is the empty set, we say that they do not intersect. *Caution:* Every pair of sets has an intersection, but not all sets intersect.

Example 3   Suppose we let $U = R$. Then $\sim H = Q$ and $\sim Q = H$, and $\sim(\sim H) = H$ and $\sim(\sim Q) = Q$.

PROBLEM SET 1.7

1. Let U, the universal set, be J, the set of integers. Suppose $A = \{1, 2, 3\}$, $B = \{0, 1, 2, 3, 4\}$, $C = \{x \in J : x \text{ is even}\}$, $D = \{x \in J : x \text{ is odd}\}$, and $E = \{-1, 0, 1\}$. Find:

(a) $A \cup E$ (b) $A \cap E$ (c) $C \cup D$ (d) $B \cap D$
(e) $\sim C$ (f) $(\sim A) \cap B$ (g) $C \cap D$ (h) $(\sim A) \cap (\sim E)$
(i) $B \cup (\sim B)$

2. Analogous to the truth tables of logic is a device called the *membership table*, which can be used to decide whether two sets have the same elements, that is, whether they are equal. The membership tables for $A \cup B$ and $A \cap B$ are as follows:

A	B	$A \cup B$
$\in$	$\in$	$\in$
$\in$	$\notin$	$\in$
$\notin$	$\in$	$\in$
$\notin$	$\notin$	$\notin$

A	B	$A \cap B$
$\in$	$\in$	$\in$
$\in$	$\notin$	$\notin$
$\notin$	$\in$	$\notin$
$\notin$	$\notin$	$\notin$

You should observe here that the condition "$\in$" or "$\notin$" in the right-hand column follows from the conditions on A and B in the same way that T or F was obtained in the construction of truth tables.

(a) The above membership tables are similar to a pair of truth tables. Which truth tables are they?

(b) Rewrite the definitions of union and intersection using the words "or" and "and."

(c) Construct membership tables for $(\sim A) \cap (\sim B)$ and $\sim(A \cup B)$. What do you conclude?

3. Suppose A, B, and C are subsets of some universal set U. Using Fig. 1.6 as a guide, arrange the following sets in sequential order so that each set in the sequence is a subset of the succeeding set: $A \cup B$; $A \cap B$; U; $(A \cup B) \cup C$; $\emptyset$; $(A \cap B) \cap C$; $A \cup (B \cup C)$.

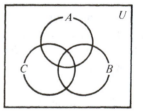

FIGURE 1.6

4. (a) Complete the following table:

SET	NUMBER OF ELEMENTS	NUMBER OF SUBSETS	NUMBER OF SUBSETS WITH n ELEMENTS IF:				
			$n = 0$	$n = 1$	$n = 2$	$n = 3$	$n = 4$
$\emptyset$				0			
$\{a\}$		2					0
$\{a, b\}$	2		1				
$\{a, b, c\}$				3			
$\{a, b, c, d\}$			1				

(b) Can you discern a pattern in the "number of subsets" column? A set of five elements has how many subsets? A set of n elements has how many subsets?

(c) Consider, for example, the fourth entry in the column below $n = 2$. This is the number of subsets of two elements that can be taken from a set of three elements. This is often referred to as the number of combinations of three things taken two at a time. Carefully examine the triangular array of nonzero entries in the five right-most columns. What pattern do you discover? (*Hint:* Each row is related to the one above it.) Use this pattern to write down the sixth row (through $n = 5$) of the table.

(d) On the basis of the result in (c), solve the following problem. During a recent basketball game, two players on the same team collided, lost their contact lenses, and had to be helped from the floor. The team's coach had five eager substitutes sitting on the bench. From these five men he had to choose two replacements. How many possibilities were there?

5. Suppose X, Y, and Z are subsets of some universal set U. Which of the following statements are always true? If you are doubtful in any instance, a Venn diagram should prove helpful.

(a) $X \cup Y = Y \cup X$ (b) $(X \cap Y) \subseteq X$ (c) $\varnothing \subseteq Z$
(d) $Y \subseteq Y$
(e) $X \cup (Y \cap Z) = (X \cup Y) \cap Z$ (f) $\sim(\sim X) = \varnothing$
(g) $(\sim X) \cup (\sim Z) = \sim(X \cup Z)$
(h) If $X \subseteq Y$ and $Y \subseteq Z$, then $X \subseteq Z$ (i) $(X \cup Y) \cap X = X$
(j) $X \cup (\sim X) = U$
(k) If $X \subseteq Y$ and $Y \subseteq X$, then $X = Y$

6. If A has six elements and B has five elements, find:

(a) The minimum number of elements in $A \cup B$ and in $A \cap B$.
(b) The maximum number of elements in $A \cup B$ and in $A \cap B$.

7. In a group of 300 persons, it was found that each one owned a dog or a cat (or both). If 175 owned a dog and 200 owned a cat, how many owned both a cat and a dog?

8. In a certain class, there are 15 veterans on the G.I. bill, 12 of whom are men. There are also 15 men who are not on the G.I. bill. There are 30 women in the class. How many students are in the class?

CHAPTER 1: REVIEW EXERCISES

1. Decide which of the following are true.

    (a) A proposition must be either true or false.
    (b) The sentence "$p \wedge q$" is a disjunction.
    (c) If p is false, then the sentence "$p \wedge q$" is false.
    (d) If the conclusion of an implication is true, the implication is true.
    (e) An argument is valid if the conclusion is not false and all the assumptions are true.
    (f) The number of elements in a set is always a counting number.
    (g) The set of protons in the solar system is a finite set.
    (h) $5 \in \{3, 4, 5\}$
    (i) $\{0, 2, 4, 6, \ldots\}$ and $\{0, 1, 2, 3, \ldots\}$ are matching sets.
    (j) $\{\varnothing\} = \varnothing$
    (k) $X \subseteq (X \cap Y)$ for any choice of X and Y.
    (l) $(X \cap Y) \cup X \subseteq X$ for any choice of X and Y.
    (m) If $A \subseteq B$, then $A \cup B = B$.
    (n) If $X \subseteq Y$, then whenever $y \in Y$, it follows that $y \in X$.
    (o) If $X \notin (A \cap B)$, then $(X \in A)$ or $(X \in B)$.

2. Indicate which of the basic argument forms you are using to supply (if possible) a valid conclusion for each of the following sets of assumptions. If no conclusion follows using one of these forms, so indicate.

    (a) All men are mortal.
         Socrates was a man.
    (b) Some students are lazy.
         Boris is not a student.
    (c) Every implication has a conclusion.
         This sentence has no conclusion.
    (d) No emperors are dentists.
         All dentists are dreaded by children. (From Lewis Carroll)
    (e) All jokes are amusing.
         No act of Congress is a joke.
    (f) $A \subseteq B$
         $b \notin B$
    (g) If an integer is odd, it is divisible by 2. Four is not divisible by 2. (*Be careful:* Do not confuse truth and validity.)

(h) Monks have good habits.
   Kathy has good habits.
(i) If Republicans are elected, taxes decrease.
   If taxes decrease, unemployment increases.

3. Find three nonempty sets A, B, and C such that

(a) $A \cap (B \cup C) = (A \cap B) \cup C$
(b) $A \cup (B \cap C) = (A \cup B) \cap C$

4. Construct truth tables for the following sentences:

(a) $p \vee q \vee (\sim p)$ (b) $(p \wedge (\sim q)) \vee ((\sim p) \wedge q)$
(c) $p \rightarrow (p \vee (\sim q))$

5. Use truth tables to decide whether each of the arguments that follow is valid.

(a)	$p \wedge q$	(b)	p	(c)	$\sim p$	(d)	p
	$\sim p \rightarrow q$		q		$p \vee \sim q$		$(\sim q) \vee r$
	$\overline{\sim q}$		$\overline{p \wedge q}$		$\overline{\sim q}$		$(\sim p) \rightarrow q$
							$\overline{r}$

6. Use membership tables (see Problem Set 1.7, problem 2) to show that for any two sets A and B, $A \cup (A \cap B) = A \cap (A \cup B)$.

7. If $A \subseteq B$ but $A \neq B$, then A is said to be a proper subset of B. Give an example of two matching sets A and B, such that A is a proper subset of B.

8. Write a short paragraph distinguishing between the validity of an argument and the truth of its conclusion.

9. Using the symbols of logic and the rule method of set notation, rewrite the definitions of union and intersection.

10. Let $U = \{1, 2, 3, 4, 5, 6, 7, 8, 9, 10\}$
        $A = \{1, 3, 5, 7, 9\}$
        $B = \{2, 4, 6, 8, 10\}$
        $C = \{1, 2, 3, 4, 5\}$
        $D = \{6, 7, 8, 9, 10\}$
        $E = \{4, 5, 6\}$
        $F = \{2, 4, 6\}$

Find the following sets:

(a) $A \cup B$ (b) $A \cap B$ (c) $E \cap D$
(d) $E \cap \sim B$ (e) $\sim(F \cap B)$ (f) $B \cap (D \cup E)$
(g) $(E \cup F) \cap B$

11. The language of sets can be used to convert verbal descriptions to symbolic descriptions.

Suppose $A = \{$all college students$\}$, $B = \{$all beer drinkers$\}$, $C = \{$all college teachers$\}$, and $D = \{$all women college teachers$\}$. Then, for example, the set of all college students who do not drink beer is $A \cap \sim B$.

Substitute such symbolic descriptions for each of the following verbal descriptions:

(a) All college teachers and students.
(b) All beer-drinking college teachers.
(c) All college teachers who are men.
(d) All beer drinkers who are not college students.
(e) All college teachers who are neither women nor beer drinkers.

12. Suppose we define the *relative complement* of one set with respect to another as follows:

The *relative complement* of A with respect to B, denoted $B - A$, is the set of all elements of B that are not elements of A; that is, $B - A = \{x : (x \in B) \wedge (x \notin A)\}$. For example, if $A = \{2, 4, 6\}$ and $B = \{1, 2\}$, then $B - A = \{1\}$ and $A - B = \{4, 6\}$.

If $X = \{1, 2, 3, \ldots, 10\}$, $Y = \{2, 4, 6, 8, 10\}$, and $Z = \{1, 2, 3, 4, 5\}$, find:

(a) $Y - Z$ (b) $(\sim Y) - (\sim Z)$ (c) $(X - Y) \cap (\sim Z)$
(d) $Y \cap (\sim Z)$ (e) $X - \varnothing$

REFERENCES

Allendoerfer, C. B., and Oakley, C. O. *Principles of Mathematics.* 3d ed. New York: McGraw-Hill Book Co., 1963. pp. 1–47.
    First published in 1955, yet already a classic in its field, it contains in its first chapter many clearly written examples of the types of problems we have been developing in the previous pages.

Carroll, Lewis. *Symbolic Logic and the Game of Logic.* New York: Dover Publications, 1958. pp. 101–106.
    A delightful list of pairs of concrete assumptions for which conclusions are to be found.

Gamow, G. *One, Two, Three . . . Infinity.* New York: The New American Library, Mentor Books, 1947. pp. 25–34.
    This immensely successful popularization of the ideas of modern science and mathematics explores, in these pages, some of the problems posed by infinite sets.

Meserve, B. E., and Sobel, M. A. *Introduction to Mathematics*. 3d ed. Englewood Cliffs, N.J.: Prentice-Hall, 1973. pp. 33–100.
> An elementary and clear exposition of the ideas of set theory and logic. Venn diagrams are used for logic as well as the algebra of sets.

Stoll, Robert R. *Sets, Logic, and Axiomatic Theories*. San Francisco: W. H. Freeman and Co., Publishers, 1961.
> The first two chapters give a much more rigorous introduction to sets and logic, for those who have found the ideas of the preceding chapter too transparent.

Wilder, R. L. *Evolution of Mathematical Concepts*. New York: John Wiley & Sons, 1968. pp. 103–110.
> A discussion of the causes and effects of the early Greek introduction of logic to the methodology of mathematics.

All the operations of our intellect tend to geometry, as to the goal where they find their perfect fulfillment.

HENRI BERGSON (1859–1941)

CHAPTER TWO

GEOMETRIC SETS OF POINTS:

Lines, Planes, and Distance

2.1 INTRODUCTION

After the death of Alexander the Great in 323 B.C., his empire underwent a series of divisions. One of the three separate empires that emerged from these changes was the Egyptian empire under the rule of Ptolemy,[1] one of Alexander's generals and a trusted confidant. Ptolemy selected Alexandria, a city at the mouth of the Nile River, as his capital and ordered that a university be established there. The university opened sometime around 300 B.C. and remained until the Middle Ages a center for the preservation of the Greek heritage.

One of the intellectual giants attracted to this fine university shortly after it opened was Euclid, who became professor of mathematics there. Euclid's reputation has come down to us mainly through a text he wrote called *The Elements*.[2] It is probably the most

[1] Not to be confused with a later Ptolemy (150 B.C.), who provided us with a fairly good model of the mechanics of the universe in his book *The Almagest*.

[2] See T. L. Heath, ed., *The Thirteen Books of Euclid's Elements* (New York: Dover Publications, 1956); a three-volume paperback edition.

important and widely used textbook ever written. *The Elements* is composed of thirteen books, and most of the material covered in the following pages can be found in the first six books of the work.

It would be ridiculous to assume that Euclid's work was so perfect that no improvements have been made on it in the intervening millennia. Euclid purported to derive all of the 465 propositions in his 13 volumes from just 10 assumptions. Until the nineteenth century there were not many mathematicians willing to do battle with this Goliath, but a David or two has subsequently appeared to point out some other important assumptions underlying Euclid's work. One of these men, David Hilbert (1862–1943), formulated an alternative set of postulates that seem less open to criticism. The geometry in this text book is based essentially on the work of Euclid but has been, in a mathematically significant way, somewhat perfected along the lines suggested by Hilbert.

2.2 OUR FIRST ASSUMPTIONS

In the following development of Euclidean geometry, as in any logical system, we adopt primitive or undefined terms. Here, they are the words *point*, *line*, *plane*, and *set*.

Before attempting to verify any conjectures we might wish to make, we need some assumptions to tell us something about the nature of our primitive terms. Such underlying assumptions we shall call *postulates*. They are sometimes called *axioms* as well.

POSTULATE 1 There exist at least two lines.

POSTULATE 2 Every line is a set of points and contains at least two distinct points.

The following postulate we will call the Line Postulate. Some of the more important general statements of our study will be given such names in order to make it easier to refer to them.

POSTULATE 3 *The Line Postulate:* If P and Q are distinct points, then there is exactly one line that contains both of them.

An alternative form of this postulate is the statement "Two points determine a line."

NOTATION RULE 2.1   Let P and Q be distinct points. The line determined by them can be denoted by either $\overleftrightarrow{PQ}$ or $\overleftrightarrow{QP}$. We will also use single lowercase letters to denote lines as follows: $\overleftrightarrow{l}, \overleftrightarrow{m}, \overleftrightarrow{t}$, and so on.

Example 1   The diagram shows several representations of lines. We will represent them in this manner throughout the book.

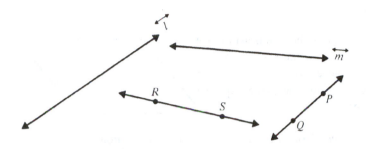

Suppose $\overleftrightarrow{l}$ and $\overleftrightarrow{m}$ are two lines that intersect.[3] How would you describe their intersection? Draw a picture of this situation. Your picture should help you answer the question and lead you to make the following conjecture.

THEOREM 1   If two lines intersect, their intersection contains exactly one point.

We can verify this conjecture on the basis of our first three postulates; but because, at this juncture, the logic involved is fairly complicated, we will postpone the task.

DEFINITION 2.1   The points of a set are said to be *collinear* if there is a line which contains all of them. Otherwise, they are said to be *noncollinear*.

If $P \in \overleftrightarrow{l}$, it is common to say that "P is on $\overleftrightarrow{l}$." Obviously, then, A and B are on $\overleftrightarrow{AB}$.

---

[3] Hereafter, when we refer to "*two* lines," "*two* points," or "*two* planes," we shall mean two *distinct* lines, points, or planes. However, if we just refer to "points A and B," "lines $\overleftrightarrow{l}$ and $\overleftrightarrow{m}$," and so forth, we allow the possibility that A and B are the same point, $\overleftrightarrow{l}$ and $\overleftrightarrow{m}$ are the same line, etc.

Example 2 In the figure, *A*, *B*, and *C* are collinear, whereas *A*, *B*, and *D* are noncollinear.

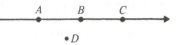
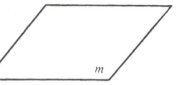

The next postulate introduces the undefined term *plane* into the
discussion. Notice that this postulate and its successor are analogous
to postulates P2 and P3.[4]

POSTULATE 4 A plane is a set of points and contains at least three
noncollinear points.

Example 3 We will adopt the common practice
of using a figure such as the one shown here to
represent a plane. We generally name the plane
using a lowercase letter. Thus the plane depicted
would be called "plane *m*."

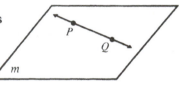

POSTULATE 5 *The Plane Postulate:* If *P*, *Q*, and *R* are three noncollinear points, then there is exactly one plane that contains all of
them.

It is common usage to say that sets of points "lie in" or "lie on" a
plane if the plane contains them, as in the following postulate.

POSTULATE 6 If two points of a line lie in a plane, then the line
lies in that plane.

Example 4 The situation described by P6 is
depicted in the accompanying figure. Since *P* and
Q are elements of plane *m*, then $\overset{\leftrightarrow}{PQ}$ lies in plane *m*.

Thus far we have been referring to numerous sets of points without mentioning their source. Definition 2.2 names our universe.

DEFINITION 2.2 *Space* is the set of all points.

We employ this term in the next postulate.

POSTULATE 7 Space contains at least four points which are noncollinear and do not lie in the same plane.

Analogous to Definition 2.1 we have the following definition.

---

[4] We will use such abbreviations to refer to postulates whenever it is convenient.
Similarly, theorems will be abbreviated T1, T2, T3, etc.

DEFINITION 2.3 The points of a set are said to be *coplanar* if there is a single plane containing them. Otherwise, they are said to be *noncoplanar*.

Example 5 In the above figure, B, C, and D are coplanar, whereas A, B, C, and D are noncoplanar.

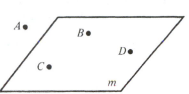

A remark about preconceived notions and axiomatic systems is called for here. Strictly speaking, the axioms or postulates are the only sources of information about the undefined terms that we have at our disposal. Thus, it would be argued by many, we should put pre-conceived notions and intuition aside while we view the relationships between different sets of points in the bright light furnished by the axiomatic method. Aesthetically and mathematically, this *would* be more satisfying. However, such a conscious and contrived disregard of our previous experience would only serve to make us suspicious of our well-founded discoveries, squelch our spontaneity, and impede progress toward the object of our study here. Therefore, we will not be reluctant to use our intuition as an aid.

PROBLEM SET 2.2

$A \bullet$

1. (a) How many lines are determined by the points A and B?

$\bullet B$

$P \bullet$

   (b) How many lines are determined by the points P, Q, and R?

$Q \bullet$

$\bullet R$

$H \bullet$ $\bullet J$

   (c) How many lines are determined by the points H, J, K, and L?

$K \bullet$ $\bullet L$

   (d) How many lines are determined by five points, no three of which are collinear?

2. Which postulate asserts that three noncollinear points determine a plane?

3. If $Q \in \overleftrightarrow{AB}$, in how many distinguishable ways may the name of this line be written, using the names of the points A, B, and Q? Assume $A \neq Q$ and $B \neq Q$.

4. Criticize the following restatement of T1:

The intersection of two lines contains exactly one point.

5. Which postulate assures us that lines must be "straight"? Use a diagram to explain this fact.

6. Assume that P1, P2, and P3 are the only information we have. On the basis of this information, how many points can we be sure exist? Explain.

7. Postulates 7 and 4 assure us that at least six lines exist. Draw a diagram illustrating this.

8. Postulate 7 assures us that at least n planes exist. Find n. (*Hint:* The table you completed in problem 4(a) of Problem Set 1.7 should make this a cinch.)

9. To which of the other postulates is P7 analogous?

10. Which one of the postulates can be used to argue that planes must be "flat"? Explain.

2.3   ABSOLUTE VALUE

In Section 2.2 we are assured that the objects of geometry (points, lines, planes, etc.) do indeed exist. In our discussion of the relationships between such sets of points, the questions as to how "big" a set is or how far one set is from another will become important to us. One device which helps to clarify these relationships is called *absolute value.*

DEFINITION 2.4   Let x be any real number. The *absolute value of x,* denoted $|x|$, is defined as follows:

$$|x| = x \quad \text{if } x \text{ is positive}$$
$$|x| = 0 \quad \text{if } x \text{ is zero}$$
$$|x| = -x \quad \text{if } x \text{ is negative}^5$$

---

[5] Remember that if x is negative, then $(-x)$ is positive.

It follows from this definition that the absolute value of every real number is nonnegative.

Example 1

$$|3| = 3 \quad \text{since 3 is positive}$$

$$|-3| = -(-3) = 3 \quad \text{since } -3 \text{ is negative}$$

Similarly,

$$|44| = 44, \quad |-44| = -(-44) = 44$$

There is, of course, nothing magic in our choice of 3 or of 44 in the above examples. These examples should lead you to conclude more generally that $|x| = |-x|$ for any real number x. In fact you will be given the opportunity to verify this result in Problem Set 2.3.

Sometimes the expression inside the absolute value signs can be very complicated. In this course you will be dealing with sentences involving expressions such as $|x - 3|$, $|x + 4|$, $|x^2 - y^2|$, and so on, and will be asked to evaluate such expressions for specific values of the variables.

Example 2   If $x = 2$ and $y = 3$, then

$$|x - 3| = |2 - 3| = |-1| = 1$$
$$|x + 4| = |2 + 4| = |6| = 6$$
$$|x^2 - y^2| = |2^2 - 3^2| = |4 - 9| = |-5| = 5$$

PROBLEM SET 2.3

1. Find the absolute value of each of the following:

(a) 3   (b) -6 (c) π (d) $\frac{1}{3}$

2. (a) Evaluate: (i) $|2 - 3|$; (ii) $|3 - 2|$
   (b) Evaluate: (i) $|14 - 6|$; (ii) $|6 - 14|$
   (c) Evaluate: (i) $|5 - 1|$; (ii) $|1 - 5|$

3. Do the results of problem 2 suggest any general result to you? If so, what?

4. (a) Evaluate $|x + 3|$ for all $x \in \{-4, -3, -1, 2, 3\}$.

   (b) Evaluate $|x| + |3|$ for all $x \in \{-4, -3, -1, 2, 3\}$.

   (c) How do you think $|x + y|$ and $|x| + |y|$ are related?

5. Evaluate:

   (a) $|x - y|$ if (i) $x = 2$ and $y = -3$

   　　　　　　　　(ii) $x = -1$ and $y = -5$

   　　　　　　　　(iii) $x = -\frac{1}{2}$ and $y = 7$

   (b) $|x| - |y|$ if (i) $x = 2$ and $y = -3$

   　　　　　　　　(ii) $x = -1$ and $y = -5$

   　　　　　　　　(iii) $x = -\frac{1}{2}$ and $y = 7$

   (c) How do you think $|x - y|$ and $|x| - |y|$ are related for all values of x and y?

6. Show that $|x| = |-x|$ for all real numbers x. (*Hint:* Use the definition of absolute value and divide the problem into three different problems.)

7. For which integers is each of the following true?

   (a) $|x|$ is less than 2

   (b) $|x|$ is greater than x

8. For what values of x is $|x - 1| = x - 1$?

9. For what values of x is $|x - 2| = 2 - x$?

2.4 COORDINATES AND DISTANCE

In Problem Set 2.2 we were able to ascertain the existence of a very limited number of points. In fact, with the postulates we have stated thus far, we cannot prove that a line contains more than two points. Our intuition tells us that there is an unlimited number of such points. Postulate 8 affirms this.

POSTULATE 8 The set of points on a line and the set of real numbers are matching sets.

This statement assures us that space contains an unlimited number of points, and can also be used to show that a plane contains an unlimited or infinite number of lines.

One might well ask whether or not there is a natural matching suggested by P8. The reader's experience in algebra should help here.

The type of matching illustrated by what is sometimes called a *line graph* or a *number line* is part of every algebra student's experience. Such a matching will be most useful to us and is depicted in Fig. 2.1 on line $\overleftrightarrow{l}$.

FIGURE 2.1

DEFINITION 2.5 Such a correspondence as that suggested by Fig. 2.1 is called a *coordinate system*. The numbers associated with the points are called their *coordinates*. The point with coordinate 0 is called the *origin*.

POSTULATE 9 *The Ruler Postulate:* Every line has a coordinate system.

This postulate is indeed obvious, but nevertheless it must be stated.

NOTATION RULE 2.2 Suppose $\overleftrightarrow{l}$ is a line, and a coordinate system is defined on $\overleftrightarrow{l}$. If $P \in \overleftrightarrow{l}$ and the coordinate of P is x, we will write

$$c(P) = x$$

Thus, using Fig. 2.1 we may write $c(S) = -2$, $c(R) = (-1)$, $c(Q) = 0$, $c(P) = 1$, and so forth.

The former algebra student will notice that the notation we have used is that used in algebra for functions. A coordinate system is indeed a function.

Figure 2.1 and P9 suggest that we may view a number line as sort of an infinite ruler. How this ruler is placed on the line is purely arbitrary. Also arbitrary is the choice of scale—that is, how far apart we choose to have points which correspond to consecutive integers. Thus, as the following postulate affirms, a line may have many different coordinate systems.

POSTULATE 10 *The Ruler Placement Postulate:* Let $\overleftrightarrow{l}$ be a line with $P \in \overleftrightarrow{l}$, $Q \in \overleftrightarrow{l}$ and $P \neq Q$. Then there exists a coordinate system for $\overleftrightarrow{l}$ such that $c(P) = 0$ and $c(Q)$ is positive.

Having introduced numbers and the idea of a ruler into our geometry, we can employ this ruler to measure distances.

DEFINITION 2.6 Let $\overleftrightarrow{l}$ be a line; let a coordinate system be defined on $\overleftrightarrow{l}$. If $P \in \overleftrightarrow{l}$, $Q \in \overleftrightarrow{l}$, and $c(P) = x$ and $c(Q) = y$, then the *distance from P to Q*, denoted PQ, is defined by

$$PQ = |y - x|$$

In Section 2.3 we saw that $|x| = |-x|$. Therefore,

$$|y - x| = |-(y - x)|$$

However, in elementary algebra one learns that

$$-(y - x) = (x - y)$$

Thus we may conclude that

$$|y - x| = |x - y|$$

This fact makes calculations of distances easier for us, since it tells us that we can subtract the coordinates in either order, take the absolute value, and get the same result.

Example 1 In Fig. 2.2,

$$QT = |1 - (-2)| = |1 + 2| = 3$$
$$TQ = |-2 - 1| = |-2 + -1| = 3$$

Hence, as one would hope, the distance from Q to T is the same as the distance from T to Q.

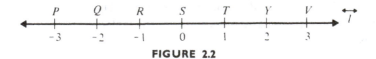

FIGURE 2.2

Example 2 Suppose $X \in \overleftrightarrow{l}$ in Fig. 2.2 and $TX = 2$. Then there are two possibilities for the coordinate of X. Note that $|1 - 3| = |-2| = 2$ and $|1 - (-1)| = |2| = 2$. Therefore, $c(X) = 3$ or

$c(X) = -1$. The same result could be obtained more intuitively by placing your finger at T and moving it two units to the left, and then placing it at T and moving it two units to the right.

What we have done in this section is to move from a pure consideration of geometric sets of points as such to their relationships with the set of real numbers. It was the French philosopher René Descartes (1596–1650) who, in an appendix to a more philosophical work, published the first successful wedding of geometry to algebra. This marriage is today called *analytic geometry*. The methods of analytic geometry are powerful and pervasive.

In Section 2.5 we shall see how a rather difficult algebra problem becomes much simpler when viewed geometrically.

PROBLEM SET 2.4

1. State the coordinate of each of the points shown below. The vertical markings are one unit apart.

(a)   (b)

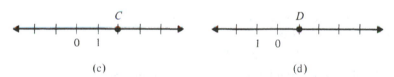

(c)   (d)

2. Is a line a finite set or an infinite set? Explain.

3. Consider line $\overset{\leftrightarrow}{l}$.

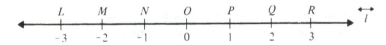

Calculate

(a) MO (b) NP (c) RP (d) PN
(e) $RP + PN$ (f) RN (g) LP (h) PR
(i) $LP + PR$ (j) LR

4. Consider the line shown.

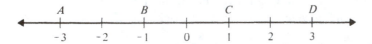

(a) What is the coordinate of a point at a distance 3 to the left of B?

(b) What is the coordinate of a point at a distance 7 to the right of C?

(c) What is the coordinate of P if $AP = PD$?

(d) What is the name of the point that is a distance 6 to the left of D?

5. On the line shown, $AB = BC$.

(a) If B is the origin and $c(A) = -2$, what is $c(C)$?

(b) If $c(A) = 5$ and $c(B) = 10$, find $c(C)$

(c) If $c(A) = -\frac{1}{2}$ and $c(C) = 6$, find $c(B)$

(d) If $c(A) = -3$ and $c(B) = 4\frac{1}{2}$, find AC

6. (a) Suppose $X \in l$. If $OX = 1.5$, what are the possible values for $c(X)$? O is the origin.

(b) If $X \in l$ and $MX = 1.5$, what are the possible values for $c(X)$? Use l of problem 3.

(c) If $X \in l$ and $c(X) = -c(Q)$, calculate XQ. Use l of problem 3.

7. (a) If $X \in l$, PX is never negative. Why not? (*Hint:* Carefully examine the definition of distance.)

(b) If $PX = -(PY)$, what can you say about P, X, and Y?

8. Which of the following do you think is (are) true for all values of a and b?

(a) $|a + b| = |a| + |b|$ (b) $|a - b| = |a| - |b|$
(c) $|a \cdot b| = |a| \cdot |b|$

9. Look closely at problems 3(e) and 3(f). Then look at 3(i) and 3(j). These should suggest some general result to you. State this result as well as you can.

10. Use P3, P4, and P8 to argue that a plane contains an infinite number of lines.

2.5 INEQUALITIES, ORDER, AND SOLUTION SETS

In algebra classes you focused attention primarily on mathematical sentences called *equations*. An ability to handle similar sentences, called *inequalities*, should also be part of the equipment of every student studying geometry. To this end, further definitions and laws are developed.

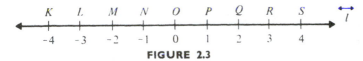

FIGURE 2.3

Consider the coordinate system suggested by Fig. 2.3. Those real numbers corresponding to the points to the right of the origin are the positive real numbers. Some of these are 1, $\frac{5}{4}$, π, and $\sqrt{7}$. The points corresponding to these numbers are not always designated by name, but they *are* there. If, on that number line, a point A, with coordinate a, is to the right of point B, with coordinate b, we say that b *is less than a*. However, since not all lines "point the same way," this does not adequately define the term "less than." Furthermore, "less than" is a numerical concept and thus is better defined independently of geometric intervention. The following pattern suggests a definition:

$$3 \text{ is less than } 5, \text{ and } 3 + 2 = 5$$

$$-2 \text{ is less than } 1, \text{ and } -2 + 3 = 1$$

$$14 \text{ is less than } 100, \text{ and } 14 + 86 = 100$$

$$-6 \text{ is less than } -2, \text{ and } -6 + 4 = -2$$

$$4 \text{ is not less than } 1, \text{ and } 4 + (-3) = 1$$

$$3 \text{ is not less than } 3, \text{ and } 3 + 0 = 3$$

Before going on to the following definition of *less than*, try to invent your own.

DEFINITION 2.7 Let a and b be real numbers. *a is less than b*, denoted $a < b$, if there exists a positive number c such that $a + c = b$; or, in words, *a is less than b* if there is a positive number that can be added to a to obtain b.

DEFINITION 2.8 *a is greater than b*, denoted $a > b$, if and only if $b < a$.

49

Thus we may write $3 < 5$, $5 > 3$, $-1 < 7$, $-7 > -9$, and so forth. It also should be clear that the expression "x is a positive real number" is equivalent to "$x > 0$." Likewise, if x is negative, we may write $x < 0$.

We now state three fairly obvious laws of inequalities, which will be used later.

1. *The trichotomy law:* If a and b are real numbers, exactly one of the following is true.

   (a) $a < b$ (b) $a = b$ (c) $a > b$

2. *The addition law for inequalities:* If $a < b$, and c is any real number, then

   $$a + c < b + c$$

   and

   $$a - c < b - c$$

   For example, $3 < 5$ implies that $3 + (-2) < 5 + (-2)$ or $1 < 3$.

3. *The multiplication law for inequalities:* If $a < b$ and $c > 0$, then

   $$ac < bc$$

   [*Note:* The stipulation that $c > 0$ is a necessary one. If $3 < 5$, $3(-2)$ is not *less* than $5(-2)$.]

4. *The transitive law for inequalities:* Let a, b, and c be real numbers. If $a < b$ and $b < c$, then

   $$a < c$$

   For example, if Patrick is younger than Mark, and Mark is younger than Kathy, then Patrick is younger than Kathy.

The reader should note that 2, 3, and 4 are analogous to three of the properties for equality, given in Appendix A.

Now we are able to talk about pictures or graphical representations of mathematical sentences in a manner that is widely used and understood. The following definitions should clarify matters.

DEFINITION 2.9   Let S be a sentence involving one variable. The set of replacements for the variable which makes S true is called the *solution set* of S.

Thus when you are asked to solve an equation or inequality, you are actually being asked to find its solution set.

Example 1 $x + 1 = 5$ is a sentence S involving the variable x. Obviously, 4 is the only replacement for x which makes $x + 1 = 5$ a true sentence. Hence we say that $\{4\}$ is the solution set of S. In general, if S is an equation, its solution set is made up of what in your algebra course you no doubt called the "roots" of S. Therefore, 4 is a root or a solution of S.

Example 2 $x + 5 < 7$. Applying the addition law for inequalities, we discover that $x + 5 - 5 < 7 - 5$ or $x < 2$. Thus any real number less than 2 will make the sentence true, and we say that its solution set is the set $\{x : x < 2\}$, which we read, "the set of all numbers x such that x is less than 2."

DEFINITION 2.10   The *graph of the solution set of a sentence* S is the set of points on a number line that corresponds to the elements of the solution set.

Example 3   The graph of the solution set for Example 1 is indicated in the accompanying figure. Note that the circle at 4 is filled in.

The graph of the solution set for Example 2 is

Notice that the circle at 2 is empty, indicating that 2 is not in the solution set.

It seems somewhat long-winded to say, "Draw the graph of the solution set of the sentence $x + 5 < 7$." Therefore, we will adopt the rather common and less cumbersome usage "graph $x + 5 < 7$" throughout this book.

NOTATION RULE 2.3   The disjunction $a < b$ or $a = b$ is denoted by $a \leq b$, and is read "a is less than or equal to b." The meaning of $a \geq b$ is analogous.

Example 4   Recalling the truth table defining the disjunction (Section 1.3), it should be clear that each of the following sentences is true:

$$3 \leq 5$$
$$5 \geq 3$$
$$-1.2 \leq 1.2$$
$$0 \leq 6$$
$$3 \leq 3$$
$$5 \geq 5$$

Example 5   Find the solution set and sketch the graph of

$$S: 3x + 5 \leq 8.$$

Here again we use the addition law for inequalities, together with the addition law in Appendix A:

$$\text{If} \quad 3x + 5 \leq 8$$
$$3x + 5 - 5 \leq 8 - 5$$
$$\text{Hence,} \quad 3x \leq 3$$

Applying the multiplication laws, we obtain

$$\tfrac{1}{3}(3x) \leq \tfrac{1}{3}(3)$$

Thus $x \leq 1$ and the solution set of S is

$$\{x : x \leq 1\}$$

The graph of this set is as shown in the figure. Notice that the circle at 1 is filled in, meaning, of course, that 1 is in the solution set.

NOTATION RULE 2.4   The conjunction $a < b$ and $b < c$ is denoted by $a < b < c$. The meaning of $a > b > c$ is analogous. The conjunction $a \leq b$ and $b \leq c$ is denoted by $a \leq b \leq c$.

Example 6   Let $A = \{x: -1 < x < 3\}$. Since A is the set of all numbers between -1 and 3, the graph of A looks as shown here.

PROBLEM SET 2.5

1. Graph the following coordinates on a number line: 3, -2, $\frac{2}{3}$, and $-\sqrt{2}$.

2. On a number line, what is the coordinate of the origin?

3. On a number line, arrange the points corresponding to each of the following sets of numbers, in increasing order, and label them with their coordinates:

   (a) $\{-5, -2, 0, 1, 6\}$ (b) $\{1, -3, -7, 5, 2\}$

   (c) $\{2, \dfrac{1}{2}, \dfrac{-13}{3}, 4, 15\}$

4. Furnish whichever member of $\{<, >, =\}$ would correctly replace "?" in each of the following statements:

   (a) 5 ? 6
   (b) 2.1 ? 1.2
   (c) $-(-3)$? 3
   (d) 3(5) ? 4(5)
   (e) $3(-5)$? $4(-5)$
   (f) $|-7|$? $|7|$
   (g) $|-5|$? 0
   (h) If $a < b$ and $c \in R$, then $c - a$? $c - b$.

5. At the top of page 54 is a graph of the solution set of a sentence involving an inequality. What is this sentence?

6. Solve (find the solution sets):

   (a) $x + 5 = 11$ (b) $3x - 2 = 19$ (c) $x - 5 < 2$
   (d) $3x + 5 \geq 11$ (e) $3x + 2 \leq 2x$

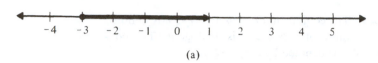

(a)

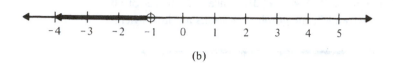

(b)

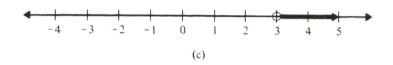

(c)

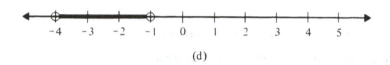

(d)

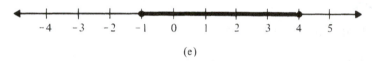

(e)

7. Graph:

   (a) $3x + 1 = 10$ (b) $3x + 1 < 10$ (c) $x \geq 5$
   (d) $|x| > 0$ (*Hint:* Consider the positive numbers, the negative
   numbers, and zero, separately.)
   (e) $-2 \leq x < 1$

8. Rewrite the definition of absolute value using inequality notation.

9. Using the word "and," rewrite the conjunction $3 < x < 5$.

10. On a number line, sketch the graph of $\{x : -3 \leq x \leq 1\}$.

11. The graph of $|x| \leq 2$ is a segment.

    (a) Sketch this graph.
    (b) State a conjunction that is equivalent to this sentence.

12. Suppose $a \neq 0$ and $b \neq 0$.

    (a) If $a < b$, how do you think $(-a)$ is related to $(-b)$?
    (b) If $a < b$, how do you think $1/a$ is related to $1/b$?

2.6 BETWEENNESS AND SEGMENTS

The concept of betweenness is rather intuitive. We can talk about a number being between two other numbers, as when we talk about the ages of three children. We can also talk about a certain location being between two others. This interplay between points and numbers was illustrated in problem 9 of Problem Set 2.4 and is formalized in the following definition.

DEFINITION 2.11 Let $\overleftrightarrow{l}$ be a line. Let A, B, and C be three elements of $\overleftrightarrow{l}$. Suppose a coordinate system is defined on $\overleftrightarrow{l}$. Then B is *between* A and C, denoted A-B-C, if and only if either of the following conditions holds:

1. $AB + BC = AC$

2. $c(A) < c(B) < c(C)$ or $c(A) > c(B) > c(C)$

Actually conditions (1) and (2) in this definition can be shown to be equivalent, so Definition 2.11 is really saying that a point is between two other points if and only if the coordinate of that point is between the coordinates of the two other points (see Notation Rule 2.4).

The statement of condition (1) has a geometry all its own. B, which is between A and C, is written between them on the left-hand side of the statement. We might have written

$$BA + BC = AC \quad \text{or} \quad BA + CB = AC$$

but $AB + BC = AC$ gives us a visual clue as to the relationship between A, B, and C.

Example 1 Suppose P, Q, and R are collinear and $QR + QP = PR$.

Since $QR = RQ$, and $PR = RP$, we may write

$$RQ + QP = RP$$

and conclude that Q is between R and P.

A similar rearrangement leads us to rewrite $AB + BC = AC$ as

$$CB + BA = CA$$

and conclude that A-B-C iff C-B-A or, in words, B is between A and C if and only if B is between C and A.

If two points, A and C, are given, do you think that there is always a third point between them? Suppose $c(A) = x$ and $c(C) = y$ and $x < y$. Then the above question is reduced to discovering whether there exists a number between x and y. One might guess that what is called "the average" of x and y would be between them. We will prove this result, and from it the related conjecture about A and C will follow.

Mathematicians call a result used primarily as a "helping theorem" a *lemma* (from the Greek word for branch).

LEMMA 2.1 If $x < y$, then

$$x < \frac{x + y}{2} < y$$

Proof: If $x < y$, then $x + y < y + y$, by the addition law. Hence $x + y < 2y$; and $\frac{1}{2}(x + y) < \frac{1}{2}(2y)$, by the multiplication law. Thus

$$\frac{x + y}{2} < y$$

By adding x to both sides of $x < y$, we can show that

$$x < \frac{x + y}{2}$$

Combining these two results, we obtain

$$x < \frac{x + y}{2} < y$$

Combining this result with P8 and the definition of *between*, our original question can now be answered in the affirmative. It follows as a theorem.

THEOREM 2 If A and C are two points, then there is a point B such that A-B-C.

Now that we know that such points exist, the following definition is meaningful.

DEFINITION 2.12 Suppose $\overleftrightarrow{AB}$ is given. The union of A and B with the set of all points between A and B is called a *line segment* (or more briefly, segment) and is denoted by $\overline{AB}$. A and B are called the *endpoints* of $\overline{AB}$. All other points of the segment are called *interior points*. The *length of $\overline{AB}$* is defined as the distance from A to B and is therefore denoted by AB. AB is sometimes referred to as the *distance between* A and B.

Example 2   Consider $\overleftrightarrow{l}$ and the graphs of $\overline{BD}$ and $\overline{GJ}$.

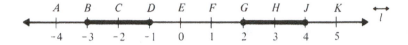

> C is an interior point of $\overline{BD}$, and $BD = 2$
> H is an interior point of $\overline{GJ}$, and $GJ = 2$

Example 3   Consider the algebraic inequality

$$|x - 1| < 2$$

The algebraic method of solving this sentence is quite long and tedious. A geometric approach is more manageable.

Since the absolute value of the difference of two numbers represents the distance between the points that have those numbers as coordinates (see Definition 2.6), x in the above inequality is the coordinate of any point which is less than two units from the point with coordinate 1.

On the line graph shown, place your index fingers at E, and move them in opposite directions along $\overleftrightarrow{l}$ up to (but not including) two units in each direction. What you have traced out is the interior of $\overline{CG}$, and the graph of the solution set of

$$|x - 1| < 2$$

The solution set is therefore

$$\{x : -1 < x < 3\}$$

or all real numbers between -1 and 3.

Since betweenness for points is equivalent to betweenness for numbers, it is common to speak of the coordinates of points as if they were the points. We will use this language when it is convenient. Thus

$$|x - 1| < 2$$

may be spoken of as a statement about all the numbers that are less than two units away from 1.

Example 4 Consider $|x - 2| > 1$. Using the language of Example 3, we see that this is a sentence about all numbers that are more than one unit from 2. The graph is as shown.

The solution set is

$$\{x : x < 1 \ or \ x > 3\}$$

Do you see why the italicized connective is "or" rather than "and"?

Example 5 Consider $|x - 1| < 3$. Since distance is represented in terms of the absolute value of a difference, we rewrite the inequality as

$$|x - (-1)| \leq 3$$

Here then, we are interested in all the numbers less than or equal to three units from -1. The graph is as shown.

The solution set is

$$\{x : -4 \leq x \leq 2\}$$

PROBLEM SET 2.6

1. Suppose A-B-C. If $c(B) = 3$, $c(C) = 6$, and $AB = 2$, find AC. Draw a picture.

2. Suppose A-B-C. If $AC = 6$, $c(A) = -2$, and $AB = 2\frac{1}{2}$, find BC and $c(B)$.

3. If $BC = 6$, $AC = 4$, and $AB = 10$, which of A, B, and C is between the other two?

4. Consider $\overset{\leftrightarrow}{l}$.

   Find a simpler name for the following:

   (a) $\overline{AB} \cup \overline{BC}$ (b) $\overline{AD} \cap \overline{BC}$

   (c) $\{A, B\} \cup \{K \in \overset{\leftrightarrow}{l} : A\text{-}K\text{-}B\}$

5. Given $x < y$ as in Lemma 2.1, complete the proof of the lemma by showing that

$$x < \frac{x + y}{2}$$

   Give a reason for each step.

6. Use Lemma 2.1 to find a number between $\frac{4}{7}$ and $\frac{5}{8}$.

7. Given $\overline{AC}$. Does T2 tell us something about the number of points on $\overline{AC}$? Discuss.

8. Use Lemma 2.1 to show that there is no "smallest" positive real number. [*Hint:* Assume there is one (call it x) and apply the lemma.]

9. Graph and give the solution sets.

   (a) $|x - 3| < 2$ (b) $|x - 3| = 2$
   (c) $|x - 3| \geq 2$ (d) $|x - 3| = 0$
   (e) $|x + 2| \leq 2$ (f) $|x| < 2$ (*Hint:* $x = x - 0$.)

10. Graph and solve the inequality $|2x - 1| < 6$. (*Hint:* $|2x - 1| < 6$ implies $|2(x - \frac{1}{2})| < 6$, or $|2|\,|(x - \frac{1}{2})| < 6$, etc.)

11. Shown is the graph of a disjunction and an absolute value inequality

(a) State the disjunction.
(b) State the absolute value inequality.

12. Shown is the graph of a conjunction and an absolute value inequality.

(a) State the conjunction.
(b) State the absolute value inequality.

13. Assume that *A-B-C* and *B-C-D*. (a) Which postulate assures us that *A*, *B*, *C* and *D* are collinear? (b) Use the definition of *between* to prove *A-B-D*.

2.7 MORE ABOUT COLLINEAR SETS

We generally conceive of a light ray as something that begins at a given point and continues outward in a straight line path. This idea motivates the following definition.

DEFINITION 2.13 Consider $\overleftrightarrow{AB}$. *Ray AB*, denoted $\overrightarrow{AB}$, is the set of all points C, $C \in \overleftrightarrow{AB}$, such that A is not between B and C. We call A the *endpoint* of this ray. Any other point of the ray is called an *interior point*.

The definition of ray is a bit tricky, but looking at Fig. 2.4, we see that A and all the points to the

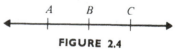

FIGURE 2.4

right of it constitute $\overrightarrow{AB}$, whereas B and all the points to the left of it make up $\overrightarrow{BA}$.

We have sketched graphs of sentences such as

$$x \le 2$$

$$x \ge -3$$

and

$$x \leq -5$$

in Section 2.5. These can now be seen to be rays.

It should be obvious that given the line of Fig. 2.4, we may write

$$\overrightarrow{BA} \cup \overrightarrow{BC} = \overleftrightarrow{AB}$$

We call such a pair of rays *opposite rays*. You should notice that $\overrightarrow{BA}$ and $\overrightarrow{BC}$ intersect only at their endpoints. This idea is thinly veiled in the following definition.

DEFINITION 2.14 Let $\overleftrightarrow{AC}$ be a line. If B is between A and C, then $\overrightarrow{BA}$ and $\overrightarrow{BC}$ are *opposite rays*.

Very much like T2 (with its implications for segments) is Theorem 3.

THEOREM 3 If A and B are two points, then there is a point C such that B is between A and C.

We will omit the proof of this theorem. It strongly resembles the proof of T2.

The length of a segment $\overline{AB}$ is simply AB. What about the length of ray? Your intuition should lead you to conclude that although a ray has an endpoint, its length is infinite.

If line $\overleftrightarrow{AB}$ is given, how many points on $\overleftrightarrow{AB}$ are six units from A? How many points on $\overrightarrow{AB}$ are six units from A? The following postulate underlies the answers to these questions.

POSTULATE 11 *The Point-Plotting Postulate* (PPP): Suppose $\overrightarrow{AB}$ and real number $r, r > 0$, are given. Then there is exactly one point C, $C \in \overrightarrow{AB}$, such that $AC = r$.

Actually P11 can be proven as a theorem from the first 10 postulates, but its verification is tedious.[6] The significance of the PPP will

[6] See, for example, C. R. Wylie, Jr., *Foundations of Geometry* (New York: McGraw-Hill Book Co., 1964), pp. 61–62.

become apparent later. It is a key to the proof of the famous Pythagorean Theorem and many others.

Consider the following definition.

DEFINITION 2.15   Let A-B-C. B is called the *midpoint* of $\overline{AC}$ if and only if $AB = BC$. B is said to *bisect* $\overline{AC}$. More generally, any set S such that $S \cap \overline{AC} = \{B\}$ is said to bisect $\overline{AC}$.

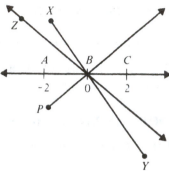

FIGURE 2.5

Example 1   In Fig. 2.5, B, $\overleftrightarrow{XY}$, $\overline{ZB}$, and $\overrightarrow{PB}$ all bisect AC. You should be able to name some other sets in Fig. 2.5 that bisect $\overline{AC}$.

Many students fail to recognize that Definition 2.15 is not an assurance that a midpoint B of $\overline{AC}$ exists.[7] The definition merely affirms that if there is a point $B \in \overline{AC}$ such that $AB = BC$, then B is *the* midpoint of $\overline{AC}$. Implicit in the use of the definite article "*the*" is the added assertion that B is the only such point. To make our structure more tidy we can state the following theorem.

THEOREM 4   *The Midpoint Theorem:* Every segment has one and only one midpoint.

To many readers all this concern about a midpoint is much ado about nothing. But as the famous philosopher and mathematician Bertrand Russell (1872–1969) once remarked: "The obvious is the most difficult to prove and the most often wrong." Although its proof is omitted, the Midpoint Theorem is *not* wrong, and only moderately difficult to prove, requiring some logical techniques yet to come.

Suppose we encounter the situation depicted by Fig. 2.6, and M is the midpoint of $\overline{AB}$. Then $AM = MB$. If $a < b$, then

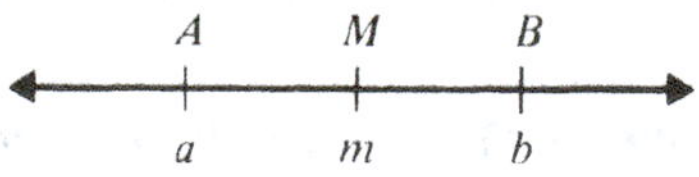

$$m - a = b - m$$

FIGURE 2.6

Applying the addition law[8] (twice), we obtain

$$2m = a + b$$

---

[7] One *can* define nonexistent points. For example, we might define the "globpoint" of AC as follows. Let A-D-C. D is called the *globpoint* of AC iff $AD = 2AC$.

[8] See Appendix A. First we added a to both sides of the equation, and then we added m to both sides.

Thus

$$m = \frac{a + b}{2}$$

The same result is obtained if we assume $b < a$, which gives us the following theorem.

THEOREM 5 Suppose M is the midpoint of $\overline{AB}$ and $c(A) = a$, $c(M) = m$, and $c(B) = b$. Then

$$m = \frac{a + b}{2}$$

This may be stated in words as "The coordinate of the midpoint of a segment is one half the sum of the coordinates of its endpoints."

Example 2 Suppose we want to find m, the coordinate of the midpoint of the segment shown in the figure:

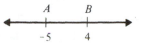

$$m = \frac{-5 + 4}{2} = \frac{-1}{2}$$

Example 3 If H is the midpoint of $\overline{AB}$ and $c(M) = \frac{1}{2}$, $c(B) = 4\frac{1}{3}$, we can calculate a, the coordinate of A.

$$\frac{1}{2} = \frac{a + 4\frac{1}{3}}{2}$$

or

$$\frac{1}{2} = \frac{a}{2} + \frac{13}{6}$$

Thus

$$a = 2\left(\frac{1}{2} - \frac{13}{6}\right) = \frac{-10}{3}$$

PROBLEM SET 2.7

1. Sketch an illustration of two rays:

   (a) If their intersection is a segment.
   (b) If their intersection contains only an interior point of both of them.

(c) If their intersection contains only the endpoint of each of them.

(d) If their intersection is a ray.

(e) If they are collinear and their intersection is the empty set.

2. Graph:

(a) $x \geq -1$ (b) $x \geq -1$ and $x \leq 1$

(c) $|x| \geq 1$ (d) $x \leq 0$ and $x \geq 0$

3. Given $\overrightarrow{AB}$. Does the PPP tell us something about the length of $\overrightarrow{AB}$? Discuss. What implications does this have for $\overleftrightarrow{AB}$?

4. Find a simpler name for each of the following, using the accompanying figure:

(a) $\overrightarrow{AB} \cup \overrightarrow{BA}$ (b) $\overline{AB} \cup \overrightarrow{BC}$

(c) $\overrightarrow{CD} \cap \overrightarrow{BA}$ (d) $\overrightarrow{BC} \cap \overrightarrow{CB}$

(e) $\overline{AB} \cup \{X : A\text{-}B\text{-}X\}$

$$\begin{array}{cccc} A & B & C & D \\ \end{array}$$

5. Suppose A-B-C and $AB = BC$. Find b if

(a) $c(A) = -1$, $c(B) = b$, and $c(C) = 3$.

(b) $c(A) = -\frac{1}{2}$, $c(B) = b$, and $c(C) = \frac{1}{3}$.

(c) $\overline{AC} = 6$ and $c(A) = -2$. (Be careful!)

6. The point A in the accompanying figure can be said to separate $\overleftrightarrow{AB}$ into two parts, and B belongs to one of these parts. We call this part of $\overleftrightarrow{AB}$, the "B-side of A." Use this phrase to write an alternative definition of *ray*.

$$\begin{array}{cc} A & B \\ \end{array}$$

7. Sketch rays $\overrightarrow{AP}$ and $\overrightarrow{AQ}$ such that $\overrightarrow{AP}$ and $\overrightarrow{AQ}$ are noncollinear. Choose what you think is an appropriate name for this set of points.

8. A, B, and C are collinear. B is the midpoint of $\overline{AC}$. If $c(A) = -3$ and $c(C) = 6$, what is $c(B)$?

9. On $\overleftrightarrow{AB}$, the coordinates of A and B are $\frac{1}{2}$ and -5, respectively. If M is the midpoint of $\overline{AB}$, find $c(M)$.

10. Suppose A-B-C and B is the midpoint of $\overline{AC}$. If $c(A) = 4$ and $c(B) = -\frac{1}{3}$, find $c(C)$.

11. Suppose A-B-C, $c(A) = \frac{1}{3}$, $c(B) = -5\frac{1}{3}$, and $c(C) = -11$. Show that B is the midpoint of $\overline{AC}$.

12. True or false? If $AB = BC$, then B is the midpoint of $\overline{AB}$. Explain.

2.8 CONGRUENCE FOR SEGMENTS, FORMAL PROOFS

In early Greek geometry there arose a formalization of a particular kind of equivalence. Given two geometric figures, if one can be moved so as to coincide with the other, we say that they are *congruent*. This rather imprecise description is formulated in sufficiently exact form for segments in the following definition.

DEFINITION 2.16 Let $\overline{AB}$ and $\overline{CD}$ be segments. $\overline{AB}$ and $\overline{CD}$ are said to be *congruent* if $AB = CD$. We say $\overline{AB}$ *is congruent to* $\overline{CD}$, and we write $\overline{AB} \cong \overline{CD}$.

You should notice that we are not saying that $\overline{AB}$ and $\overline{CD}$ are collinear. They may be any pair of segments in space, or they may be the same segment. If they have the same length, they are congruent.

There are many theorems which are obviously true and which can be proven readily from Definition 2.16 together with some of the fundamental properties of the real number system. (If the names of these properties are unfamiliar to you, you will find them in Appendix A.) Next, we will derive some of these simple results while focusing our attention on the method of proof.

Such proofs as the ones that follow are called *formal proofs*. Heretofore, we have proven theorems *informally*.

The forms that we are going to follow are two-column form and three-column form.

THEOREM 6 Segment congruence is reflexive. Symbolically, if $\overline{AB}$ is a segment, then $\overline{AB} \cong \overline{AB}$.

Proof:

STATEMENTS	REASONS	LOGIC
1. $\overline{AB}$ is a segment	1. Hypothesis	1. Hypothesis
2. The length of $\overline{AB}$ is AB	2. Definition of length	2. Reason 2, Statement 1, and modus ponens
3. $AB = AB$	3. Reflexive law (see Appendix A)	3. Reason 3, Statement 2, and modus ponens
4. $\overline{AB} \cong \overline{AB}$	4. Definition of $\cong$	4. Reason 4, Statement 3, and modus ponens

A close examination of this proof should lead you to the discovery that it is really a sequence of three modus ponens (m.p.) arguments. Our first statement is merely the assumption or hypothesis. Each following statement is a conclusion of a syllogism having "established truths" as its assumptions. These "established truths" may come from previous material (e.g., Reason 4), or they may be results established as part of the proof (e.g., Statement 3). Each entry in the "Reasons" column is either a general statement such as a definition, theorem, postulate, algebraic law, or the word "hypothesis," signifying that the corresponding statement is taken from the hypothesis of the theorem. You will see these patterns again and again.

Very often we reinforce the idea that we are trying to derive a certain conclusion from a given assumption by rewriting both of them above the proof.

THEOREM 7 Segment congruence is symmetric; that is, if $\overline{AB} \cong \overline{CD}$, then $\overline{CD} \cong \overline{AB}$. (It is common to provide a diagram to illustrate the theorem.)

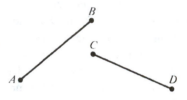

Hypothesis: $\overline{AB} \cong \overline{CD}$
Conclusion: $\overline{CD} \cong \overline{AB}$

Proof:

STATEMENTS	REASONS	LOGIC
1. $\overline{AB} \cong \overline{CD}$	1. Hypothesis	1. Hypothesis
2. $AB = CD$	2. Definition of $\cong$	2. Reason 2, Statement 1, and modus ponens
3. $CD = AB$	3. Symmetric law (see Appendix A)	3. R3, S2, m.p.[9]
4. $\overline{CD} \cong \overline{AB}$	4. Definition of $\cong$	4. R4, S3, m.p.

THEOREM 8 Segment congruence is transitive; that is, if $\overline{AB} \cong \overline{CD}$ and $\overline{CD} \cong \overline{EF}$, then $\overline{AB} \cong \overline{EF}$.

The proof is similar to the previous two proofs and is left as an exercise.

The following theorems are derived from analogous algebraic facts.

---

[9] We shall use abbreviations in the proofs wherever possible.

THEOREM 9   *Segment Addition Theorem* (SAT): If A-B-C, R-S-T, $\overline{AB} \cong \overline{RS}$, and $\overline{BC} \cong \overline{ST}$, then $\overline{AC} \cong \overline{RT}$.

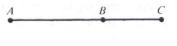

Hypothesis: A-B-C; R-S-T;
$\overline{AB} \cong \overline{RS}$; and $\overline{BC} \cong \overline{ST}$
Conclusion: $\overline{AC} \cong \overline{RT}$

Proof:

STATEMENTS	REASONS	LOGIC
1. $\overline{AB} \cong \overline{RS}$; $\overline{BC} \cong \overline{ST}$	1. Hypothesis	1. Hypothesis
2. $AB = RS$; $BC = ST$	2. Definition of $\cong$	2. R2, S1, m.p.
3. $AB + BC = RS + ST$	3. Addition law (see Appendix A)	3. R3, S2, m.p.
4. A-B-C; R-S-T	4. Hypothesis	4. Hypothesis
5. $AB + BC = AC$; $RS + ST = RT$	5. Definition of between	5. R5, S4, m.p.
6. $AC = RT$	6. Substitution law (Appendix A)	6. R6, S5, S3, m.p.
7. $\overline{AC} \cong \overline{RT}$	7. Definition of $\cong$	7. R7, S6, m.p.

You should notice that we delayed the introduction of part of the hypothesis until it was needed.

Example 1   In the figure shown, if $\overline{AB} \cong \overline{CD}$, then $\overline{AC} \cong \overline{BD}$.

Proof:

STATEMENTS	REASONS	LOGIC
1. $\overline{AB} \cong \overline{CD}$	1. Hypothesis	1. Hypothesis
2. $\overline{BC} \cong \overline{BC}$	2. $\cong$ is reflexive	2. R2, $\overline{BC}$ is a segment, m.p.
3. $\overline{AC} \cong \overline{BD}$	3. SAT	3. R3, S1, S2, m.p.

THEOREM 10   *Segment Subtraction Theorem* (SST): If A-B-C, R-S-T, $\overline{AB} \cong \overline{RS}$, and $\overline{AC} \cong \overline{RT}$, then $\overline{BC} \cong \overline{ST}$.

The proof mimics that of T9. The SST says, in effect, that if we delete congruent segments from the ends of (larger) congruent segments, the parts that remain are also congruent.

Example 2 In Example 1, we took the information depicted in the figure shown here as part of our hypothesis. Using the same figure again, we can prove the converse of Example 1. If $\overline{AC} \cong \overline{BD}$, then $\overline{AB} \cong \overline{CD}$.

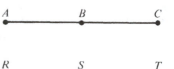

Proof:

STATEMENTS	REASONS
1. $\overline{AC} \cong \overline{BD}$	1. Hypothesis
2. $\overline{BC} \cong \overline{BC}$	2. $\cong$ is reflexive
3. $\overline{AB} \cong \overline{CD}$	3. SST, S1, S2

In this example we have derived the result in two-column form. In the ensuing chapters, three-column form will be abandoned.

THEOREM 11 *Segment Bisection Theorem:* If $\overline{AC} \cong \overline{RT}$, B is the midpoint of $\overline{AC}$, and S is the midpoint of $\overline{RT}$, then $\overline{AB} \cong \overline{RS}$.

Proof:

STATEMENTS	REASONS	LOGIC
1. $\overline{AC} \cong \overline{RT}$; B the midpoint of $\overline{AC}$; S the midpoint of $\overline{RT}$	1. Hypothesis	1. Hypothesis
2. $AB = BC$; $RS = ST$; B is between A and C; S is between R and T	2. Definition of midpoint	2. R2, S1, m.p.
3. $AB + BC = AC$; $RS + ST = RT$	3. Definition of between	3. R3, S2, m.p.
4. $AC = RT$	4. Definition of $\cong$	4. R4, S1, m.p.
5. $AB + BC = RS + ST$	5. Substitution law (see Appendix A)	5. R5, S3, S4, m.p.
6. $AB + BC = AB + AB = 2(AB)$; $RS + ST = RS + RS = 2(RS)$	6. Substitution law	6. R6, S2, m.p.
7. $2(AB) = 2(RS)$	7. Substitution law	7. R7, S6, S5, m.p.
8. $AB = RS$	8. Multiplication law (see Appendix A)	8. R8, S7, m.p.

Example 3 Syllogisms other than m.p., can play
a role in the proof of a theorem, as can be seen in
the following three-column proof. Given the configuration shown,
if $AB > CD$, then $AC \not\cong BD$.

$$A \qquad\qquad B \quad C \quad D$$

Proof:

STATEMENTS	REASONS	LOGIC
1. $AB > CD$	1. Hypothesis	1. Hypothesis
2. $AB \neq CD$	2. Trichotomy law	2. R2, S1, m.p.
3. $\overline{AB} \not\cong \overline{CD}$	3. Definition of $\cong$	3. R3, S2, m.t.
4. $\overline{BC} \cong \overline{BC}$	4. Congruence is reflexive	4. R4, $\overline{BC}$, m.p.
5. $\overline{AC} \not\cong \overline{BD}$	5. SST	5. R5, S3, S4, m.t.

If you worked hard trying to absorb the idea of the Point-Plotting
Postulate (PPP), you have earned the following corollary.[10]

COROLLARY P11 *The Segment Construction Theorem* (SCT): Let
$\overline{AB}$ be any segment and $\overrightarrow{PQ}$ any ray. Then there is exactly one point
$S, S \in \overrightarrow{PQ}$, such that $\overline{AB} \cong \overline{PS}$.

Proof: Let $AB = r = PS$ in the PPP. Then
$AB = PS$, or $\overline{AB} \cong \overline{PS}$.

We will have many opportunities to apply this theorem.

PROBLEM SET 2.8

1. Given $\overleftrightarrow{l}$, state the postulate, definition, or theorem
   that justifies each of the following statements:

$$A \quad B \quad C \quad D \quad E \quad F \qquad \overleftrightarrow{l}$$

   (a) $\overline{BC} \cong \overline{BC}$
   (b) If $\overline{AB} \cong \overline{CD}$, and $\overline{BC} \cong \overline{DE}$, then $\overline{AC} \cong \overline{CE}$
   (c) If $\overline{AB} \cong \overline{EF}$, then $\overline{EF} \cong \overline{AB}$
   (d) If $\overline{AC} \cong \overline{DF}$, then $\overline{AB} \not\cong \overline{DF}$

---

[10] Mathematicians use the word "corollary" to describe a theorem that follows
easily from a previously established fact. The word "corollary" comes from the
Latin word for a gratuity or a tip.

(e) If $\overline{CE} \cong \overline{DF}$, then $\overline{CD} \cong \overline{EF}$

(f) If $AB = CD$, then $\overline{AB} \cong \overline{CD}$

(g) If $c(A) = a$, $c(B) = b$, $c(D) = d$, and $|b - a| + |d - b| = |d - a|$, then B is between A and D. (Two reasons must be given here.)

2. (a) On $\overleftrightarrow{l}$ in problem 1, if D is the midpoint of both $\overline{AF}$ and $\overline{BE}$, what can be said about $\overline{AB}$ and $\overline{EF}$?

(b) On $\overleftrightarrow{l}$ in problem 1, if D is the midpoint of $\overline{AF}$ and $\overline{BE}$, what can be said about $\overline{AE}$ and $\overline{BF}$?

(c) Given $\overleftrightarrow{l}$, if D is the midpoint of $\overline{AF}$ and $\overline{BE}$, then $\overline{AB} \cong \overline{EF}$. Supply the missing parts of the proof.

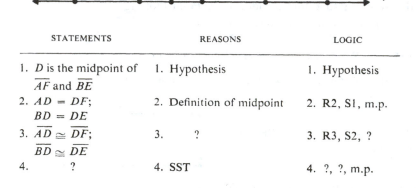

STATEMENTS	REASONS	LOGIC
1. D is the midpoint of $\overline{AF}$ and $\overline{BE}$	1. Hypothesis	1. Hypothesis
2. $AD = DF$; $BD = DE$	2. Definition of midpoint	2. R2, S1, m.p.
3. $\overline{AD} \cong \overline{DF}$; $\overline{BD} \cong \overline{DE}$	3. ?	3. R3, S2, ?
4. ?	4. SST	4. ?, ?, m.p.

(d) Given $\overleftrightarrow{l}$, if D is the midpoint of $\overline{AF}$ and $\overline{BE}$, it follows that $\overline{AE} \cong \overline{BF}$. Write out the three-column proof of this conjecture. The proof is nearly identical to that of problem 2(c).

3. Given $\overleftrightarrow{l}$ as in problem 1, if $\overline{BC} \cong \overline{EF}$, $c(B) = 0$, $c(C) = 5$, and $c(E) = 9$, find $c(F)$.

4. Give a two-column proof of the SST. You have already been told that the proof mimics that of the SAT.

5. Supply the missing reasons for the following proof.

Hypothesis: Given $\overleftrightarrow{l}$;
$\overline{AB} \cong \overline{CD}$
Conclusion: $\overline{AC} \cong \overline{BD}$

Proof:

STATEMENTS	REASONS
1. $\overline{BC} \cong \overline{BC}$	1.
2. If $\overline{AC} \cong \overline{BD}$, then $\overline{AB} \cong \overline{CD}$	2.
3. $\overline{AB} \not\cong \overline{CD}$	3.
4. $\overline{AC} \not\cong \overline{BD}$	4. SST (Modus tollens was used here.)

6. Why is the following implication false?

$$\text{If } \overline{AB} \cong \overline{CD}, \text{ then } \overline{AC} \cong \overline{BD}$$

CHAPTER 2: REVIEW EXERCISES

1. For each of the following statements indicate whether it is *always* true, *sometimes* but not always true, or *never* true.

(a) The intersection of two lines contains a single point.

(b) If two points of a line lie in a plane, then the entire line lies in the plane.

(c) $|x + y| < |x| + |y|$

(d) $|-x| = |x|$

(e) If $a < b$, then $b > a$

(f) If $a < b$, then $ac < bc$ (a, b, and c are real numbers.)

(g) Suppose $X \in \overline{AB}$. Then X is between A and B.

(h) Let $\overline{AB}$ be given; then $AB = BA$.

(i) The union of two rays is a segment.

(j) The intersection of two rays is a point.

(k) If B is between A and C, then $AB = BC$

(l) If B is between A and C, and if $c(A) = x$, $c(B) = y$, $c(C) = z$, then $x < y < z$

2. Graph and give the solution sets:

(a) $x = 2$ (b) $|x - 2| = 3$

(c) $x - 3 < 2$ (d) $|x + 2| \geq 5$

(e) $3 \leq x \leq 7$ (f) $|x| \geq 2$

3. Given $\overleftrightarrow{l}$, find a simpler name for each of the following:

(a) $\overline{AB} \cap \overline{BC}$ (b) $\overline{AB} \cup \overline{BC}$

(c) $\overrightarrow{BA} \cap \overrightarrow{CD}$ (d) $\overleftrightarrow{AB} \cap \overrightarrow{BC}$

(e) $\overrightarrow{BC} \cup \overrightarrow{BA}$ (f) $AC + DC$

(g) $AC - AB$

4. If B is the midpoint of $\overline{AC}$ and the coordinates of A, B, and C are a, b, and c, respectively, find:

(a) b, if $a = -2$ and $c = \frac{4}{3}$

(b) a, if $b = -2$ and $c = \frac{3}{4}$

(c) c, if $a = -2$ and $b = \frac{3}{4}$

5. Name each postulate, definition, or theorem stated below:

(a) $\overline{AB} \cong \overline{AB}$, for any two points A and B

(b) If $A\text{-}B\text{-}C$, $R\text{-}S\text{-}T$, $\overline{AB} \cong \overline{RS}$, and $\overline{AC} \cong \overline{RT}$, then $\overline{BC} \cong \overline{ST}$

(c) $AB = CD$ iff $\overline{AB} \cong \overline{CD}$

(d) Every line has a coordinate system.

(e) Two points determine a line.

REFERENCES

Birkhoff, G. D., and Beatley, Ralph. *Basic Geometry*. Glenview, Ill.: Scott, Foresman and Co., 1940.

The first geometry book at this level to simplify Euclidean geometry by introducing the real number system.

Heath, T. L., ed. *The Thirteen Books of Euclid's Elements*. New York: Dover Publications, 1956.

The "granddaddy" of them all. For over two thousand years, educated men in all fields looked upon its study as an indispensable aid to logical reasoning.

Hilbert, David. *The Foundations of Geometry*. Translated by E. J. Townsend. La Salle, Ill.: Open Court Publishing Company, 1902.

In his now-famous lectures at Edinburgh, Hilbert set forth a system of axioms that enabled him to correct errors found in Euclid's *Elements*.

Moise, E. E. *Elementary Geometry from an Advanced Standpoint*. Reading, Mass.: Addison-Wesley Publishing Co., 1963.

A text that is widely used for the training of geometry teachers, it applies the simplifications introduced by Birkhoff and Beatley (in the reference given above) to the system instituted by Hilbert.

Wylie, C. R., Jr. *Foundations of Geometry.* New York: McGraw-Hill Book Co., 1964.

    This text covers much of the material in the Moise text, with the advantage that the reader is given a better sense of what the mathematics *means.*

The knowledge at which geometry aims is knowledge of the eternal and not of anything perishing or transient.

SOCRATES (*ca.* 400 B.C.) in Plato's *Republic*

CHAPTER THREE

SEPARATION
ANGLES

3.1 INTRODUCTION

In Chapter 2 we saw that we could assign measures to segments. In this chapter we will introduce angles and develop a theory of measure for them. You no doubt believe that you know what angles are. In Problem Set 3.1, you will be asked to act on that belief and to define the term "angle."

Before embarking on the study of angles and their measures, we will lay some groundwork by discussing the concept of convexity. A little of the theory of convex sets is used here to develop some clear statements about angles and their measures. This theory has other more prosaic applications in such diverse fields as the social and management sciences and engineering.

3.2 CONVEX SETS

Figures 3.1(a), (b), and (c) suggest a property that parts (d) and (e) do not possess. In the first three cases, $\overline{PQ}$ is a subset of the region

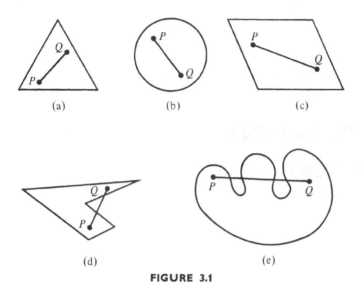

(a)   (b)   (c)

(d)   (e)

FIGURE 3.1

bounded by the figure, whereas in (d) and (e) $\overline{PQ}$ leaves the region. In fact, for *any* choice of P and Q in one of the first three regions, $\overline{PQ}$ is a subset of that region. Such a region is called a *convex set*.

DEFINITION 3.1 Let A be a set of points. A is said to be *convex* if, for all choices of P and Q such that $P \in A$ and $Q \in A$, it follows that $\overline{PQ} \subseteq A$.

Hereafter in definitions, the word "if" will be used to mean "iff."

Example 1 Every line is a convex set, as is every segment and every ray.

Example 2 Let $\overline{AB}$ and $\overline{CD}$ belong to line $\overleftrightarrow{l}$. If $\overline{AB}$ and CD are disjoint, it follows that $\overline{AB} \cup \overline{CD}$ is not convex, even though $\overline{AB}$ and $\overline{CD}$ are convex.

Example 3 Consider $\overleftrightarrow{AB}$ as shown in the diagram. $\overline{AB}$ is a convex set but its complement

is not convex. This follows from the fact that although both D and C are in $\sim(\overline{AB})$, $\overline{DC} \nsubseteq \sim(\overline{AB})$.

From these examples we can conclude that two of the following three statements are false.

1. The union of two convex sets is a convex set.
2. The intersection of two convex sets is a convex set.
3. The complement of a convex set is a convex set.

We state, as a theorem, the conjecture that is true.

THEOREM 12   The intersection of two convex sets is a convex set.

Proof (an informal proof will do the job): Let A and B be two convex sets. It must be shown that for any two points P and Q, in their intersection, it follows that $\overline{PQ}$ is in their intersection.

Suppose $P \in (A \cap B)$, $Q \in (A \cap B)$. Then $P \in A$ and $Q \in A$, which implies that $\overline{PQ} \subseteq A$ (A is convex). Likewise, $P \in B$ and $Q \in B$, and therefore $\overline{PQ} \subseteq B$ (B is convex). Since $\overline{PQ} \subseteq B$ and $\overline{PQ} \subseteq A$, it follows that $\overline{PQ} \subseteq (A \cap B)$.

Convex sets may be very large or they may be very small. For example, all of space is a convex set since if P and Q are points of space, $\overline{PQ}$ is a subset of space. On the other hand, the shortest segment you can imagine is also a convex set, as is the empty set.

Consider the plane m, depicted in the diagram. The subsets of the plane that lie on each of the sides of $\overleftrightarrow{l}$ (denoted by $\mathscr{H}_1$ and $\mathscr{H}_2$) are easily seen to be convex. In addition, it should be clear from the diagram that $\overline{PQ}$ intersects $\overleftrightarrow{l}$, whereas $\overline{PR}$ and $\overleftrightarrow{l}$ are disjoint.

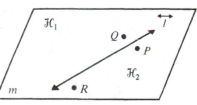

Although these statements are "obvious," they cannot be proven from the facts we have at our disposal. The only other means by which we can establish such statements as facts is to absorb them into our general theory as a postulate.

POSTULATE 12   *The Plane Separation Postulate* (PSP): Given a line and a plane containing it, the set of all points of that plane that do not lie on the line is the union of two disjoint sets such that

1. Each of the sets is convex.
2. If P belongs to one of the sets and Q belongs to the other, then $\overline{PQ}$ intersects the line.

DEFINITION 3.2 The pair of disjoint sets mentioned in P12 (and denoted by $\mathscr{H}_1$ and $\mathscr{H}_2$ in the preceding diagram) are called the *half planes* determined by $\overleftrightarrow{l}$. If $P \in \mathscr{H}_1$ and $Q \in \mathscr{H}_2$, then P and Q are on *opposite sides* of $\overleftrightarrow{l}$. $\mathscr{H}_1$ is called the *P-side* of $\overleftrightarrow{l}$. $\mathscr{H}_2$ is called the *Q-side* of $\overleftrightarrow{l}$.

In the preceding diagram, R and Q are on opposite sides of $\overleftrightarrow{l}$, and R is on the *P-side* of $\overleftrightarrow{l}$.

Until fairly recently, the PSP never appeared in geometry text-books. The German mathematician Moritz Pasch (1843–1931) appears to have been among the first to recognize that Euclid tacitly assumed this postulate (in Proposition 21, Book I of *The Elements*, for example). Pasch formulated a postulate that is equivalent to the PSP.[1] That Euclid himself was aware of this assumption and chose to ignore it seems doubtful. In any event, the PSP is really little more than a sophisticated form of the children's riddle: Why did the chicken cross the road?

PROBLEM SET 3.2

1. Use the line shown to decide whether the following sets are convex or not convex.

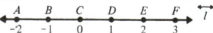

(a) $\overline{AB} \cap \overline{AC}$
(b) $\overline{AB} \cap \overline{BC}$
(c) The graph of $\{x : -1 \le x \le 5\}$
(d) The graph of $\{x : |x| \ge 5\}$
    $\overline{AB} \cup \overline{DC}$
(f) $\overrightarrow{AB} \cap \overrightarrow{DC}$

[1] A statement of the postulate of Pasch and its derivation from the PSP can be found in E. E. Moise, *Elementary Geometry from an Advanced Standpoint* (Reading, Mass.: Addison-Wesley Publishing Co., 1963), p. 53; and in C. R. Wylie, Jr., *Foundations of Geometry* (New York: McGraw-Hill Book Co., 1964), p. 100.

(g) $\overrightarrow{CD} \cup \overrightarrow{BA}$

(h) The graph of $\{x : |x| < 0\}$

(i) $\{P \in \overleftrightarrow{l} : CP = 2\}$

2. (a) Give an example to show that the complement of a convex set is not necessarily convex.

(b) Give an example of a set (not $\varnothing$) and its complement, both of which are convex.

(c) Give two examples to show that the union of two convex sets is not necessarily convex. Give one example of each case.

3. Which postulate of Chapter 2 assures us that a plane is a convex set?

4. May two lines separate a plane into three regions? Four regions?

5. May three lines separate a plane into four regions? Five regions? Six regions? Seven regions?

6. If the answer is "no" to any of the parts of problems 4 or 5, try to explain why.

7. (a) State a Line Separation Postulate analogous to the PSP. Illustrate with a diagram.

(b) State a Space Separation Postulate. Illustrate with a diagram.

(c) Find the maximum number of nonintersecting regions into which three planes may separate space.

8. Sketch your conception of an angle. Give a definition of "angle." (*Hint:* An angle is *not* a convex set.)

9. Copy the accompanying diagram, and illustrate the following with shading:

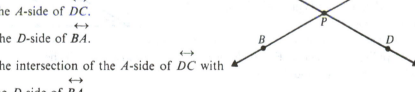

(a) The A-side of $\overleftrightarrow{DC}$.

(b) The D-side of $\overleftrightarrow{BA}$.

(c) The intersection of the A-side of $\overleftrightarrow{DC}$ with the D-side of $\overleftrightarrow{BA}$.

10. If a point is removed from a plane, is the resulting set a convex set?

11. Is the union of two half planes a plane? Discuss the several different possibilities. A sketch or two would probably help.

12. Problems 4, 5, and 7(c) quite naturally lead to the following:

What is the maximum number of nonintersecting regions into which a plane can be separated by the following numbers of lines?

(a) One line.
(b) Two lines.
(c) Three lines.
(d) Four lines.

3.3 ANGLES AND THEIR MEASURES

In problem 7 of Problem Set 2.7, you were asked to sketch noncollinear rays $\overrightarrow{AP}$ and $\overrightarrow{AQ}$, and then to choose a name for this set of points. An appropriate name for our purposes would have been "angle."

DEFINITION 3.3 The union of a pair of noncollinear rays with a common endpoint is called an *angle*. The common endpoint is called the *vertex* of the angle, and the rays are called the *sides* of the angle.

If, as in the preceding figure, the rays are $\overrightarrow{AP}$ and $\overrightarrow{AQ}$, the angle is denoted by either of the symbols $\angle PAQ$ and $\angle QAP$. Notice the position of the vertex A in each of these representations; it is sandwiched between interior points on each of the rays. At times when there is no possibility of confusion with another angle having the same vertex, the notation is shortened further to $\angle A$. When more than one angle does have the same vertex, we sometimes find it convenient to shorten the notation as suggested in the diagram shown. $\angle CBD$ can be called $\angle 1$, and $\angle CBA$ can be called $\angle 2$. Another piece of shorthand we will employ is the symbol " $\underline{/s}$ " to mean "angles."

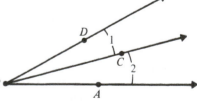

We have already discussed the notion of measure as it applies to segments. For angles we begin the discussion with the following postulate and definition.

POSTULATE 13 *The Angle Measurement Postulate:* To every angle there corresponds a unique real number k, $0 < k < 180$.

DEFINITION 3.4 Let $\angle A$ be given; the *measure of* $\angle A$, denoted $m\angle A$, is that real number specified by the Angle Measurement Postulate.

In these pages, the units of measure we will be using will be degrees, so that if we write $m \angle B = 60$, we mean that $\angle B$ is a 60-degree (60°) angle.

Postulate 13 is introduced here to simplify many of the following definitions and theorems. Although it is possible to develop the subject in such a manner as to allow any real number to correspond to some angle, many difficulties are encountered if we permit this to be the case (e.g., in trying to give a precise definition of the term "adjacent angles"). Postulate 13 skillfully skirts these difficulties and the problems growing out of them.

Analogous to the Point-Plotting Postulate, we have a similar postulate for angles.

POSTULATE 14   *The Angle Construction Postulate:* Let $\overrightarrow{AB}$ be a ray on the edge of half plane $\mathcal{H}$, and let k be a real number such that $0 < k < 180$; then there is exactly one ray $\overrightarrow{AP}$ with $P \in \mathcal{H}$ and such that $m \angle BAP = k$.

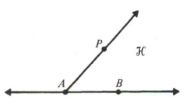

Just as a line separates a plane, $\angle C$ in some sense separates the plane into regions that allow us to distinguish between X and Y. These regions are called the *interior* and *exterior* of the angle.

DEFINITION 3.5   Let $\angle ABC$ be given in plane m. The intersection of the set of all points on the A-side of $\overleftrightarrow{BC}$ with the set of all points on the C-side of $\overleftrightarrow{AB}$ is called the *interior of $\angle ABC$*, and is denoted by $\mathscr{I} \angle ABC$.

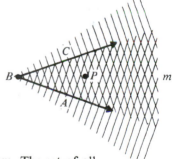

Thus in the diagram, $P \in \mathscr{I} \angle ABC$, whereas A, B, and C are not.

DEFINITION 3.6   Let $\angle ABC$ be given in plane m. The set of all points which are neither in the interior of the angle, nor points of the angle, is called *the exterior of $\angle ABC$*, and is denoted by $\mathscr{E} \angle ABC$.

It should be fairly obvious that if $\angle ABC \subseteq$ plane m, $\mathscr{E} \angle ABC$, $\mathscr{I} \angle ABC$, and $\angle ABC$ are mutually disjoint and their union is m.

If the sketch of an angle is given, its degree measure can be approximated with the aid of a protractor (Fig. 3.2). The protractor can

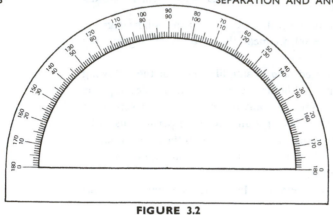

FIGURE 3.2

also be employed to sketch an angle of a given measure. Each student of geometry should possess such an instrument as well as a compass and a ruler.

PROBLEM SET 3.3

1. Which of the following statements is (are) true? Use the accompanying figure.

   (a) $A \in \angle BAC$
   (b) $B \in \mathscr{I} \angle BAC$
   (c) $C \in \mathscr{E} \angle BAC$
   (d) $D \in \mathscr{I} \angle BAC$
   (e) $E \in \mathscr{E} \angle BAC$

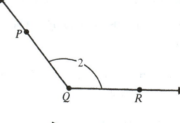

2. Give four different names for the angle shown.

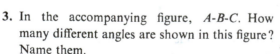

3. In the accompanying figure, *A-B-C*. How many different angles are shown in this figure? Name them.

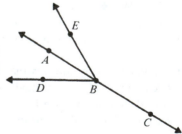

4. Use the figure for problem 3 and give a simpler name for each of the following:

   (a) $\angle ABD \cap \angle ABE$ (b) $\angle DBC \cap \angle EBA$
   (c) $\mathscr{I} \angle DBA \cap \mathscr{I} \angle DBE$ (d) $\mathscr{I} \angle EBC \cap \mathscr{E} \angle EBC$
   (e) $\mathscr{I} \angle DBE \cap \mathscr{E} \angle ABE$

5. (a) The measure of an angle is a number. Is the converse true? Explain.
   (b) Discuss the statement "If x and y are the measures of $\angle X$ and $\angle Y$, then there exists some angle, $\angle Z$, such that $m\angle Z = (x + y)$."
   (c) For what values of x do the following make sense?
      (i) $m\angle A = 36 - x$
      (ii) $m\angle B = 2x - 10$

6. Prove: The interior of an angle is a convex set. [*Hint:* See T12 (Section 3.2).]

7. Using a protractor, draw $\angle BAC$ such that $m\angle BAC = 50$. If $P \in \mathscr{I} \angle BAC$ and $m\angle BAP = 15$, sketch $\overrightarrow{AP}$. Let $\overrightarrow{AP}$ be on the edge of the half plane containing C and then use the protractor to find $m\angle PAC$.

8. Let $\overrightarrow{AB}$ be on the edge of half plane $\mathscr{H}$. Using a protractor, sketch $\angle BAC$, $C \in \mathscr{H}$, such that $m\angle BAC = 60$. Find D, if $D \in \overrightarrow{AC}$, and $AD = 2$ inches (in.). Find E, if $E \in \overrightarrow{AB}$, and $AE = 1$ in. Use a protractor to measure $\angle ADE$. What is this measure?

9. Sketch three rays, no two of which are collinear, with a common endpoint. Label your sketch.

   (a) How many angles are formed?
   (b) Which angle do you think has the largest measure?
   (c) How do you think this measure is related to the measures of the other angles? Be as specific as you can.

10. Problem 9 should suggest a statement to you that is analogous to the definition of betweenness. Express your candidate for this statement.

11. Using the sketch for problem 7, name two angles that have non-intersecting interiors.

3.4 CONGRUENCE FOR ANGLES

Just as congruence for segments is defined in terms of equality of measures (lengths), we can develop a dual notion for angles.

DEFINITION 3.7 Let $\angle A$ and $\angle B$ be given. $\angle A$ and $\angle B$ are said to be congruent if $m\angle A = m\angle B$. We write $\angle A \cong \angle B$.

As with congruence for segments, the following theorems are useful and easy to prove.

THEOREM 13 Let $\angle A$ be given, then $\angle A \cong \angle A$ (reflexive property).

THEOREM 14 If $\angle A \cong \angle B$, then $\angle B \cong \angle A$ (symmetric property).

THEOREM 15 If $\angle A \cong \angle B$ and $\angle B \cong \angle C$, then $\angle A \cong \angle C$ (transitive property).

The proofs of these theorems are left for Problem Set 3.4.

Problems 9 and 10 of Problem Set 3.3 were designed to help you see the necessity of the next postulate.

POSTULATE 15 *The Angle Addition Postulate* (AAP): Suppose $\angle BAC$ is given. If $P \in \mathscr{I}\angle BAC$, then $m\angle BAP + m\angle PAC = m\angle BAC$.

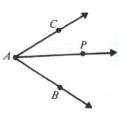

A note of caution: The name "Angle Addition Postulate" is somewhat of a misnomer. We are not adding angles; we are adding their measures.

Example 1 In the accompanying figure, each number written by a ray is the measure of the angle formed by that ray and $\overrightarrow{AB}$. Thus, for example, $m\angle BAD = 110$. From the AAP we know that

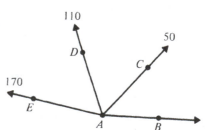

$$m\angle BAC + m\angle CAD = m\angle BAD$$

Thus

$$50 + m\angle CAD = 110$$

so that we can conclude that $m\angle CAD = 60$. Likewise, $m\angle DAE$
$= 170 - m\angle BAD = 60$.

You should now be able to prove the following theorems about angles. The proofs are analogous to those of the Segment Addition Theorem and the Segment Subtraction Theorem, with the AAP playing the role of the definition of *between*.

THEOREM 16(a)   *The Angle Addition Theorem:*
Let $\angle ABC$ and $\angle DEF$ be given. Suppose $P \in \mathscr{I}\angle ABC$ and $Q \in \mathscr{I}\angle DEF$. If $\angle ABP \cong \angle DEQ$ and $\angle PBC \cong \angle QEF$, then $\angle ABC \cong \angle DEF$.

THEOREM 16(b)   *The   Angle   Subtraction Theorem:*   Let $\angle ABC \cong \angle DEF$.   Suppose $P \in \mathscr{I}\angle ABC, Q \in \mathscr{I}\angle DEF$, and $\angle PBC \cong \angle QEF$; then $\angle PBA \cong \angle QED$.

Note that in the diagrams we have marked the congruent angles with similar scratches. We will adopt this device in future diagrams. We will vary the markings by adding more "dashes" if more than one congruence is present. For segment congruence we will adopt the practice used in the adjacent figure, where $\overline{AC} \cong \overline{BC}$.

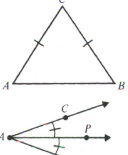

Study the accompanying diagram. Can you think of a good name for $\overrightarrow{AP}$? If you chose "angle bisector," you have chosen the name we will use.

DEFINITION 3.8   Let $\angle BAC$ be given. If $P \in \mathscr{I}\angle BAC$ and $\angle BAP \cong \angle PAC$, then $\overrightarrow{AP}$ is said to *bisect* $\angle BAC$. $\overrightarrow{AP}$ is called the *angle bisector* of $\angle BAC$.

Example 2   Assume, as in Example 1, that each number by a ray is the measure of the angle formed by that ray and AB.

Suppose that $\overrightarrow{AD}$ bisects $\angle CAE$. Then $m\angle CAD = m\angle DAE$. But $m\angle CAE = 120 - 70 = 50$. By AAP it also follows that $m\angle CAD + m\angle DAE = 50$, and since these two measures are equal, we may conclude that $m\angle CAD = m\angle DAE = 25$.

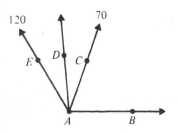

In the accompanying configuration, suppose that $\angle ABC \cong$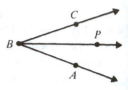
$\angle DEF$, that $\overrightarrow{BP}$ bisects $\angle ABC$, and that $\overrightarrow{EQ}$ bisects $\angle DEF$. Then it
easily follows that $m\angle PBC = \frac{1}{2}m\angle ABC$ and $m\angle QEF = \frac{1}{2}m\angle DEF$.
But $\angle ABC$ and $\angle DEF$ are congruent. Thus we may conclude that
$\angle PBC \cong \angle QEF$, and we have the following theorem.

THEOREM 17 *The Angle Bisection Theorem:* Suppose that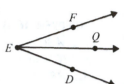
$\angle ABC \cong \angle DEF$; $\overrightarrow{BP}$ bisects $\angle ABC$, and $\overrightarrow{EQ}$ bisects $\angle DEF$. Then
$\angle PBC \cong \angle QEF$.

Just as a careful study of the Point-Plotting Postulate led you to the
Segment Construction Theorem in Chapter 2, here your efforts to
digest the Angle Construction Postulate and the definition of congruence for angles should make the proof of the following corollary
almost obvious.

COROLLARY P14 *The Angle Construction Theorem* (ACT): Let
$\angle A$ be given. Let $\overrightarrow{PB}$ be a ray on the edge of half plane $\mathcal{H}$. Then
there exists exactly one ray $\overrightarrow{PQ}$, $Q \in \mathcal{H}$ such that $\angle BPQ \cong \angle A$.

PROBLEM SET 3.4

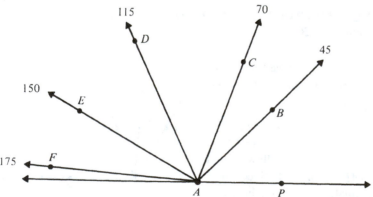

1. In the accompanying diagram, the number by each ray is the measure of the angle formed by that ray and $\overrightarrow{AP}$. Find the measure of each of the named angles:

(a) $\angle PAB$ (b) $\angle BAC$ (c) $\angle CAE$
(d) $\angle FAE$ (e) $\angle PAD$ (f) $\angle EAB$
(g) $\angle BAE$ (h) $\angle CAF$

2. Using only a pencil and a straightedge, draw angles you estimate to have measures of 30, 45, and 120. Check your guesses with a protractor.

3. When measuring an angle with a protractor, one can obtain the desired result without placing the line at the bottom of the protractor along one of the sides of the angle. Explain.

4. Prove the Angle Addition Theorem in two-column form. (*Hint:* Mimic the proof of the SAT.)

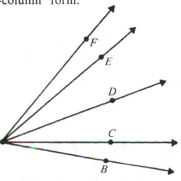

5. Given the figure shown, name the postulate definition or theorem that justifies each of the following statements:

(a) If $\angle BAC \cong \angle EFA$, then $\angle EFA \cong \angle BAC$
(b) If $\angle FAD \cong \angle EAC$, then $\angle FAE \cong \angle DAC$
(c) If $\angle EAD \cong \angle BAC$, then $\angle EAC \cong \angle BAD$
(d) If $\angle FAE \cong \angle BAC$ and $\angle EAD \cong \angle CAD$, then $\angle FAD \cong \angle BAD$
(e) If $\overrightarrow{AC}$ bisects $\angle BAD$, then $\angle BAC \cong \angle CAD$
(f) $m\angle BAE = m\angle BAD + m\angle DAE$

6. Suppose $\overrightarrow{AP}$ bisects $\angle BAC$.

(a) If $m\angle BAC = 40$, find $m\angle BAP$
(b) If $m\angle PAC = 40$, find $m\angle BAC$
(c) If $\overrightarrow{AQ}$ bisects $\angle BAP$, and $m\angle BAC = 40$, find $m\angle QAC$

7. Let $\overrightarrow{AG}$ bisect $\angle DAF$ in the diagram for problem 1. Find:

(a) $m\angle PAG$ (b) $m\angle CAG$

8. Which theorem assures us that an angle has only one angle bisector? Explain.

9. Give an informal proof of ACT.

10. Given: $\angle BAC \cong \angle DAE$, in the figure
    Prove: $\angle BAD \cong \angle CAE$ (Use two-column form.)

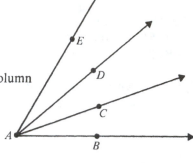

11. Given: $\overrightarrow{AD}$ bisects $\angle BAF$ and $\angle CAE$
    Prove: $\angle EAF \cong \angle CAB$

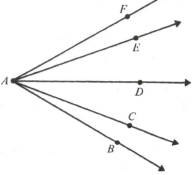

3.5 COMPLEMENTS AND PERPENDICULARITY

The trichotomy law (Section 2.5) assures us that the set of order relations between pairs of real numbers may be divided into three parts. If, for example, we pick the number 90 and consider $\angle A$, exactly one of the following statements is true:

1. $m\angle A < 90$
2. $m\angle A = 90$
3. $m\angle A > 90$

It suits our purposes to give each of these possibilities a name.

DEFINITION 3.9 Let $\angle A$ be given. If $m\angle A < 90$, $\angle A$ is an *acute* angle. If $m\angle A = 90$, $\angle A$ is a *right* angle. If $m\angle A > 90$, $\angle A$ is an *obtuse* angle.

The following theorem is an obvious consequence of the foregoing definition.

THEOREM 18   Right angles are congruent.

In each of the diagrams shown, $\angle BPA$ and $\angle APC$ are related in a like manner. They share a vertex and a side; and moreover their

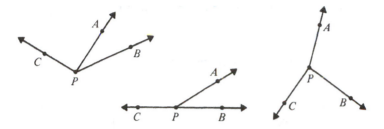

interiors do not intersect. This situation occurs so frequently that it, too, is given a name.

DEFINITION 3.10   A pair of coplanar angles are said to be *adjacent* if they have a common vertex and a common side and their interiors do not intersect. The sides that are not the common side are called the *exterior* sides. If the exterior sides are collinear, the adjacent angles are called a *linear pair*.

In the second of the preceding three illustrations, $\angle BPA$ and $\angle APC$ are a linear pair. Further details about the linear pair will be given in Section 3.6.

DEFINITION 3.11   If $m\angle BAC = 90$ or, equivalently, if $\angle BAC$ is a right angle, then $\overrightarrow{AB}$ is said to be *perpendicular to* $\overrightarrow{AC}$. In such a case we also say that $\overleftrightarrow{AB}$ is perpendicular to $\overleftrightarrow{AC}$, or that any other pair of lines, rays, or segments that are subsets of $\overleftrightarrow{AB}$ and $\overleftrightarrow{AC}$ are perpendicular.

NOTATION RULE 3.1   Suppose that $\overleftrightarrow{l}$ is perpendicular to m in the accompanying figure. We then write

$$\overleftrightarrow{l} \perp \overleftrightarrow{m}$$

As Definition 3.11 indicates, we may also write $\overline{AB} \perp \overrightarrow{PD}$, $\overline{DE} \perp \overrightarrow{PB}$, $\overrightarrow{DE} \perp \overline{AP}$, and so on.

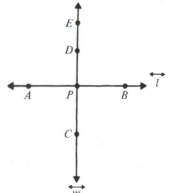

NOTATION RULE 3.2 If $\overrightarrow{AB} \perp \overrightarrow{AC}$, we will depict such a situation by the device shown in the diagram.

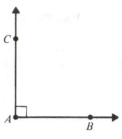

DEFINITION 3.12 Let $\angle A$ and $\angle B$ be given. If $m\angle A + m\angle B = 90$, then $\angle A$ and $\angle B$ are *complementary angles*, and each is said to be the *complement* of the other.

Example 1 In the angles shown, the degree measures are given.

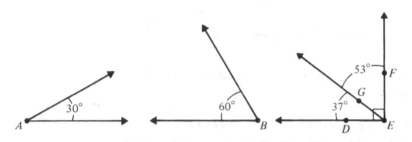

$\angle A$ is the complement of $\angle B$, and $\angle DEG$ is the complement of $\angle DEF$. The Angle Addition Postulate assures us that $m\angle DEF = 90$, and hence that $\overrightarrow{DE} \perp \overrightarrow{EF}$.

Example 2 If the measure of an acute angle is x, then the measure of its complement is $(90 - x)$. Thus, if an angle has measure 40, the measure of its complement is $(90 - 40) = 50$.

Example 3 Suppose the measure of an angle is $\frac{4}{5}$ the measure of its complement. Then, if the measure of its complement is x, we have $x + \frac{4}{5}x = 90$ or $\frac{9}{5}x = 90$. Multiplying both sides by $\frac{5}{9}$, we obtain $x = 50$. Since $\frac{4}{5}(50) = 40$, the measures of the two angles are 40 and 50.

Given angles A, B, C, and D, as shown, suppose $\angle A \cong \angle B$.

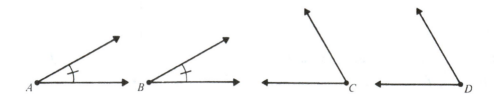

Furthermore, suppose that $\angle A$ and $\angle C$ are complementary and that $\angle B$ and $\angle D$ are complementary. Then

$$m\angle A + m\angle C = 90 = m\angle B + m\angle D$$

or,

$$m\angle A + m\angle C = m\angle B + m\angle D$$

But

$$m\angle A \qquad\qquad = m\angle B \quad (\angle A \cong \angle B)$$

so we may conclude that $m\angle C = m\angle D$.

We can summarize these ideas with a theorem.

THEOREM 19   Complements of congruent angles are congruent.

A word of caution: Like the animals on Noah's Ark, complementary angles come in pairs. Statements such as "This is a complementary angle" and "$\angle A$ and $\angle B$ and $\angle C$ are three complementary angles" have no meaning in the geometric context.

PROBLEM SET 3.5

1. Find the measure of the complement of $\angle A$ if $m\angle A$ is

    (a) 35   (b) 21.2   (c) x

    (d) $(90 - x)$ (e) $x + 12$

2. (a) Let $m\angle B = 15 + x$. For what values of x does $\angle B$ have a complement?

    (b) Can an obtuse angle have a complement? Explain.

3. Suppose $\overrightarrow{AB} \perp \overrightarrow{AC}$. Is there a third ray $\overrightarrow{AD}$ in the plane of $\overrightarrow{AB}$ and $\overrightarrow{AC}$ such that $\overrightarrow{AB} \perp \overrightarrow{AD}$? Sketch and discuss.

4. The measure of an angle is 20 less than the measure of its complement. Find the measures of both angles.

5. If $\angle A$ is the complement of $\angle B$, and $m\angle A = \frac{2}{3}m\angle B$ find $m\angle A$.

6. If $\angle A$ and $\angle B$ are complementary and the measure of $\angle A$ is four times the measure of $\angle B$, find the measure of each angle.

7. Use the accompanying diagram to prove: If two angles are complementary and adjacent, then their exterior sides are perpendicular. Use two-column form.

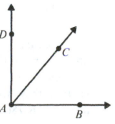

8. In a plane $m\angle BAC = 22$ and $m\angle CAD = 78$.
    (a) Are $\angle BAC$ and $\angle CAD$ complementary?
    (b) Is $\overrightarrow{AB} \perp \overrightarrow{AD}$? Explain.

9. Prove: If $m\angle A + m\angle B = 90$, then $\angle A$ is acute (*Hint:* Recall the definition of "less than.")

10. Prove T18

3.6 SUPPLEMENTS AND VERTICAL ANGLES

Just as we defined complementary angles in terms of the number 90, we use the number 180 to define *supplementary angles*.

DEFINITION 3.13   Let $\angle A$ and $\angle B$ be given. If $m\angle A + m\angle B = 180$, then $\angle A$ and $\angle B$ are called *supplementary angles*. Each of them is said to be the *supplement* of the other.

Example 1   In the figure shown, the degree measures of the angles are given. $\angle A$ and $\angle B$ are supplementary, as are $\angle EPC$ and

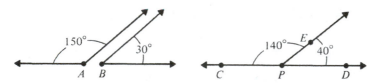

$\angle DPE$. In addition, $\angle EPC$ and $\angle DPE$ are adjacent. From the diagram, it appears that $\angle EPC$ and $\angle DPE$ are a linear pair. Is a linear pair always supplementary? We cannot prove such a conjecture from the previous postulates. Therefore, we will adopt it as a postulate.

POSTULATE 16   *The Supplement Postulate:* If two angles are a linear pair, then they are supplementary.

Of course, two angles may be supplementary without being a linear pair (for that matter, one may be in Kalamazoo and the other in Honolulu); if they are a linear pair, the Supplement Postulate assures us that they are supplementary, and, therefore, the sum of their measures is 180.

Example 2 In the accompanying figure, $\angle A$ and $\angle B$ are

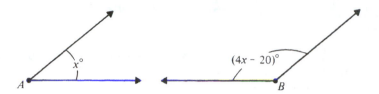

supplementary. The degree measures are given. To find $m\angle A$, we note that

$$m\angle A + m\angle B = 180$$

Therefore,

$$x + 4x - 20 = 180$$
$$5x = 200$$

which implies that

$$x = 40 \quad \text{or} \quad m\angle A = 40$$

Suppose $\angle C$ and $\angle D$ are supplementary and congruent. Then, letting $m\angle C = m\angle D = x$, we have

$$x + x = 180 \quad \text{or} \quad x = 90$$

Thus $\angle C$ and $\angle D$ are right angles, and we may state Theorem 20.

THEOREM 20 If two angles are supplementary and congruent, then each is a right angle.

In Section 3.5 we proved that complements of congruent angles are congruent. The following theorem may be proved in similar fashion.

THEOREM 21 Supplements of congruent angles are congruent.

Example 3 A special case of T21 is illustrated in the diagram, where we are given that $\angle ABE$ is the supplement of $\angle BED$. Then, since

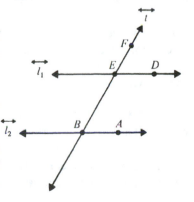

$\angle BED$ and $\angle DEF$ are a linear pair, it follows that $\angle BED$ is the supplement of $\angle DEF$. But $\angle BED \cong \angle BED$. Therefore, by T21, $\angle ABE \cong \angle DEF$.

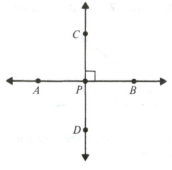

Suppose $\overleftrightarrow{AB} \perp \overleftrightarrow{CD}$ and $\overleftrightarrow{AB} \cap \overleftrightarrow{CD} = \{P\}$, as in the figure shown. Then, the definition of *perpendicular* assures us that one of the angles in the figure, $\angle BPC$ say, is a right angle. $\angle DPB$ and $\angle BPC$ are a linear pair and thus, by the Supplement Postulate, are supplementary. Then, it easily follows that $\angle DPB$ is *also* a right angle. Considering the two linear pairs, $\angle BPC$ and $\angle CPA$, and $\angle APD$ and $\angle DPB$, it can be established in a similar fashion that both $\angle CPA$ and $\angle APD$ are right angles. We can summarize these ideas as a theorem.

THEOREM 22 If two lines are perpendicular, then they form four right angles.

When a pair of lines intersect as in the figure given, four linear pairs of angles are formed. $\angle APC$ and $\angle CPB$ are one of these pairs. Other pairs of angles that are *not* linear pairs are also formed: $\angle CPB$ and $\angle DPA$, for example. Such angles are called *vertical angles*. The formal definition of this concept follows.

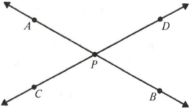

DEFINITION 3.14 Let $\overleftrightarrow{AB} \cap \overleftrightarrow{CD} = \{P\}$. Suppose $A\text{-}P\text{-}B$ and $C\text{-}P\text{-}D$. Then $\angle CPB$ and $\angle DPA$ are called *vertical angles*. $\angle APC$ and $\angle BPD$ are also vertical angles.

Like complementary angles and supplementary angles, vertical angles, too, come only in pairs.

One of the most useful theorems you will meet in this book concerns vertical angles.

THEOREM 23 *The Vertical Angle Theorem* (VAT): If two angles are vertical angles, then they are congruent.

The proof of this theorem is an easy consequence of the Supplement Postulate and T21 and is left as an exercise.

The VAT has been attributed by many to Thales of Miletus (*ca.* 624–548 B.C.), who, some scholars believe, was the first mathematician to apply the deductive method to geometry.

PROBLEM SET 3.6

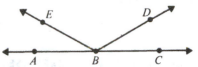

1. In the figure, *A-B-C*. Name as many pairs of supplementary angles as you can.

2. Fill in the blanks:

    (a) The complement of an acute angle is a(n) _____ angle.
    (b) The supplement of an acute angle is a(n) _____ angle.

3. $\angle A$ and $\angle B$ are supplementary. Find $m\angle A$ if $m\angle B$ is

    (a) 135
    (b) 88.7
    (c) x

    (d) $180 - x$
    (e) $90 - x$
    (f) $x + 100$

4. (a) Let $m\angle A = x - 10$. For what values of x does $\angle A$ have a supplement?
    (b) Can the measure of the supplement of an angle ever be twice the measure of its complement? Explain.

5. If the measure of an angle is 40 greater than the measure of its supplement, find the measure of both angles.

6. $\angle A$ and $\angle B$ are supplementary. $m\angle A = \frac{2}{3}m\angle B$. Find $m\angle B$.

7. Prove T21

8. Given: $\overleftrightarrow{l}$, $\overrightarrow{PA}$, and $\overrightarrow{PB}$, as in the accompanying figure; and $\angle 3 \cong \angle 4$
    Prove: $\angle 1 \cong \angle 2$

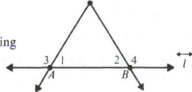

9. Given: $\overleftrightarrow{l}$, $\overrightarrow{PQ}$; and $\angle 1 \cong \angle 2$
    Prove: $\angle 1$ is a right angle

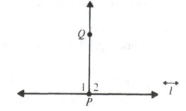

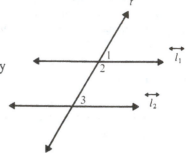

10. Given: $\overset{\leftrightarrow}{l_1}$, $\overset{\leftrightarrow}{l_2}$, $\overset{\leftrightarrow}{t}$; and $\angle 1 \cong \angle 3$

Prove: $\angle 3$ and $\angle 2$ are supplementary

11. Prove T23

CHAPTER 3: REVIEW EXERCISES

1. For each of the following statements indicate whether it is *always* true, *sometimes* but not always true, or *never* true.

   (a) The union of two half planes is a plane.

   (b) $\overrightarrow{AB} \cap \overrightarrow{AC}$ is an angle.

   (c) The complement of a convex set is a convex set.

   (d) Let $\overrightarrow{AP} \perp \overrightarrow{AQ}$. If $Q \in \mathscr{I} \angle PAR$, then $\angle QAR$ is acute.

   (e) If $\angle A$ and $\angle B$ are a linear pair, then $m\angle A + m\angle B = 180$

   (f) If two angles are supplementary, then they are a linear pair.

   (g) If $\angle A$ is an obtuse angle, then $\angle A$ has a complementary angle.

   (h) Suppose $P \in \mathscr{I} \angle BAC$; if $\angle BAP \cong \angle PAC$, then $\overset{\leftrightarrow}{AP} \perp \overset{\leftrightarrow}{BC}$

   (i) $m\angle BAC + m\angle CAD = m\angle BAD$

   (j) Two coplanar angles are adjacent if they have a common vertex and a common side.

   (k) The interior of an angle is a subset of the angle.

   (l) If two angles are a linear pair and congruent, then each is a right angle.

2. In the diagram, name:

   (a) A pair of complementary angles.

   (b) Two pairs of supplementary angles.

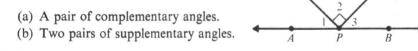

3. Sketch a figure showing that the union of two convex sets is not necessarily a convex set.

4. Draw a diagram to show that the sum of the measures of a pair of adjacent angles (not a linear pair) is not necessarily the measure of an angle.

5. Given: $\angle BAD \cong \angle CAE$

   Prove: $\angle BAC \cong \angle DAE$ (Do not use AAP!)

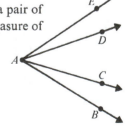

6. Given: Four rays (no two of which are collinear) with a common endpoint. How many angles are formed by these four rays? [You might want to consult the fifth row of the chart you constructed in problem 4(a) of Problem Set 1.6.]

7. Prove: If two angles are congruent and supplementary, then each is a right angle.

8. Find the measures of the complements and supplements of angles with the following measures:

    (a) 45   (b) 60
    (c) 75   (d) $90 + x$

9. $m\angle A = 2x + 14$. For what values of x will $\angle A$ have each of the following?

    (a) A complement.
    (b) A supplement.

10. The measures of an angle and its supplement differ by 18. What are the measures of these two angles?

11. Given: $\angle 1 \cong \angle 2$
    Prove: $\angle 1 \cong \angle 3$

12. Prove the Angle Subtraction Theorem.

My own suspicion is that the universe is not only queerer than we suppose, but queerer than we can suppose.

J. B. S. HALDANE (1892–1964)

CHAPTER FOUR

TRIANGLES

4.1 INTRODUCTION

In a dialogue entitled the *Timaeus*, Plato developed a theory of matter that held that everything was composed of triangles. Plato depicted the process of metabolism as a battle in which the triangles of the body cut up into their own likenesses the triangles of the food taken in. As we grow older, our triangles begin to lose this battle to these ingested invaders and we gradually waste away. This very early example of applied mathematics would hardly satisfy a modern physiologist; but then the physiologist, as J. B. S. Haldane suggests, would probably view the mathematical models he himself uses as simply tentative.

This perception of the triangle as some kind of primordial figure was not unique with Plato. For example, the figure takes on considerable religious significance in early Hindu meditation ornaments as well as in Christian representations of the Trinity. Even in a mathematical sense, the triangle can be considered to be the most primitive of all the plane figures that enclose convex sets. In the following pages, we will explore the *mathematical properties* of triangles and the relationships between them. We will also examine some applications of these properties and relationships.

4.2 TRIANGLES AND CONGRUENCE

We begin this section by putting together some of the vocabulary and notation we have learned in previous chapters in order to formulate a definition of "triangle."

DEFINITION 4.1   Let A, B, and C be three noncollinear points. *Triangle ABC*, denoted $\triangle ABC$, is the union of the segments $\overline{AB}$, $\overline{BC}$, and $\overline{CA}$ determined by the three noncollinear points. Each of the points A, B, and C is called a *vertex* (plural vertices) of $\angle ABC$, and each of $\overline{AB}$, $\overline{BC}$, and $\overline{CA}$ is called a *side* of $\triangle ABC$. The *angles* of $\triangle ABC$ are $\angle A$, $\angle B$, and $\angle C$. The angles and the sides of $\triangle ABC$ are called the *parts* of $\triangle ABC$.

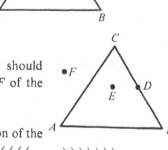

A word of caution: Although $\angle B$, for example, is called an angle *of* the triangle, it is not a subset of the triangle. It is comprised of the points illustrated in the accompanying diagram by the "dashed portion" of the rays $\overrightarrow{BA}$ and $\overrightarrow{BC}$ as well as two sides of the triangle, $\overline{AB}$ and $\overline{BC}$. When referring to the angles of a triangle, it is customary not to include these "dashed portions" in the illustration of the triangle.

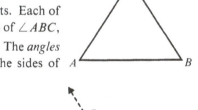

In Chapter 3 we defined the interior of an angle. You should have little trouble guessing which of the points D, E, and F of the adjacent figure are in the *interior* of $\triangle ABC$.

DEFINITION 4.2   The *interior of a triangle* is the intersection of the interiors of any two of its angles. The points that are neither elements of the triangle nor its interior comprise the *exterior of the triangle*.

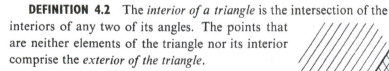

NOTATION RULE 4.1   If P is in the interior of $\triangle ABC$, we write $P \in \mathscr{I} \triangle ABC$. If Q is in the exterior of $\triangle ABC$, we write $Q \in \mathscr{E} \triangle ABC$.

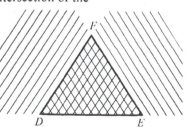

Example 1 Given $\triangle ABC$ as in the figure, and points P, Q, and R, the following statements are all true:

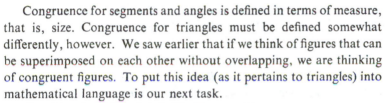

A is a vertex of $\triangle ABC$

$R \in \triangle ABC$

$P \in \mathscr{I}\,\triangle ABC$

$Q \in \mathscr{E}\,\triangle ABC$

$\overline{AB}$ is a side of $\triangle ABC$

$\angle BAC$ is an angle of $\triangle ABC$

Congruence for segments and angles is defined in terms of measure, that is, size. Congruence for triangles must be defined somewhat differently, however. We saw earlier that if we think of figures that can be superimposed on each other without overlapping, we are thinking of congruent figures. To put this idea (as it pertains to triangles) into mathematical language is our next task.

Consider $\triangle ABC$ and $\triangle DEF$. The matching

$$A \leftrightarrow D$$

$$B \leftrightarrow E$$

$$C \leftrightarrow F$$

is only one of a number of matchings between their vertices. Once we mention such a correspondence between the vertices of two triangles, there is a natural correspondence induced between their sides. If $A \leftrightarrow D$ and $B \leftrightarrow E$, it seems natural to let $\overline{AB} \leftrightarrow \overline{DE}$, since these are the sides which have the matching vertices as endpoints. Thus, given the foregoing correspondence between the vertices of $\triangle ABC$ and $\triangle DEF$ (which we denote $ABC \leftrightarrow DEF$), the following correspondences are induced between the sides:

$$\overline{AB} \leftrightarrow \overline{DE}$$

$$\overline{BC} \leftrightarrow \overline{EF}$$

$$\overline{AC} \leftrightarrow \overline{DF}$$

In an obvious way, the one-to-one correspondence

$$\angle A \leftrightarrow \angle D$$

$$\angle B \leftrightarrow \angle E$$

$$\angle C \leftrightarrow \angle F$$

between angles can also be established.

When we mention "corresponding angles" or "corresponding sides," we will always be speaking of such induced correspondences. Furthermore, if the corresponding angles and corresponding sides are congruent, we say that *the triangles are congruent.*

DEFINITION 4.3   Given two triangles, or a triangle and itself, and a one-to-one correspondence between their vertices; if the corresponding angles and corresponding sides are congruent, then the *triangles are congruent.*

If we are given $\triangle ABC$, $\triangle DEF$, and $ABC \leftrightarrow DEF$, the statement "$\triangle ABC$ is congruent to $\triangle DEF$," denoted $\triangle ABC \cong \triangle DEF$, means that all the following conditions hold:

$$\angle A \cong \angle D \qquad \overline{AB} \cong \overline{DE}$$

$$\angle B \cong \angle E \qquad \overline{BC} \cong \overline{EF}$$

$$\angle C \cong \angle F \qquad \overline{AC} \cong \overline{DF}$$

Notice that in the statement "$\triangle ABC \cong \triangle DEF$" the vertices of the corresponding angles are written in the corresponding positions in the

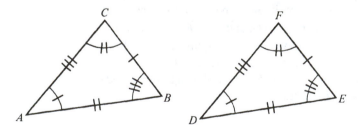

names of the two triangles. Such a situation is depicted in the figure shown. The angles or segments with identical markings are the corresponding parts.

THEOREM 24 (a) Triangle congruence is reflexive.

(b) Triangle congruence is symmetric.

(c) Triangle congruence is transitive.

The following expansion of our vocabulary will facilitate future discussions.

DEFINITION 4.4 A *side* of a triangle is *opposite an angle* of that triangle if the vertex of the angle is not on that side. *Two sides* of a triangle *include an angle* of that triangle if the vertex of the angle is the common endpoint of those two sides. *Two angles* of a triangle *include a side* of that triangle if the vertices of the two angles are the endpoints of the side.

Example 2   Given $\triangle ABC$, we may say that

$\angle A$ is opposite $\overline{BC}$

$\overline{AB}$ is opposite $\angle C$

$\overline{AB}$ is included by $\angle A$ and $\angle B$

$\angle A$ and $\angle C$ include $\overline{AC}$

$\angle A$ is included by $\overline{AC}$ and $\overline{AB}$

$\overline{AB}$ and $\overline{BC}$ include $\angle B$

PROBLEM SET 4.2

1. Given the figure, complete each of the following statements.

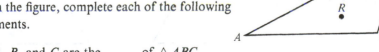

(a) A, B, and C are the _____ of $\triangle ABC$.

(b) $R \in$ _____ whereas $P \in$ _____.

(c) Q is _____ A and C.

(d) $\angle A$ is included by sides _____ and _____.

(e) $\overline{BC}$ is opposite _____.

(f) P and R are on _____ of $\overset{\leftrightarrow}{AC}$.

(g) _____ and _____ include the side $\overline{AC}$.

2. Given $\triangle ABC$, it follows that $AB + BC \neq AC$.

    (a) If the above sum were equal to AC, two of our definitions would conflict. Which two? Why?

    (b) The trichotomy law tells us that, in this case, either $AB + BC < AC$, or $AB + BC > AC$. Which of these inequalities do you think holds?

3. Do you think a triangle is a convex set? Why or why not?

4. Given $\triangle ABC$ and $\triangle DEF$, in how many distinguishably different ways might these triangles be congruent? (*Hint:* you have already done this problem in a different context in Chapter 1.)

5. Supply the missing reasons in the following proof of T24(a): Triangle congruence is reflexive.

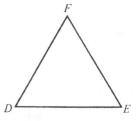

        Given: $\triangle ABC$

        Prove: $\triangle ABC \cong \triangle ABC$

STATEMENTS	REASONS
1. $\overline{AB} \cong \overline{AB}$; $\overline{BC} \cong \overline{BC}$; $\overline{AC} \cong \overline{AC}$	1.
2. $\angle A \cong \angle A$; $\angle B \cong \angle B$; $\angle C \cong \angle C$	2.
3. $\triangle ABC \cong \triangle ABC$	3.

6. (a) Prove T24(b)

    (b) Prove T24(c)

7. Given the following pairs of congruent triangles, state the pairs of corresponding parts:

(a)

$\triangle ABC \cong \triangle DEF$

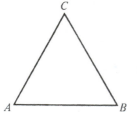

(b)

$\triangle PRU \cong \triangle SQT$

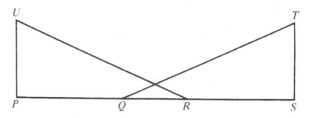

(c)

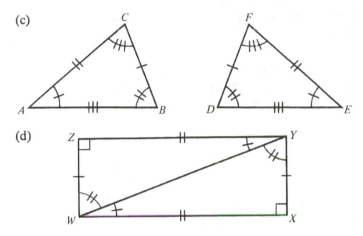

(d)

(e) △ *HKL* ≅ △ *EFG* (The congruence statement has a geometry of its own.)

(f) △ *ABC* ≅ △ *BAC*

4.3 TRIANGLE CONGRUENCE SIMPLIFIED, I

In order for two triangles to be congruent, six simpler congruence relations must hold. The verification of the existence of the six congruences can be a tedious matter, if indeed it is possible at all. Look at the pair of triangles shown, which have $\angle A \cong \angle A'$ (read *A*-prime), $\overline{AC} \cong \overline{A'C'}$, and $\overline{AB} \cong \overline{A'B'}$. Experience tells us that $\overline{BC}$ *should be* congruent to $\overline{B'C'}$, and that $\angle B$ and $\angle C$ *should be* congruent

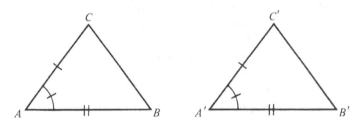

to $\angle B'$ and $\angle C'$, respectively, and that there is no other possibility. Make some sketches of your own if you have doubts. Then the following postulate will seem more natural.

POSTULATE 17 *The Side-Angle-Side Postulate* (SAS): Given two triangles, or one triangle and itself, and a one-to-one correspondence between their vertices such that two sides and the included angle in one triangle are congruent to the corresponding parts of the other triangle, then the triangles are congruent.

It is important to stress that in order for the SAS postulate to apply, the angle must be included by the sides. In the accompanying diagrams, one of the three triangles is not congruent to either of the others.

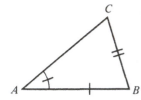 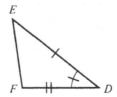 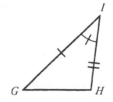

Example Suppose $\overleftrightarrow{CD}$ is perpendicular to $\overline{AB}$ at D. Furthermore, suppose $\overleftrightarrow{CD}$ bisects $\overline{AB}$. We can show that $\overline{AC} \cong \overline{BC}$ in the following manner.

Proof:

STATEMENTS	REASONS
1. $\overleftrightarrow{CD} \perp \overline{AB}$ at D; $\overleftrightarrow{CD}$ bisects $\overline{AB}$	1. Given
2. $\angle CDA$ and $\angle CDB$ are right $\angle$s	2. Perpendicular lines form four right $\underline{\angle}$s [T22 (Section 3.6)]
3. $\angle CDA \cong \angle CDB$	3. Why?
4. $\overline{AD} \cong \overline{BD}$	4. Definition of bisect
5. $\overline{CD} \cong \overline{CD}$	5. Why?
6. $\triangle ADC \cong \triangle BDC$	6. SAS
7. $\overline{AC} \cong \overline{BC}$	7. CPCTC[1]

You will find the SAS postulate to be very useful. Before considering some easy but important applications of the SAS postulate, we need some further definitions. Sketch these sets of points as you first read their definitions.

---

[1] CPCTC is an abbreviation for the sentence "Corresponding parts of congruent triangles are congruent." We will employ it often.

DEFINITION 4.5   A triangle is a *right triangle* if one of its angles is a right angle. The sides which include the right angle are called the *legs* of the triangle. The side opposite the right angle is called the *hypotenuse*.

DEFINITION 4.6   A triangle is an *acute triangle* if each of its angles is an acute angle. If a triangle has an obtuse angle, it is an *obtuse triangle*. If all three angles of a triangle are congruent, it is said to be *equiangular*.

DEFINITION 4.7   A triangle is *isosceles* if it has a pair of congruent sides. The angle included by these sides is called the *vertex angle*; its vertex is called *the vertex* of the triangle. The side opposite the vertex angle is called the *base*. The angles that include the base are called the *base angles*. A triangle is *scalene* if it is not isosceles. If all three sides of a triangle are congruent, it is said to be *equilateral*.[2]

From Definition 4.7 it should be clear that every equilateral triangle is isosceles, though not every isosceles triangle is equilateral.

The accompanying diagram portrays a situation that leads to our first corollary of the SAS postulate.

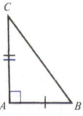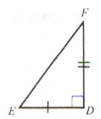

COROLLARY P17   If two legs of one right triangle are congruent to the corresponding legs of another right triangle, then the triangles are congruent.

Another corollary to P17 has been credited to Thales, the first Greek mathematician, and is known as the *Isosceles Triangle Theorem*. The proof given here is essentially that of Pappus of Alexandria (*c.* 300 A.D.), the last great Greek mathematician of the ancient period. It involves a congruence of a triangle with itself that is *not* reflexive. That there is more than one matching between the set of vertices of a triangle and itself should be clear to you. (See problem 4 of Problem Set 4.2.)

THEOREM 25   *The Isoceles Triangle Theorem* (ITT): If two sides of a triangle are congruent, then the angles opposite them are congruent.

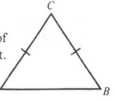

Hypothesis: $\overline{AC} \cong \overline{BC}$
Conclusion: $\angle B \cong \angle A$

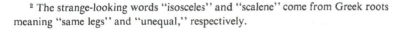

[2] The strange-looking words "isosceles" and "scalene" come from Greek roots meaning "same legs" and "unequal," respectively.

Proof:

STATEMENTS	REASONS
1. $\overline{AC} \cong \overline{BC}$	1. Given
2. $\overline{BC} \cong \overline{AC}$	2. Why?
3. $\angle C \cong \angle C$	3. Why?
4. $\triangle ACB \cong \triangle BCA$	4. SAS
5. $\angle B \cong \angle A$	5. Why?

The ITT is often stated in the following way:

The base angles of an isosceles triangle are congruent.

Our next theorem can be proven using the SAS postulate and T20 (Section 3.6).

THEOREM 26   The bisector of the vertex angle of an isosceles triangle bisects and is perpendicular to the base of the triangle.

The adjacent diagram for the proof of this theorem is based on the assumption that the angle bisector intersects the base. That this *is* the case does follow from our postulates.[3] Students are usually unaware that they are making such an assumption, and we will proceed without raising the question again.

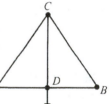

PROBLEM SET 4.3

1. In which of the following diagrams can the SAS postulate be used to prove pairs of triangles are congruent? State the triangle congruence in each case where it exists.

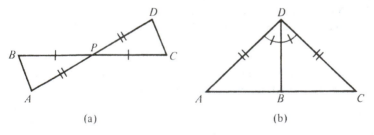

(a)   (b)

[3] See E. E. Moise, *Elementary Geometry from an Advanced Standpoint* (Reading, Mass.: Addison-Wesley Publishing Co., 1963), p. 69, for a proof of what is often referred to as the "Cross-bar Theorem."

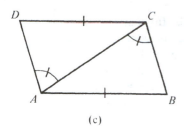

(c)

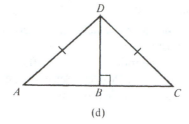

(d)

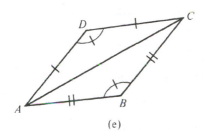

(e)

2. The set of acute triangles and the set of obtuse triangles are disjoint sets. Use Definition 4.6 to explain why.

In problems 3–5, use the accompanying diagram.

3. If $\overline{AE} \cong \overline{DE}$, prove that $\angle 1 \cong \angle 2$.

4. If $\triangle BEC$ is isosceles with vertex E, and $\overline{AB} \cong \overline{DC}$, prove that $\angle AEB \cong \angle DEC$.

5. If $\overline{AB} \cong \overline{DC}$, $\overline{BE} \cong \overline{DE}$, prove that $\triangle AED$ is isosceles.

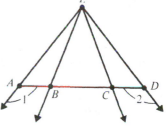

6. If $\overline{AB}$ and $\overline{CD}$ bisect each other at P, prove that $\triangle APC \cong \triangle BPD$. Provide your own diagram.

7. By means of successive applications of the ITT, prove that every equilateral triangle is equiangular.

8. Given: $\triangle ABC$; $\overline{AC} \cong \overline{BC}$; $\overline{AN}$ bisects $\overline{BC}$; and $\overline{BM}$ bisects $\overline{BC}$
Prove: $\overline{AN} \cong \overline{BM}$

9. Prove Corollary P17

10. Prove T26

11. State the converse of the ITT. Do you think such a statement is true?

12. Supply the missing reasons in the proof of T25.

4.4 TRIANGLE CONGRUENCE SIMPLIFIED, II

The SAS postulate provided a subset of the six fundamental congruence relations necessary to define a congruence between two triangles. The SAS postulate itself can be used to obtain other such subsets. Two of these (one follows from the other) are introduced in this section without proof. In Problem Set 4.4, you will be asked to provide some missing details in the proof of the first of these theorems.

THEOREM 27 *The Angle-Side-Angle Theorem* (ASA): Given two triangles, or a triangle and itself, and a one-to-one correspondence between their vertices. If two angles and the included side of the first triangle are congruent to the corresponding parts of the second triangle, then the triangles are congruent.

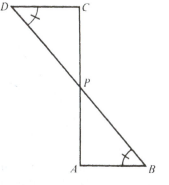

Example 1 In the figure, $\overline{AC}$ bisects $\overline{BD}$ at P, and $\angle D \cong \angle B$. Prove that $\triangle PDC \cong \triangle PBA$.

Proof:

STATEMENTS	REASONS
1. $\overline{AC}$ bisects $\overline{BD}$ at P; $\angle D \cong \angle B$	1. Given
2. $\overline{DP} \cong \overline{BP}$	2. Definition of bisect
3. $\angle DPC \cong \angle BPA$	3. Vertical Angle Theorem (VAT)
4. $\triangle PDC \cong \triangle PBA$	4. ASA

In problem 11 of Problem Set 4.3, you stated the converse of the Isosceles Triangle Theorem (ITT) and gave your opinion as to the legitimacy of such a conjecture. The ASA theorem can be used to substantiate this conjecture, which we now state as a theorem.

THEOREM 28 If two angles of a triangle are congruent, then the sides opposite those angles are congruent. In terms of the adjacent diagram, we have:

Hypothesis: $\triangle ABC$ with $\angle A \cong \angle B$
Conclusion: $\overline{AC} \cong \overline{BC}$

The proof of this theorem, like the ITT, involves a congruence of $\triangle ABC$ with itself that is not reflexive. We leave this proof for an exercise.

More difficult to prove than the ASA theorem is another basic triangle congruence proposition.

THEOREM 29   *The Side-Side-Side Theorem* (SSS): Given two triangles, or a triangle and itself, and a one-to-one correspondence between their vertices. If the three sides of one triangle are congruent to the corresponding sides of the other triangle, then the triangles are congruent.[4]

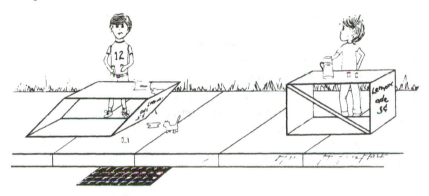

Example 2   In the figure, the salesman on the right is smiling because he knows (and the SSS theorem affirms) that the triangles supporting his lemonade will not let him down the way the four-sided support did his competitor on the left.

Example 3   Given $\overline{AB} \cong \overline{CD}$, and $\overline{AD} \cong \overline{CB}$
Prove: $\triangle ABD \cong \triangle CDB$

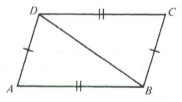

Proof:

STATEMENTS	REASONS
1. $\overline{AB} \cong \overline{CD}$; $\overline{AD} \cong \overline{CB}$	1. Given
2. $\overline{DB} \cong \overline{BD}$	2. Reflexive property
3. $\triangle ABD \cong \triangle CDB$	3. SSS

Often, the diagram accompanying a problem must be augmented before a clear solution comes into view. This is usually done by

[4] A proof of this theorem can be found in Moise, *Elementary Geometry*, pp. 87–89.

sketching into the figure what are called *auxiliary sets* of points. One should always be able to give the justification for such additions, as in the following example.

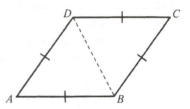

Example 4   Given $\overline{AB} \cong \overline{BC} \cong \overline{CD} \cong \overline{DA}$
Prove: $\angle A \cong \angle C$

Proof:

STATEMENTS	REASONS
1. $\overline{AB} \cong \overline{BC} \cong \overline{CD} \cong \overline{DA}$	1. Given
2. Introduce $\overline{DB}$	2. The Line Postulate, definition of segment
3. $\overline{DB} \cong \overline{BD}$	3. Why?
4. $\triangle ABD \cong \triangle CDB$	4. Why?
5. $\angle A \cong \angle C$	5. Why?

Lines, segments, and rays are frequently introduced as *auxiliary sets* through application of the Line Postulate, the Segment Construction Theorem, or the Angle Construction Theorem.

PROBLEM SET 4.4

1. In each of the following diagrams can SAS, ASA, or SSS be used to prove pairs of triangles are congruent? State the triangle congruence in each case it exists. If none of these applies, so indicate.

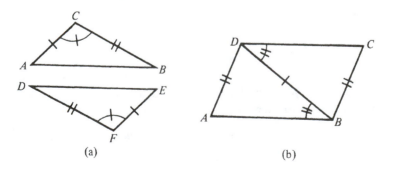

(a)   (b)

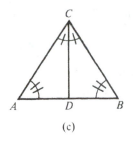

(c)

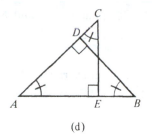

(d)

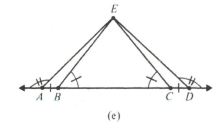

(e)

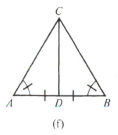

(f)

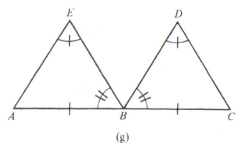

(g)

2. Use the converse of the ITT to prove that if a triangle is equi-angular, it is equilateral.

3. Given: $\triangle ADE$; $\angle BAE \cong \angle CDE$; and $\overline{AC} \cong \overline{BD}$
   Prove: $\triangle BEC$ is isosceles

4. If $\overline{AB}$ and $\overline{CD}$ bisect each other at P, and $\overline{AC} \cong \overline{BD}$, prove (*without using VAT*) that $\triangle APC \cong \triangle BPD$. Provide your own diagram.

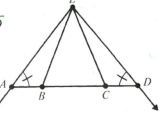

5. Given: $\triangle ABC$ and $\overline{AC} \cong \overline{BC}$; let M be the midpoint of $\overline{AB}$.
   Prove: $\angle ACM \cong \angle BCM$

6. Given: $\triangle ABC$; $\overline{AC} \cong \overline{BC}$; $\overrightarrow{AN}$ and $\overrightarrow{BM}$ bisect $\angle A$ and $\angle B$, respectively; A-M-C; and B-M-C
   Prove: $\overline{AN} \cong \overline{BM}$

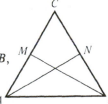

7. Given: $\triangle ABC$; $\overline{AC} \cong \overline{BC}$; and D, M, and N are the midpoints of $\overline{AB}$, $\overline{AC}$, and $\overline{BC}$, respectively
   Prove: $\angle ADM \cong \angle BDN$

8. Given: $\triangle ABC$ is isosceles with vertex C. If D, M, and N are the midpoints of $\overline{AB}$, $\overline{AC}$, and $\overline{BC}$, respectively, do you think that $\overline{MD} \cong \overline{CN}$? Can you prove your assertion?

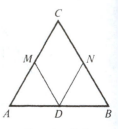

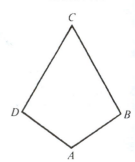

Exercises 7, 8

9. Given: In the figure, $\overline{DC} \cong \overline{BC}$, and $\overline{AD} \cong \overline{AB}$

   (a) Prove: $\angle D \cong \angle B$
   (b) Why is it not necessarily true that $\angle A \cong \angle C$?

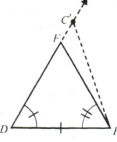

10. In terms of the accompanying diagrams, we can restate the ASA theorem.

   Given: $\triangle ABC$, $\triangle DEF$, and $ABC \leftrightarrow DEF$.
   If $\angle A \cong \angle D$, $\angle B \cong \angle E$, and $\overline{AB} \cong \overline{DE}$, then $\triangle ABC \cong \triangle DEF$.

   Supply the missing details in the proof below.

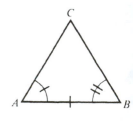

STATEMENTS	REASONS
1. $\angle A \cong \angle D$; $\overline{AB} \cong \overline{DE}$	1. Given
2. There is a point C' on $\overrightarrow{DF}$ such that $\overline{DC'} \cong \overline{AC}$	2.
3. $\triangle ABC \cong \triangle DEC'$	3.
4. $\angle ABC \cong \angle DEC'$	4.
5. $\overrightarrow{EF} = \overrightarrow{EC'}$	5. Angle Construction Theorem
6. $F = C'$	6.
7. Therefore, $\overline{EF} \cong \overline{BC}$	7. CPCTC and S6
8.	8. SAS

4.5 INDIRECT PROOF AND SOME NEW VOCABULARY

Some of the theorems introduced in previous sections were written as implications. We could write all our theorems in this form, but too often a loss of clarity and elegance would ensue. To keep our language from becoming too stilted and contrived, we write implications in other forms: forms that require some care on our part if we are to distinguish between hypothesis and conclusion.

Consider, now, the "theorem"

$$p \rightarrow q$$

In many cases, the hypothesis, p, is actually a long list of assumptions, that is, a conjunction. Such a proposition becomes false when one or more of its parts is false (recall the table defining "$\wedge$"). In Chapter 1, we saw that

$$p \rightarrow q$$
and
$$\sim q \rightarrow \sim p$$

are equivalent sentences. Therefore, if by denying q, we can deduce a denial of one of our list of assumptions (and hence of p), we will have proven the theorem

$$p \rightarrow q$$

To employ this tactic is to use the most common technique of indirect proof.[5] We begin by assuming that the conclusion we wish to derive is false, and proceed to deduce a contradiction of the hypothesis.

NOTATION RULE 4.2 In an indirect proof of a theorem, we will place the symbol $\rightarrow$/$\leftarrow$ following the derived contradiction, thus indicating that the theorem has been proven.

All methods of indirect proof are based on the *law of the excluded middle*, which states that a proposition is true or false. Symbolically this law may be written

$$p \vee \sim p$$

The validity of arguments based on this law has been denied by a school of twentieth-century mathematicians called the *intuitionists*.

---

[5] See Alfred Tarski, *Introduction to Logic* (New York: Oxford University Press, 1965), pp. 157–159.

Most mathematicians, however, continue to use this assumption and feel at ease with it. Hopefully, the following examples will leave you feeling just as comfortable.

Example 1 In the figure, $\angle A \cong \angle B$. Prove that $\overline{AC} \not\cong \overline{BC}$.

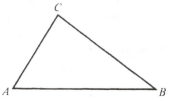

Proof: Assume $\overline{AC} \cong \overline{BC}$. Then, by the Isosceles Triangle Theorem, it follows that $\angle A \cong \angle B$. →|←

Our second example is the long-delayed proof of our very first theorem.

THEOREM 1 If two lines intersect, their intersection contains exactly one point.

Proof: Assume their intersection does *not* contain exactly one point. Then, by the trichotomy law, there are two possibilities. Each of these possibilities leads us to a contradiction, as follows:

(i) Assume their intersection contains no points. Then by the definition of intersect, they do not intersect. →|←

(ii) Assume that their intersection contains more than one point. Then it contains two distinct points. Call these two points P and Q. Since P and Q are in the intersection, P and Q are in *both* of the *two* lines. But the Line Postulate informs us that there is *only one* line that contains P and Q. →|←

The first part of the proof of T1 established the *existence* of a point in the intersection of two lines; the second part affirmed the *uniqueness* of such a point. Mathematical statements like T1 are referred to as *existence and uniqueness* statements. The technique of *indirect proof* is frequently applied to the verification of the uniqueness part of such statements.

In Chapter 3, the Angle Construction Postulate and its corollary, the ACT, were established. Both of these are existence and uniqueness statements. The Angle Construction Postulate is helpful in establishing the following result.

THEOREM 30 Let $\overleftrightarrow{l}$ be given, $\overleftrightarrow{l}$ in plane m. Let $P \in \overleftrightarrow{l}$. Then there is one and only one line in plane m that is perpendicular to $\overleftrightarrow{l}$ at P.

Proof: Let $\mathcal{H}$ be either of the half planes of *m*

determined by $\overleftrightarrow{l}$. Let $A \in \overleftrightarrow{l}$, $A \neq P$. Then, by the
Angle Construction Postulate, there exists *exactly*

one ray, $\overrightarrow{PQ}$, $Q \in \mathcal{H}$, such that $m\angle APQ = 90$.

Thus $\overrightarrow{PQ} \perp \overleftrightarrow{l}$, from which it follows that $\overleftrightarrow{PQ} \perp \overleftrightarrow{l}$.

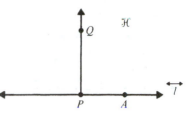

If the other half plane determined by $\overleftrightarrow{l}$ were selected, the same line

$\overleftrightarrow{PQ}$ would be obtained. The proof of this is left as an exercise.

DEFINITION 4.8 A *median of a triangle* is a segment joining a
vertex of that triangle to the midpoint of the opposite side.

An *altitude of a triangle* is a segment which joins a vertex of that
triangle to the line containing the opposite side and which is perpendicular to that line. The opposite side is called the *base*, with reference
to that altitude.

An *angle bisector of a triangle* is that portion of an angle bisector of
any of its angles which joins a vertex of the triangle to the opposite side.

Example 2

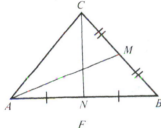

$\overline{AM}$ and $\overline{CN}$ are medians of
$\triangle ABC$.

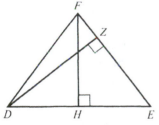

$\overline{FH}$ is an altitude to $\overline{DE}$ as a base.
$\overline{DZ}$ is an altitude to $\overline{EF}$ as a base.

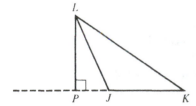

$\overline{LP}$ is an altitude of $\triangle JKL$, to $\overline{JK}$
as a base.

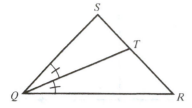

$\overline{QT}$ is an angle bisector of $\triangle QRS$.

The term "angle bisector," as just defined, is an abuse of the language, but should cause no difficulty. The context will always enable us to decide whether the term refers to a segment or a ray. Likewise, the term "altitude" is often used to mean the length of the segment of Definition 4.8. Here, too, ambiguity is no problem.

Our new vocabulary appears in the following theorem.

THEOREM 31   The median from the vertex of an isosceles triangle is an altitude and an angle bisector.

Given: Let isosceles $\triangle ABC$ be given with vertex C. Suppose $\overline{CM}$ is a median.

Prove: $\overline{CM}$ is an altitude and $\overline{CM}$ is an angle bisector.

Proof: It is easy to establish that $\overline{AC} \cong \overline{BC}$, $\angle A \cong \angle B$, and $\overline{AM} \cong \overline{BM}$. Then by the SAS postulate, it follows that $\triangle ACM \cong \triangle BCM$. Thus $\angle ACM \cong \angle BCM$, and, clearly then, $\overline{CM}$ is an angle bisector of $\triangle ABC$. Also (by CPCTC), $\angle AMC \cong \angle BMC$. But since these two angles are a linear pair, they are supplementary and, hence, each is a right angle. From this it follows that $\overline{CM} \perp \overline{AB}$, and thus $\overline{CM}$ is an altitude.

Theorems 30 and 31 can be used to obtain the following corollary.

COROLLARY T31   If a line in the plane of an isosceles triangle bisects and is perpendicular to the base, that line contains the opposite vertex.

The proof is left as an exercise.

In the proof to T31, the SAS postulate was used. Alert students should have noticed that the conditions necessary to set the SSS theorem to work were also met. We avoided the SSS theorem for a good reason: You will recall that we never *actually* proved it. Its proof is lengthy and fairly complicated. The point here is that most proofs of the SSS theorem employ Corollary T31,[6] which depended on

[6] See part (iii) of the proof outlined by Moise, *Elementary Geometry*, pp. 87–88.

T31 for its verification. To use the SSS theorem to derive T31 would put us in the position of the dog that chases its own tail.

PROBLEM SET 4.5

1. (a) Sketch an obtuse triangle and its three altitudes.
   (b) Sketch a right triangle and its three altitudes.
   (c) Sketch an acute triangle and its three altitudes.

2. Sketch an acute triangle and its three medians.

3. In Chapter 1 we discussed three basic argument forms: modus ponens, modus tollens, and hypothetical syllogism. To which of these is our technique of indirect proof most closely related? Why?

4. List four existence and uniqueness statements already established in the book but not mentioned in Section 4.5.

5. Given: $\overline{AC} \ncong \overline{BC}$
   Prove: $\angle A \ncong \angle B$ (Give an indirect proof.)

6. Given: $\triangle ABC$, $\triangle DEF$; $ABC \leftrightarrow DEF$; $\overline{AB} \cong \overline{DE}$; $\overline{AC} \cong \overline{DF}$; and $\overline{BC} \ncong \overline{EF}$

   Prove: $\angle A \ncong \angle D$ (indirect)

7. Using the vocabulary introduced in this section, restate T26 (Section 4.3).

8. (a) Suppose that in $\triangle ABC$, $m\angle A > m\angle B$. How do BC and AC appear to be related? A sketch might be helpful.
   (b) Suppose that in $\triangle ABC$, $AB > BC$. How do $m\angle C$ and $m\angle A$ appear to be related?
   (c) Sketch a few more triangles and examine them as in parts (a) and (b) above. Based on this experience, state a conjecture that generalizes these results.

9. Prove Corollary T31

10. Complete the proof of T30 by proving the following. Suppose $\mathcal{H}_1$ and $\mathcal{H}_2$ are half planes determined by l. If $\overrightarrow{PQ} \perp l$, $Q \in \mathcal{H}_1$ and $\overrightarrow{PR} \perp l$, $R \in \mathcal{H}_2$, then $\overleftrightarrow{PQ}$ and $\overleftrightarrow{PR}$ are the same line. (*Hint:* Consider $\overleftrightarrow{PQ}$, T22 (Section 3.6), and the Angle Construction Postulate.)

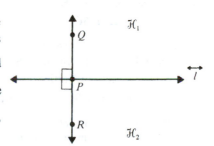

11. Give an indirect proof of the following theorem about sets:
    $[(A \subseteq B) \wedge A \cap B = \varnothing] \rightarrow A = \varnothing.$

12. In the accompanying figure, E is the midpoint
    of $\overline{BC}$, and $\overline{AE} \cong \overline{EF}$. Prove that $m \angle DBC >$
    $m \angle C$. Use two-column form. (*Hint:* Recall
    the definition of "$<$," and notice that $m \angle DBF$
    is positive.)

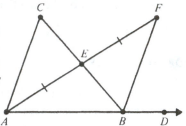

4.6 EXTERIOR ANGLES AND A VERY USEFUL THEOREM

   Associated with every triangle are three pairs of
angles that are called *exterior angles*. One of these
pairs is depicted in the diagram, where $\angle DBC$ and
$\angle ABE$ are exterior angles of $\triangle ABC$.

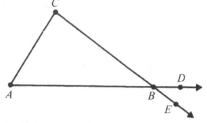

   DEFINITION 4.9 Given $\triangle ABC$. If A-B-D, then
$\angle DBC$ is called *an exterior angle* of the triangle.
The angles of the triangle that are not adjacent to the exterior angle
are called *remote interior angles* with respect to that angle.

   In the preceding diagram $\angle A$ and $\angle C$ are remote interior angles
with respect to $\angle DBC$ or $\angle ABE$.
   The following theorem defines a relation between the measures of
an exterior angle of a triangle, and those of the remote interior angles.

   THEOREM 32 *The Exterior Angle Theorem* (EAT)[7]: The measure
of an exterior angle of a triangle is greater than the measures of either
of the remote interior angles.

Hypothesis: $\triangle ABC$ with exterior angle $\angle DBC$
Conclusion: $m \angle DBC > m \angle A$
$\qquad\qquad m \angle DBC > m \angle C$

   Proof: We will prove that $m \angle DBC > m \angle C$.
The second part of this theorem has a similar proof,
using $\angle ABG$ and the Vertical Angle Theorem. Let
E be the midpoint of $\overline{BC}$. Consider the ray opposite

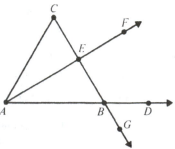

---

[7] This is a "weak" form of this theorem. In the next chapter we will strengthen
the EAT, replacing the inequality by an equality.

to $\overrightarrow{EA}$. By the Segment Construction Theorem (SCT), there exists exactly one point F on this ray such that $\overline{EF} \cong \overline{AE}$.

Now glance at the diagram for problem 12 of Problem Set 4.5. To have successfully completed *that* exercise is to have proven $m\angle DBC > m\angle C$. If you have not done the problem, use the hint given and complete this proof.

Although this is the *weak form* of the EAT, paradoxically it is a powerful tool for obtaining some very significant results. A first application is found in the proof of the following theorem. Problem 8 of Problem Set 4.5 hinted at this result.

THEOREM 33   The lengths of two sides of a triangle are unequal if and only if the measures of the angles opposite them are unequal in the same order.

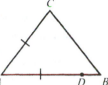

In terms of the figure, $AB > AC$ iff $m\angle C > m\angle B$. We quite naturally divide the proof of this theorem into two parts.

Part 1:

   Hypothesis: $AB > AC$
   Conclusion: $m\angle C > m\angle B$

Proof: Since $AB > AC$, it follows from the SCT and the definition of *between* that there exists a point D between A and B such that $\overline{AD} \cong \overline{AC}$. Also, by the EAT, it follows that $m\angle ADC > m\angle B$. But if $\overline{AD} \cong \overline{AC}$, we can conclude that $\angle ADC \cong \angle ACD$. Therefore, $m\angle ACD > m\angle B$. Clearly, $m\angle C > m\angle ACD$. Thus, by transitivity, we obtain the conclusion

$$m\angle C > m\angle B$$

Part 2:

   Hypothesis: $m\angle C > m\angle B$
   Conclusion: $AB > AC$

The proof to Part 2 is left as an exercise.

Example 1   Consider right $\triangle ABC$ with right angle $\angle A$. $\angle DAC$ is a right angle and an exterior angle of $\triangle ABC$; thus $\angle B$ and $\angle C$ must be acute. Therefore, we may write

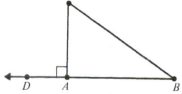

$$m\angle A > m\angle C$$

$$m\angle A > m\angle B$$

It then follows from T33 that

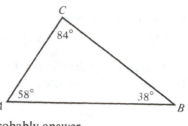

$$BC > AD \quad \text{and} \quad BC > AB$$

Example 2 In the figure, $AC < BC < AB$

Ask the celebrated man in the street, "What is the shortest path between two points?" and he will probably answer, "A straight line!" A partial verification of this assertion is given in the following theorem.

THEOREM 34 *The Triangle Inequality:* The sum of the lengths of two sides of a triangle is greater than the length of the third side.

Hypothesis: Given $\triangle ABC$
Conclusion: $AB + BC > AC$

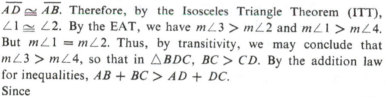

   Proof: If $\overline{AC}$ is *not* the *longest* side, the inequality is obvious. Therefore we will assume that, in the accompanying diagram, $AC > AB$ and $AC > BC$. Then, the SCT supports the assertion that there exists a point D, $D \in \overline{AC}$ such that $\overline{AD} \cong \overline{AB}$. Therefore, by the Isosceles Triangle Theorem (ITT), $\angle 1 \cong \angle 2$. By the EAT, we have $m\angle 3 > m\angle 2$ and $m\angle 1 > m\angle 4$. But $m\angle 1 = m\angle 2$. Thus, by transitivity, we may conclude that $m\angle 3 > m\angle 4$, so that in $\triangle BDC$, $BC > CD$. By the addition law for inequalities, $AB + BC > AD + DC$.
Since

$$AD + DC = AC$$

we may write

$$AB + BC > AC$$

and the theorem is proved.

Example 3 Not all the following lists of numbers can be used as lengths of the sides of a triangle:

   (a) 1, 2, 3
   (b) 3, 4, 5
   (c) 6, 8, 12

Examining the three lists we find that $1 + 2 \not> 3$; therefore, 1, 2, 3 cannot be lengths of sides of triangles.

PROBLEM SET 4.6

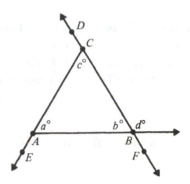

1. (a) Which exterior angle has $\angle CAB$ and
$\angle CBA$ as remote interior angles?

   (b) Name the remote interior angles associated
with exterior angle $\angle EAB$.

   (c) What is the relationship of $m\angle FBA$ to
$m\angle CAB$?

2. Fill in the blanks.

   (a) If an exterior angle of a triangle is acute, then the triangle is
$a(n)$_____triangle.

   (b) If an exterior angle of a triangle is a right angle, then the
triangle is $a(n)$_____triangle.

   (c) If all the exterior angles of a triangle are obtuse, the triangle
is $a(n)$_____triangle.

3. Use the figure for problem 1 to complete the following statements.

   (a) If $a = 30$ and $c = 90$, then $d >$ _____
   (b) If $a = 40$ and $c = 85$, then d _____
   (c) If $d = 120$, then b _____
   (d) If $d = 110$, then c _____
   (e) If $c = 90$, then a _____ and b _____ and d _____

4. Which of the following triplets of numbers can be lengths of the
sides of a triangle?

   (a) 2, 4, 5
   (b) 4, 8, 12
   (c) $\frac{1}{3}, \frac{1}{2}, \frac{1}{2}$
   (d) 140, 150, 160
   (e) $\frac{1}{6}, \frac{1}{3}, \frac{2}{3}$

5. Given $\triangle ABC$. If $AB = 4$, $BC = 6$, and $AC = 8$, name the
angles of the triangle in order of size.

6. Given: $AC > AB$
   Prove: $AC > AD$

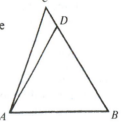

7. Let $\triangle ABC$ be given with $\angle A$ an obtuse angle. Use the accompanying diagram to prove that if $\overline{CH}$ is the altitude from C, then H is *not* between A and B. (Try an indirect proof using the EAT.)

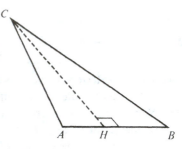

8. In each of the following figures, name the longest segment:

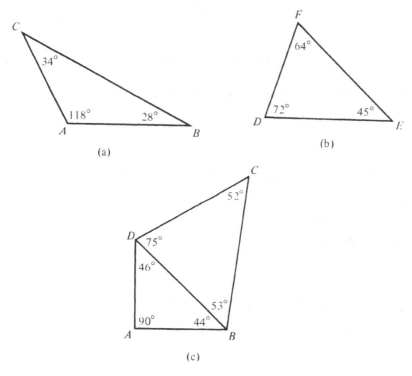

(a)

(b)

(c)

9. If $AB = 5$ and $AC = 6$, then there exist an x and a y such that it is always true that $x \leq BC \leq y$. Find x and y.

10. Suppose A, B, and C are vertices of a triangle. Furthermore suppose A and C are fixed whereas B is free to move. What happens to $m\angle B$ as $AC - (AB + BC)$ gets closer and closer to zero?

11. Write the ITT and its converse as an "iff" statement. Write the contrapositive of the statement you have just written. Comment on the relation of the ITT and its converse to T32.

12. Prove Part 2 of T33

13. Let $\triangle ABC$ be a right triangle with right angle $\angle A$. Prove that $m\angle A + m\angle B + m\angle C < 270$.

4.7 MORE THEOREMS ABOUT PERPENDICULAR LINES

We have already established the existence of a unique perpendicular (in a given plane) to a line at a given point *on that line*. Now we consider a situation in which the given point *is not* on the line.

THEOREM 35 Given a line and a point not on the line, there is one and only one line perpendicular to the given line and which contains the given point.

We first prove the *existence* of such a line.

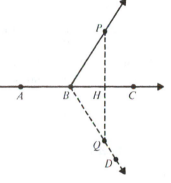

Hypothesis: Let $\overleftrightarrow{AC}$ be given; P is a point *not* on $\overleftrightarrow{AC}$

Conclusion: There is a line through P that is perpendicular to $\overleftrightarrow{AC}$

Proof: Suppose A-B-C. If $\overleftrightarrow{PB} \perp \overleftrightarrow{AC}$, the proof is complete. Suppose $\overleftrightarrow{PB}$ is *not* perpendicular to $\overleftrightarrow{AB}$. Then either $\angle ABP$ or $\angle CBP$ is acute. Assume it is $\angle CBP$. Then there exists exactly one ray $\overrightarrow{BD}$ with P and D on opposite sides of $\overleftrightarrow{AC}$ and such that $\angle CBD \cong \angle CBP$. (Why?) Also there exists exactly one point Q such that $Q \in \overrightarrow{BD}$ and such that $\overline{BQ} \cong \overline{BP}$. Therefore, $\triangle QPB$ is isosceles, with vertex B. Since P and Q are on opposite sides of $\overleftrightarrow{AC}$, $\overline{PQ}$ must intersect $\overleftrightarrow{AC}$. Call this point of intersection H. Then $\overline{BH}$ is the angle bisector of the vertex angle of an isosceles triangle. Therefore, $\overline{BH}$ is an altitude and $\overline{BH} \perp \overline{PQ}$. $\overleftrightarrow{PQ}$ is the desired line.

The proof of the "uniqueness" part of the theorem involves little more than the Exterior Angle Theorem (EAT) and is left as an exercise.

In Section 4.5 we fudged a bit when we stated the definition of the *altitude* of a triangle. It would have been difficult, at that point, to prove that such segments *always* exist. Theorem 35 assures us that they do indeed.

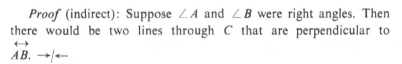

COROLLARY T35(a)　　No triangle has two right angles.

Hypothesis: $\triangle ABC$ is given
Conclusion: $\angle A$ and $\angle B$ are not both right angles

Proof (indirect): Suppose $\angle A$ and $\angle B$ were right angles. Then there would be two lines through C that are perpendicular to $\overleftrightarrow{AB}$. →|←

In Chapter 2, we defined the *distance* from one point to another. What about the distance between two *sets* of points? A single definition to cover all cases is not feasible here, so we will adopt the practice of defining distance anew in each specific instance that requires use of the concept. For example, the distance from a point to a line not containing it is defined next.

DEFINITION 4.10　　Let $\overleftrightarrow{l}$ be a line and let P be a point, such that $P \notin \overleftrightarrow{l}$. The *distance from P to $\overleftrightarrow{l}$* is the length of the perpendicular segment joining P to $\overleftrightarrow{l}$. If $P \in \overleftrightarrow{l}$, the *distance from P to $\overleftrightarrow{l}$* is defined to be zero.

That such a segment exists is assured by Theorem 35.

Example　　In the figure, AC is the distance from C to $\overleftrightarrow{AB}$. In Example 1 of the previous section we saw that $BC > AC$. We may generalize this result and conclude that the distance from a point to a line is the length of the *shortest* segment joining the point to the line.

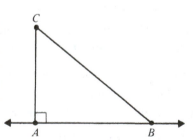

Corollary T31 and T35 can be used to obtain another corollary.

COROLLARY T35(b)　　The altitude to the base of an isosceles triangle is also a median.

The proof is left as an exercise.

Corollary T35(b) can be put to work in the proof of our next theorem.

THEOREM 36   *The Hypotenuse-Leg Theorem:* Given two right triangles, or a right triangle and itself, and a correspondence between their vertices. If one leg and the hypotenuse of one of the triangles are congruent to the corresponding parts of the second right triangle, then the triangles are congruent.

In terms of the diagrams, we may write:

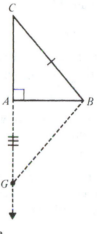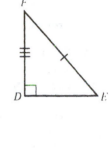

Given: Right triangles $\triangle ABC$ and $\triangle DEF$, with
       right angles $\angle A$ and $\angle D$; $\overline{BC} \cong \overline{EF}$; and
       $\overline{AB} \cong \overline{DE}$
Prove: $\triangle ABC \cong \triangle DEF$

Proof: By the Segment Construction Theorem, there exists exactly one point G on the ray opposite to $\overrightarrow{AC}$, such that $\overline{AG} \cong \overline{DF}$. Obviously, $\angle GAB$ is a right angle, and, therefore, $\angle GAB \cong \angle FDE$. It follows by the SAS postulate then, that $\triangle GAB \cong \triangle FDE$. Thus $\overline{FE} \cong \overline{GB}$, so that by transitivity we can conclude that $\overline{GB} \cong \overline{CB}$. This makes $\triangle CGB$ an isosceles triangle, with $\overline{BA}$ an altitude to the base.

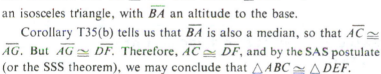

Corollary T35(b) tells us that $\overline{BA}$ is also a median, so that $\overline{AC} \cong \overline{AG}$. But $\overline{AG} \cong \overline{DF}$. Therefore, $\overline{AC} \cong \overline{DF}$, and by the SAS postulate (or the SSS theorem), we may conclude that $\triangle ABC \cong \triangle DEF$.

Half of the following corollary has an easy proof.

COROLLARY T36   *The Angle Bisector Theorem:* Let $P \in \mathscr{I} \angle BAC$. P is equidistant from the sides of $\angle BAC$ if and only if P is on the angle bisector of $\angle BAC$.

PROBLEM SET 4.7

1. In T30 (Section 4.5) the phrase "in plane m" appears. Why is such a phrase unnecessary in T35?

2. Let X and Y be sets of points. In view of the definitions of distance given thus far, how would you define the distance from X to Y?

3. Prove the "uniqueness" part of T35.

4. Prove Corollary T35(a) without using T35. (*Hint:* Use EAT.)

5. Given: $\overline{AB} \perp \overline{AD}$; $\overline{BC} \perp \overline{DC}$; and $\overline{AB} \cong \overline{BC}$
   Prove: $\overline{AD} \cong \overline{CD}$

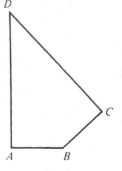

6. Prove: If two altitudes of a triangle are congruent, then the triangle is isosceles.

7. Prove: The angle bisector of the vertex angle of an isosceles triangle is also an altitude and a median.

8. Prove Corollary T35(b)

9. Given: In the figure, E is the midpoint of $\overline{AB}$;
   $\overline{DE} \cong \overline{EC}$; $\overline{AD} \perp \overline{AB}$; and $\overline{BC} \perp \overline{AB}$
   Prove: $\overline{AD} \cong \overline{BC}$

10. (a) Rewrite Corollary T36 as two "if . . . then" statements. (b) Prove the easier of the propositions given in (a).

11. From the midpoint of one side of an acute triangle, segments are drawn perpendicular to the other sides. Prove that if the segments are congruent, the triangle is isosceles.

4.8   THE PERPENDICULAR BISECTOR THEOREM AND LINE SYMMETRY

We begin this section with a definition.

DEFINITION 4.11   In a given plane, the *perpendicular bisector* of a segment is the line that is perpendicular to that segment at its midpoint.

Suppose that, in the figure, $\overset{\leftrightarrow}{PM}$ is the perpendicular bisector of $\overline{AB}$. Let $Q \in \overset{\leftrightarrow}{PM}$. Either $Q = M$ or $Q \neq M$. If $Q = M$, then $AQ = QB$. If $Q \neq M$. then $\triangle QMA \cong \triangle QMB$ (SAS) and $AQ = QB$. Summarizing, we may write: If a point is on the perpendicular bisector of a segment, then it is equidistant from the endpoints of the segment.

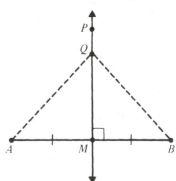

Next, let us consider a point P which is equidistant from the endpoints of $\overline{AB}$, as in the accompanying figure. If $P \in \overline{AB}$, it obviously lies on
the perpendicular bisector of $\overline{AB}$. If $P \notin \overline{AB}$, then
the fact that P is equidistant from A and B implies
that $\triangle APB$ is isosceles with vertex P. Since we have
established that the perpendicular bisector of the
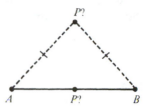
base of an isosceles triangle contains the opposite vertex [Corollary
T31 (Section 4.5)], it follows that P lies on the perpendicular bisector
of $\overline{AB}$. We combine these two results and obtain:

THEOREM 37   *The Perpendicular Bisector Theorem* (PBT): In a
plane, the perpendicular bisector of a segment is the set of all points
that are equidistant from the endpoints of the segment.

The *Line Postulate* (Section 2.2) and the PBT can be combined to
yield:

COROLLARY T37   *The Perpendicular Bisector Corollary*: In a
plane, if two points are each equidistant from the endpoints of a
segment, then the line determined by them is the perpendicular
bisector of that segment.

The PBT and its corollary are powerful theorems. There is a
tendency among students to use triangle congruence theorems to
derive results that are more swiftly obtained using these two theorems.
Be alert!

Example 1   Let D be the midpoint of $\overline{BC}$ with $\overrightarrow{DA} \perp \overline{BC}$.
We can prove that $\angle B \cong \angle C$ merely by noting that the PBT
implies that $\overline{AB} \cong \overline{AC}$, and it then follows (by the Isosceles
Triangle Theorem) that $\angle B \cong \angle C$.

The concept of symmetry is a fairly intuitive one. Moreover, a
knowledge of the symmetry properties of mathematical or physical
objects sometimes clarifies their structural characteristics. The idea of
symmetry pervades discussions in such diverse fields as physiology,
mineralogy, organic chemistry, music, and art as well as mathematics.

We have already seen that the "equals" relation and the "congruence" relations possess symmetry properties.

Symmetry with respect to a line can be defined in the following way.

DEFINITION 4.12   Let A and B be two points in a plane. A and B are said to be a *symmetric pair* with respect to a line in that plane if the line is the perpendicular bisector of $\overline{AB}$. The line is called *the axis of symmetry* of A and B. A line $\overleftrightarrow{l}$ is said to be *an axis of symmetry of a figure $\mathcal{F}$* if for every point P, such that $P \in F$ and $P \notin \overleftrightarrow{l}$, there is also some point Q such that $Q \in F$ and $\overleftrightarrow{l}$ is the axis of symmetry of P and Q.

Example 2 A and B are symmetric with respect to $\overleftrightarrow{l}$. Line $\overleftrightarrow{l}$ is also a line of symmetry of $\overline{AB}$.

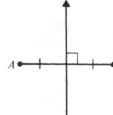

Example 3   The kite-shaped figure with vertices A, B, C, and D is symmetric with respect to $\overleftrightarrow{BD}$. It is obviously not symmetric with respect to $\overleftrightarrow{AC}$.

Example 4   Many figures have more than one line of symmetry. Each altitude of equilateral triangle $\triangle ABC$ is contained in an axis of symmetry of the figure.

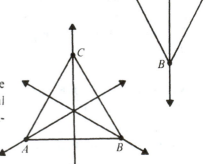

PROBLEM SET 4.8

1. State the PBT as an "iff" proposition.

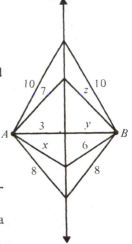

2. In the figure, if the lengths of the segments are as shown, find
   x, y, and z.

3. (a) In a plane containing a given segment, how many perpendicu-
      lar bisectors of the segment are there?
   (b) In space, how many perpendicular bisectors are there of a
      given segment?

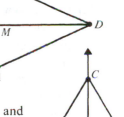

4. In the figure, if $AC = BC$ and $BD = AD$,
   prove, in two-column form, that $AM = MB$.
   Do not use congruent triangles.

5. Given: The adjacent figure; D is the midpoint of $\overline{AB}$; and
   $\overset{\leftrightarrow}{CD} \perp \overset{\leftrightarrow}{AB}$
   Prove (without using $\cong$ $\triangle$): $\triangle ABC$ is isosceles

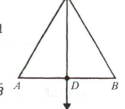

6. Given: Isosceles triangles $\triangle ABC$ and $\triangle ABD$, each with base $\overline{AB}$
   Prove: $\overset{\leftrightarrow}{CD} \perp \overline{AB}$

7. (a) Make a sketch of a square and draw in all
      of the lines of symmetry.
   (b) Which letters of the alphabet (uppercase)
      often have an axis of symmetry?
   (c) Which of the letters listed in (b) have more
      than one axis of symmetry?
   (d) List five numerals that have an axis of
      symmetry.

8. Given: The perpendicular bisector of side $\overline{AB}$
   of $\triangle ABC$ intersects $\overline{BC}$ at E
   (a) Prove: $\overline{AE} \cong \overline{BE}$
   (b) Prove: $m\angle CAB > m\angle CBA$

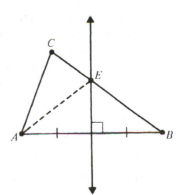

9. Given: In the accompanying figure, $\overleftrightarrow{PE} \perp$ bisector of $\overline{AB}$; and $\overleftrightarrow{PD} \perp$ bisector of $\overline{BC}$
   Prove: $AP = BP = CP$

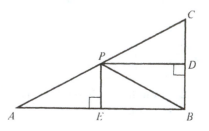

10. You probably recall from algebra that every line is the graph of an equation of the form

$$Ax + By + C = 0$$

In his textbook on the calculus, Edwin Moise begins an ingenious proof of this assertion with the statement: "Every line is the perpendicular bisector of some segment."[8] Prove this statement.

11. Suppose A, B, and $\overleftrightarrow{m}$ are coplanar and that we wish to send a light ray from A to B bouncing off a mirror containing line $\overleftrightarrow{m}$.

   (a) If we assume that the light ray will take the shortest path possible, explain how viewing $\overleftrightarrow{m}$ as an axis of symmetry of B and a second point B' helps us locate C, $C \in \overleftrightarrow{m}$ such that $AC + CB$ is a minimum.

   (b) Under these conditions, prove the well-known law of reflection: $\angle i \cong \angle r$ (the angle of incidence is congruent to the angle of reflection).

CHAPTER 4: REVIEW EXERCISES

1. For each of the following statements indicate whether it is *always* true, *sometimes* but not always true, or *never* true.

   (a) The measure of an exterior angle of a triangle is greater than the measure of every interior angle of that triangle.

   (b) The bisector of a segment is perpendicular to the segment.

   (c) If two lines are perpendicular to a third line at the same point, then those two lines are coplanar.

   (d) A median of a scalene triangle is perpendicular to a side of the triangle.

   (e) The longest side of a triangle is called the hypotenuse.

   (f) Suppose A-B-C. Then $\angle DCA$ is an exterior angle of $\triangle BCD$.

[8] Edwin E. Moise, *Calculus* (Reading, Mass.: Addison-Wesley Publishing Co., 1966), p. 27.

(g) If the median of a triangle is also an altitude, then that triangle is isosceles.

(h) $\angle BAC \cong \triangle BAC$

(i) In $\triangle BAC$, if the side included by $\angle A$ and $\angle B$ is congruent to the side included by $\angle B$ and $\angle C$, then $\angle A \cong \angle C$.

(j) If $\triangle ABC$ is a right triangle, then $\angle A$ is obtuse.

(k) In $\triangle ABC$, $AB > BC$; then $m\angle A < m\angle C$

(l) An isosceles triangle has exactly one axis of symmetry.

(m) A right triangle has at least one right angle.

(n) If $AB = DF$, $BC = FE$, $AC = DE$, then $\triangle ABC \cong \triangle DEF$

(o) Two triangles are congruent if two sides and an angle of one are congruent to the corresponding parts of the other.

2. In $\triangle PQR$, $m\angle P = 37$, $m\angle Q = 73$, and $m\angle R = 70$. Name the shortest side. Name the longest side.

3. Which of the following sets of numbers could be used as the lengths of the sides of a triangle?

(a) 2, 3, 4   (b) 1, 3, 5   (c) a-b, a, b

4. Given: In the figure, $m\angle A > m\angle B$; and $m\angle C > m\angle D$
Prove: $DB > AC$

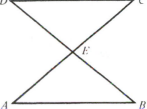

5. (a) Restate the "existence" part of T35 in "if . . . then" form.

(b) Write out the proof of the existence part of T35 in two-column form.

6. Given: $\triangle ABC$ with A-D-B; $\overline{CD} \perp \overline{AB}$; and $m\angle A = m\angle B = 45$

(a) Sketch such a situation.

(b) What conclusions do you think follow from this given information? Make a list.

(c) Are there any items on your list that you cannot prove?

7. Prove: The sum of the altitudes of any triangle is less than the sum of the lengths of the sides.

8. Given: In the figure $\overline{AB} \cong \overline{BE} \cong \overline{CE} \cong \overline{CD}$
Prove: $\triangle ADE$ is isosceles

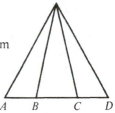

9. Given: $\overset{\leftrightarrow}{l} \perp \overset{\leftrightarrow}{m}$; $\overset{\leftrightarrow}{m} \perp \overset{\leftrightarrow}{n}$; and $\overset{\leftrightarrow}{l}$, $\overset{\leftrightarrow}{m}$, and $\overset{\leftrightarrow}{n}$ coplanar

Prove: $\overset{\leftrightarrow}{l} \cap \overset{\leftrightarrow}{n} = \varnothing$. (Give an indirect proof.)

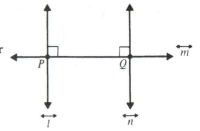

10. Sketch a pair of triangles that are not congruent, and yet have two congruent pairs of corresponding sides and a pair of congruent angles. Which of the conditions of the SAS postulate are not met?

11. Use the figure for problem 8 to prove the following statement. If $\triangle BEC$ and $\triangle AED$ are isosceles with vertex E, then $\angle AEB \cong \angle DEC$.

12. Without using congruent triangles, prove the first half of T32; that is, "The median from the vertex of an isosceles triangle is an altitude."

13. In each of the following figures, name the shortest segment.

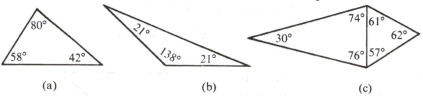

    (a)   (b)   (c)

14. In the accompanying diagram, $\overline{AC} \cong \overline{BC}$, $\angle A \cong \angle CDE$, and $\angle B \cong \angle CED$. Prove: $\triangle CDE$ is isosceles.

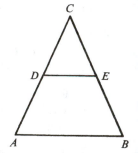

15. What is the difference between a *bisector of an angle* and an *angle bisector of a triangle*?

16. Let $\overleftrightarrow{l}$ be the edge of two half planes, $\mathscr{H}_1$ and $\mathscr{H}_2$. A and B are two points of $\overleftrightarrow{l}$, $M \in \mathscr{H}_1$ and $R \in \mathscr{H}_2$ such that $\angle MAB \cong \angle RAB$, $AM = AR$, and B is not between M and R.
    (a) Prove that $\triangle MRB$ is isosceles.
    (b) Does $\overline{MR}$ intersect $\overleftrightarrow{l}$?

17. Prove: The union of the segments joining the midpoints of the sides of an equilateral triangle is an equilateral triangle.

18. Given: $\overline{AB}$ is the perpendicular bisector of $\overline{CD}$.
    Prove: (a) $\triangle CDA$ is isosceles.
        (b) $\triangle ABD \cong \triangle ABC$.

19. Given: $\angle BAC$ such that $AB = AC$; $P \in \overrightarrow{AB}$ and $Q \in \overrightarrow{AC}$ such that $PC = QB$. Can you prove that $AP = AQ$? Explain.

20. Prove: If the altitude of a triangle is also a median, then that triangle is isosceles.

21. Complete Euclid's proof of the Isosceles Triangle Theorem.
Given: $\overline{AC} \cong \overline{BC}$.
Prove: $\angle ABC \cong \angle BAC$.

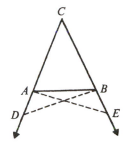

Proof: Suppose $D \in \overrightarrow{CA}$ such that C-A-D. Let C-B-E such that $\overline{CD} \cong \overline{CE}$. Complete this proof, first showing that $\triangle DBC \cong \triangle EAC$ and that $\triangle DBA \cong \triangle EAB$.

REFERENCES

Boyer, Carl B. *A History of Mathematics.* New York: John Wiley & Sons, 1968. pp. 661–663.
  The contrasting views on the nature of mathematics as held by leaders of the "formalist," "logicist," and "intuitionist" schools are spelled out.

Courant, Richard, and Robbins, Herbert. *What Is Mathematics?* New York: Oxford University Press, 1941.
  See page 22 for Euclid's beautiful indirect proof of the theorem: "There exists an infinite number of primes."

Escher, M. C. *The Graphic Work of M. C. Escher.* New York: Ballantine Books, 1967. pp. 11, 17.
  The triangle is treated here, as by Plato, as a primordial form in the evolutionary process.

Moise, Edwin E., and Downs, Floyd L., Jr. *College Geometry.* Reading, Mass.: Addison-Wesley Publishing Co., 1971.
  Chapters 5, 6, and 7 contain much of the material in this chapter. A good reference for those who have difficulty with these ideas.

Tarski, Alfred. *Introduction to Logic.* New York: Oxford University Press, 1965. pp. 157–159.
  An example of an interesting, although comparatively rare form of indirect proof called *reductio ad absurdum* is given.

The truth is rarely pure and never simple.

OSCAR WILDE (1854–1900)

CHAPTER FIVE

PARALLEL LINES

5.1 INTRODUCTION

Parallel lines do not intersect.

Lines $\overleftrightarrow{l}$ and $\overleftrightarrow{m}$ do not intersect.

These two statements appeared in an exercise in Chapter 1. You were asked to regard them as a set of assumptions and to supply (if possible) a valid conclusion using one of the basic argument forms.

Perhaps you first wanted to conclude that $\overleftrightarrow{l}$ and $\overleftrightarrow{m}$ are parallel. If so, you soon discovered that such a conclusion did not fit any of the basic argument forms.

Are there lines that do not intersect and are *not* parallel? Consider the lines suggested by the intersection of the front wall and ceiling, and the intersection of a side wall and the floor of any room. Do these lines intersect? Would you say that they are parallel? What about the lines suggested by the intersection of the front wall and ceiling, and the intersection of the front wall and the floor? The diagram shown represents such a room.

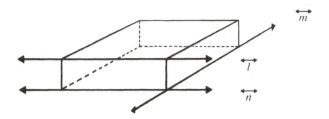

Evidently $\overleftrightarrow{l}$ and $\overleftrightarrow{m}$ are not parallel, but $\overleftrightarrow{l}$ and $\overleftrightarrow{n}$ *are* parallel. In addition to being nonintersecting, it is quite apparent that $\overleftrightarrow{l}$ and $\overleftrightarrow{n}$ are coplanar while $\overleftrightarrow{l}$ and $\overleftrightarrow{m}$ are not. Hence we make the following definition.

DEFINITION 5.1   Two lines are *parallel* if they are coplanar and they do not intersect. If two lines are not coplanar, they are *skew*.

In the preceding figure, $\overleftrightarrow{l}$ and $\overleftrightarrow{m}$ are skew lines.

Do parallel lines exist? The above definition makes no such assertion. It merely expresses what we mean by the word "parallel." The question of existence is settled in the following section.

5.2   PARALLEL LINES IN A PLANE

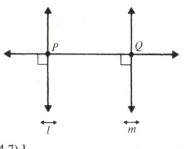

Consider two coplanar lines $\overleftrightarrow{l}$ and $\overleftrightarrow{m}$, each of which is perpendicular to a third line, as in the accompanying figure. Do $\overleftrightarrow{l}$ and $\overleftrightarrow{m}$ intersect? Suppose $\overleftrightarrow{l}$ *does* intersect $\overleftrightarrow{m}$ at a point R. Then there are two perpendiculars from R to $\overleftrightarrow{PQ}$. →/← [T35 (section 4.7).]

Summarizing these ideas as a theorem, we have Theorem 38.

THEOREM 38   In a plane, if two lines are perpendicular to the same line, then they are parallel.

A corollary of this theorem, important enough to dignify by calling it a theorem, assures us that parallels do indeed exist.

THEOREM 39   Let $\overleftrightarrow{l}$ be a line and let P be a point not on $\overleftrightarrow{l}$. Then there is a line, containing P, that is parallel to $\overleftrightarrow{l}$.

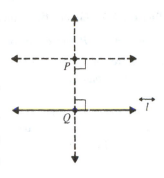

Proof: Through P there is a line perpendicular to $\overleftrightarrow{l}$ at some point Q (T35). Also through P, and in the plane determined by $\overleftrightarrow{l}$ and $\overleftrightarrow{PQ}$, there is a line perpendicular to $\overleftrightarrow{PQ}$ [T30 (Section 4.5)]. From T39 it follows that this line is parallel to $\overleftrightarrow{l}$.

NOTATION RULE 5.1   If $\overleftrightarrow{l}$ and $\overleftrightarrow{m}$ are parallel lines, we write

$$\overleftrightarrow{l} \parallel \overleftrightarrow{m}$$

One may also write $\overline{AB} \parallel \overline{CD}$, $\overrightarrow{AB} \parallel \overrightarrow{CD}$, $\overline{AB} \parallel \overrightarrow{CD}$, and so on, meaning that the lines that contain these segments or rays are parallel.

DEFINITION 5.2   A line is a *transversal* of two coplanar lines if it intersects them in two different points.

Example   In the diagrams shown, $\overleftrightarrow{t}$ is a transversal of $\overleftrightarrow{l}$ and $\overleftrightarrow{m}$. $\overleftrightarrow{r}$ is *not* a transversal of these lines.

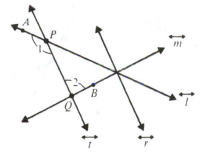
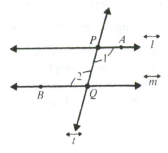

In the figures for the preceding example $\angle 1$ and $\angle 2$ are called *alternate interior angles*. Anytime two lines and a transversal of them are given, two such pairs of alternate interior angles are formed. The definition of *alternate interior angles* is fairly complicated. Refer to the above figures when reading the following definition.

DEFINITION 5.3 Given two lines $\overset{\leftrightarrow}{l}$ and $\overset{\leftrightarrow}{m}$ and a transversal $\overset{\leftrightarrow}{t}$ intersecting them at P and Q, respectively. Suppose $A \in \overset{\leftrightarrow}{l}$ and $B \in \overset{\leftrightarrow}{m}$ and A and B are on opposite sides of $\overset{\leftrightarrow}{t}$. Then $\angle APQ$ and $\angle BQP$ are *alternate interior angles*.

If we are given two lines and a transversal of those two lines, it is common to say that the two lines are *cut* by the transversal. We use this terminology in our next theorems.

THEOREM 40 If two lines are cut by a transversal so that one pair of alternate interior angles are congruent, then the other pair of alternate interior angles are congruent.

In terms of the diagram, we may write:

Given: Lines $\overset{\leftrightarrow}{l}$ and $\overset{m}{}$ with transversal $\overset{\leftrightarrow}{t}$; $\angle 1 \cong \angle 3$
Prove: $\angle 2 \cong \angle 4$

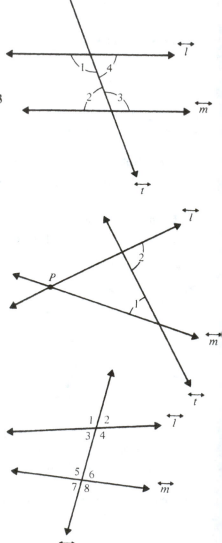

The proof is left as an exercise.

THEOREM 41 *The Alternate Interior-Parallel Theorem* (AIP): If two lines are cut by a transversal so that a pair of alternate interior angles are congruent, then the lines are parallel.

Proof (indirect): Suppose they are *not* parallel, as in the accompanying figure, where $\overset{\leftrightarrow}{t}$ is a transversal of $\overset{\leftrightarrow}{l}$ and $\overset{\leftrightarrow}{m}$. If they are not parallel, $\overset{\leftrightarrow}{l}$ and $\overset{\leftrightarrow}{m}$ intersect at some point P, and a triangle is formed. $\angle 2$ is an exterior angle of this triangle, so that $m\angle 2 > m\angle 1$. However, since $\angle 1$ and $\angle 2$ are alternate interior angles, it follows that $m\angle 2 = m\angle 1$. →/←

When two lines are cut by a transversal, eight angles are formed, as shown. $\angle 3$ and $\angle 6$ are alternate interior angles as are $\angle 4$ and $\angle 5$.

Names are also given to other pairs of these angles.

Corresponding angles:

$\angle 1$ and $\angle 5$

$\angle 2$ and $\angle 6$

$\angle 3$ and $\angle 7$

$\angle 4$ and $\angle 8$

Interior angles on the same side of the transversal:

$\angle 3$ and $\angle 5$

$\angle 4$ and $\angle 6$

The following theorems are easy corollaries of T41.

THEOREM 42 *The Corresponding-Parallel Theorem:* If two lines are cut by a transversal so that a pair of corresponding angles are congruent, then the lines are parallel.

THEOREM 43 If two lines are cut by a transversal so that a pair of interior angles on the same side of the transversal are supplementary, then the lines are parallel.

The proofs of these theorems are left as exercises.

PROBLEM SET 5.2

1. Using the plane figure shown, name all pairs of

   (a) Alternate interior angles.

   (b) Corresponding angles for l and m.

   (c) Interior angles on the same side of the transversal for l and n.

2. Using the accompanying plane figure, name all pairs of

   (a) Alternate interior angles.

   (b) Corresponding angles.

   (c) Interior angles on the same side of the transversal.

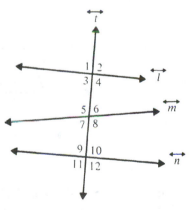

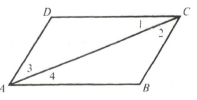

3. Using Definition 5.3 as a model, write a definition of *interior angles on the same side of the transversal.*

4. Using the diagram shown, classify the following pairs of angles:

(a) $\angle 5$ and $\angle 4$
(b) $\angle 1$ and $\angle 5$
(c) $\angle 3$ and $\angle 6$
(d) $\angle 3$ and $\angle 4$
(e) $\angle 7$ and $\angle 2$
(f) $\angle 2$ and $\angle 5$
(g) $\angle 1$ and $\angle 7$

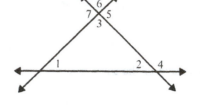

5. Prove T40

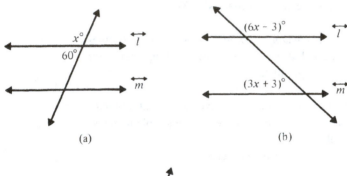

(a)   (b)

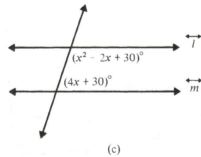

(c)

6. For each figure, find a value for x that will make $\overleftrightarrow{l} \parallel \overleftrightarrow{m}$. (*Hint:* If factoring is a chore, use the *quadratic formula.* It can be found in Appendix A.)

7. Given: $\triangle ABC$ is isosceles with vertex angle $\angle C$, and $\overrightarrow{CE}$ bisects $\angle BCD$

Prove: $\overrightarrow{CE} \parallel \overline{AB}$

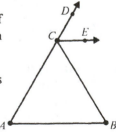

8. Prove T42

9. Given: $\angle B$ and $\angle D$ are right angles, and $\overline{AD} \cong \overline{BC}$
Prove: $\overline{AB} \parallel \overline{CD}$
(*Hint:* Introduce an auxiliary segment.)

10. Theorem 43 can be viewed as a generalization of an earlier theorem of this section. Which one?

11. Prove T43

12. Given: $\overline{ED} \perp \overleftrightarrow{m}$; $\overline{AB} \perp \overleftrightarrow{m}$; and $\overline{BC} \cong \overline{CD}$
Prove: (a) $\overleftrightarrow{l} \parallel \overleftrightarrow{n}$
(b) $\overleftrightarrow{l} \parallel \overleftrightarrow{m}$

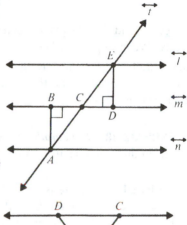

13. (a) State the converses of T41, T42, and T43.
(b) Assume the converse of T41 is true, and use that fact to prove the converse of T42.

14. *The Ironing Board Theorem, I*
Given: $\overline{AC}$ and $\overline{BD}$ bisect each other at E
Prove: $\overleftrightarrow{AB} \parallel \overleftrightarrow{DC}$

5.3 THE PARALLEL POSTULATE

In Problem Set 5.2 you were asked to state the converse of T41. You should have written something like the following:

> If two parallel lines are cut by a transversal,
> then the alternate interior angles are congruent.

The proof of this statement depends upon a postulate we have not yet stated. Without this postulate, situations that contradict our intuition can occur.

A question posed in problem 5 of Problem Set 3.2 is interesting in this regard: Can three lines separate a plane into five regions? Suppose that in the diagram $\overleftrightarrow{l} \parallel \overleftrightarrow{m}$ and $\overleftrightarrow{n} \parallel \overleftrightarrow{m}$. Note that five disjoint regions *are* indicated. However, after carefully examining the figure, you will probably decide that at least one of the lines, $\overleftrightarrow{l}$ and $\overleftrightarrow{n}$, must

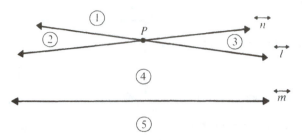

intersect $\overleftrightarrow{m}$, and that the plane is actually separated into *more* than five regions. We have previously shown that there *exists* a line containing P and parallel to $\overleftrightarrow{m}$; your unwillingness to believe that two such lines exist is an assertion that this line is unique. Our next postulate formalizes this idea.

POSTULATE 18   *The Parallel Postulate:* Through a given point not on a given line, there is at most one line parallel to the given line.

The Parallel Postulate is easily believed. Euclid included it (in a more complex form) as the fifth of the five geometric postulates on which he based *The Elements*. Euclid's Parallel Postulate is more complicated than his other four geometric postulates. For this reason, and since the converse of Euclid's fifth postulate had been shown to be provable, there was for over two thousand years considerable feeling among mathematicians that the postulate itself could be proven from the others.

By the middle of the nineteenth century, however, at least three men, Carl Friedrich Gauss, Janos Bolyai, and Nicolai Ivanovich Lobachevsky had arrived at the same conclusion: namely, that the Parallel Postulate cannot be proven and that it is but one alternative. Another alternative is Gauss's assumption that *more than one* parallel to a given line passes through a given point. Geometric systems which incorporate such an alternative have come to be called non-Euclidean geometry.

No one geometric system can be shown to be universally more applicable than the others. In his theory of relativity, for example, Einstein obtained better agreement between theoretical prediction and experimental observation using a non-Euclidean geometry than he would have obtained using Euclidean geometry.

For over two thousand years prior to the nineteenth century, the entire western world had believed that Euclid's postulates were unquestionable truths about the physical universe. The acceptance of the validity of non-Euclidean geometry brought with it the realization that man may be unable to acquire absolute truth, including that sought in various spheres of thought: economics, ethics, government, and the like. Morris Kline has written in this connection, "When the anchor of truth was lost, all bodies of knowledge were cast adrift."[1] We have learned to live with such ideas, and to recognize that "mathematical truth" merely offers us theories about how the natural world works. We have also learned to carefully reexamine our most resolutely held assumptions and to give a fair hearing to ideas that fly in the face of our intuition and common sense.

Engineers and scientists continue to use Euclidean geometry for most of their work because it is so intuitively acceptable and practically reliable. The development of the remainder of the ideas in this book will also be Euclidean.

Our first applications of the Parallel Postulate immediately follow, beginning with the converse of AIP.

THEOREM 44 (PAI): If two parallel lines are cut by a transversal, then alternate interior angles are congruent.

Proof (indirect): Assume $\overleftrightarrow{t}$ is a transversal of $\overleftrightarrow{l}$ and $\overleftrightarrow{m}$ and that $\overleftrightarrow{l} \parallel \overleftrightarrow{m}$. Suppose $\angle 1 \not\cong \angle 2$. Then by the Angle Construction Theorem, there is a unique ray $\overrightarrow{QS}$, with S on the R-side of $\overleftrightarrow{t}$ such that $\angle 1 \cong \angle PQS$. Then the Alternate-Interior Parallel Theorem implies that $\overleftrightarrow{QS} \parallel \overleftrightarrow{l}$, so that we have *two* lines through Q that are parallel to $\overleftrightarrow{l}$. This contradicts the Parallel Postulate.

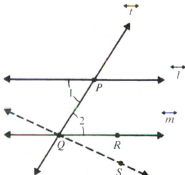

THEOREM 45 If two parallel lines are cut by a transversal, each pair of corresponding angles are congruent.

THEOREM 46 If two parallel lines are cut by a transversal, interior angles on the same side of the transversal are supplementary.

Our next theorem is obvious, but often ignored by writers of textbooks.

[1] See Morris Kline, *Mathematics: A Cultural Approach* (Reading, Mass.: Addison-Wesley Publishing Co., 1962), p. 577.

THEOREM 47   In a plane, if a line intersects one of two parallel lines in a single point, then it intersects the other.

Proof: Suppose $\overleftrightarrow{l} \parallel \overleftrightarrow{m}$ and $\overleftrightarrow{l} \cap \overleftrightarrow{n} = \{P\}$. Then the Parallel Postulate implies that $\overleftrightarrow{n} \not\parallel \overleftrightarrow{m}$, and since all these lines are coplanar, it follows from the definition of *parallel* that $\overleftrightarrow{m}$ and $\overleftrightarrow{n}$ intersect.

THEOREM 48   In a plane, if a line is perpendicular to one of two parallel lines, it is perpendicular to the other.

The proof of this theorem is left as an exercise.

THEOREM 49   In a plane, two lines parallel to the same line are parallel to each other.

Hypothesis: $\overleftrightarrow{l} \parallel \overleftrightarrow{m}$; $\overleftrightarrow{m} \parallel \overleftrightarrow{n}$; and $\overleftrightarrow{l} \neq \overleftrightarrow{m}$

Conclusion: $\overleftrightarrow{l} \parallel \overleftrightarrow{n}$

Proof (indirect): Suppose $\overleftrightarrow{l}$ and $\overleftrightarrow{n}$ are not parallel. Then they intersect at some point P. But they are both parallel to $\overleftrightarrow{m}$, and we have a contradiction of the Parallel Postulate again.

PROBLEM SET 5.3

1. In the figure, $\overleftrightarrow{t}$ is a transversal of $\overleftrightarrow{l}$ and $\overleftrightarrow{m}$ and $\overleftrightarrow{l} \parallel \overleftrightarrow{m}$. If $m\angle 2 = 130$, find the measures of angles 1, 3, 4, 5, 6, 7, and 8.

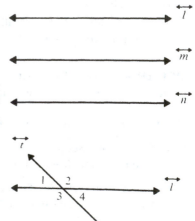

2. In the figure, $\angle 1 \cong \angle 2$, $m\angle 3 = 55$, and $\overline{AC} \parallel \overrightarrow{BD}$. Find $m\angle A$ and $m\angle C$.

3. In the figure, $m\angle 1 = 70$, $m\angle 2 = 60$, and $\overleftrightarrow{CD} \parallel \overline{AB}$.

   (a) Find $m\angle 3$, $m\angle A$, and $m\angle B$.
   (b) Compute $m\angle A + m\angle B + m\angle 2$.

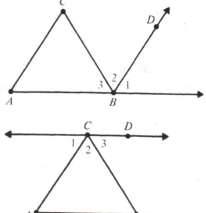

4. In the figure, $\overline{AC} \parallel \overline{FD}$, $\overline{FB} \parallel \overline{EC}$, and $\overline{BD} \parallel \overline{AE}$. If $m\angle A = 65$, $m\angle C = 55$, and $m\angle E = 60$, find:

(a) $m\angle BFD$
(b) $m\angle FDB$
(c) $m\angle FBD$

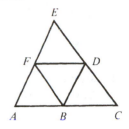

5. Given: $\overline{DE} \parallel \overline{AB}$, and $\triangle ABC$ is isosceles with vertex C
Prove: $\triangle DEC$ is isosceles

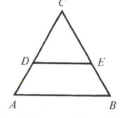

6. Given: $\triangle ABC$ is isosceles with vertex C, and $\overline{DE} \parallel \overline{BC}$
Prove: $\overline{AE} \cong \overline{DE}$

7. Prove T48

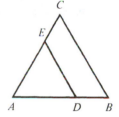

8. Given: $\overline{AB} \parallel \overline{DC}$, and $\overline{AB} \cong \overline{DC}$
Prove: $\triangle EBA \cong \triangle ECD$

9. Given: $\overline{AB} \cong \overline{DC}$; $\overline{AB} \perp \overline{BC}$; and $\overline{DC} \perp \overline{BC}$
Prove: $\overline{AD}$ and $\overline{BC}$ bisect each other

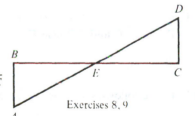

Exercises 8, 9

10. Given: $\angle 1 \cong \angle 2 \cong \angle 3 \cong \angle 4$
Prove: $\overleftrightarrow{m} \parallel \overleftrightarrow{n}$

11. (a) One of our theorems is an assertion that in a plane "parallelism" is a transitive relation. Which theorem?

(b) Is parallelism reflexive? Explain.
(c) Is parallelism symmetric? Explain.

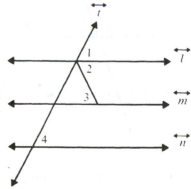

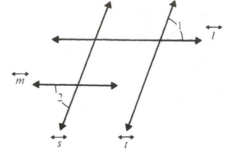

12. Given: $\overleftrightarrow{l} \parallel \overleftrightarrow{m}$, and $\overleftrightarrow{s} \parallel \overleftrightarrow{t}$
    Prove: $\angle 1 \cong \angle 2$

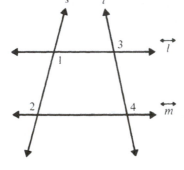

13. Given: $\angle 1 \cong \angle 2$
    Prove: $\angle 3 \cong \angle 4$

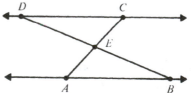

14. Given: $\overline{DC} \parallel \overline{AB}$, and $\overline{DC} \cong \overline{AB}$
    Prove: $\overline{AC}$ and $\overline{BD}$ bisect each other

15. In the following figures, $\overleftrightarrow{l} \parallel \overleftrightarrow{m}$. Find x in each case.

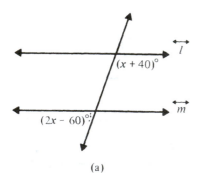

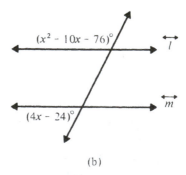

(a)   (b)

16. If two lines have a common perpendicular, are they parallel? Explain.

5.4 FURTHER FACTS ABOUT TRIANGLES

One of the most useful facts that follows from the Parallel Postulate is known as the *Angle Sum Theorem*.

THEOREM 50 *The Angle Sum Theorem* (AST): The sum of the measures of the angles of a triangle is 180.

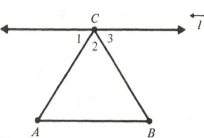

Proof: Refer to the diagram. Given $\triangle ABC$, we know there exists a line l through C and parallel to $\overleftrightarrow{AB}$. By PAI (Theorem 44), we have $\angle 1 \cong \angle A$ and $\angle 3 \cong \angle B$. Clearly $m\angle 1 + m\angle 2 + m\angle 3 = 180$. Therefore, by substitution we have $m\angle A + m\angle 2 + m\angle B = 180$, which is the desired result.

The Parallel Postulate is hidden in the proof of the AST. Moreover, if we deny the Parallel Postulate, the AST is actually false.

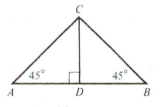

Example 1 Problem 6 in the Review Exercises for Chapter 4 states: Given $\triangle ABC$ with A-D-B, $\overline{CD} \perp \overline{AB}$ and $m\angle A = m\angle B = 45$. A sketch of this situation is given here. You were asked to make a list of the conclusions that followed from the given information. Among the items that may have been on your list and which are now provable are as follows:

1. $m\angle ACB = 90$, since $m\angle A + m\angle B + m\angle ACB = 180$
2. $m\angle ACD = 45$, which follows from the fact that $m\angle A + m\angle ADC + m\angle ACD = 180$
3. $m\angle BCD = 45$, for similar reasons
4. $\triangle ADC$ and $\triangle BDC$ are isosceles triangles, which follows from (2) and (3).

Consider $\triangle ABC$ and $\triangle DEF$ as depicted in the diagram (at top of page 150). The AST allows us to write

$$x + y + m\angle B = 180$$

and

$$x + y + m\angle E = 180$$

Therefore,

$$m\angle B = 180 - (x + y)$$

and

$$m\angle E = 180 - (x + y) \qquad \text{(Why?)}$$

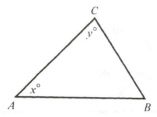

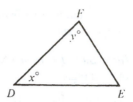

We can conclude that $m\angle B = m\angle E$, or $\angle B \cong \angle E$. These ideas can be formalized as a theorem.

COROLLARY T50(a) *The Third Angle Corollary:* Given a correspondence between two triangles. If two pairs of corresponding angles are congruent, then the third pair of corresponding angles are also congruent.

Also useful is the following corollary.

COROLLARY T50(b) The acute angles of a right angle are complementary.

The proof is left as an exercise.

In Chapter 4 we proved a weak form of the Exterior Angle Theorem. A stronger form of this theorem follows easily from the AST.

COROLLARY T50(c) *The Exterior Angle Corollary* (EAC): The measure of an exterior angle of a triangle is equal to the sum of the measures of the two remote interior angles.

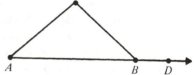

Hypothesis: $\triangle ABC$ with exterior angle $\angle CBD$

Conclusion: $m\angle CBD = m\angle A + m\angle C$

Proof:

STATEMENTS	REASONS
1. $\triangle ABC$ with exterior angle $\angle CBD$	1. Hypothesis
2. $\angle ABC$ and $\angle CBD$ are a linear pair	2. ?
3. $\angle ABC$ and $\angle CBD$ are supplementary	3. ?
4. $m\angle ABC + m\angle CBD = 180$	4. ?
5. $m\angle ABC + m\angle A + m\angle C = 180$	5. T50
6. $m\angle ABC + m\angle CBD = m\angle ABC + m\angle A + \angle C$	6. ?
7. Therefore, $m\angle CBD = m\angle A + m\angle C$	7. ?

Example 2 From the AST, it follows that $x + 39 + 35 = 180$ or $x = 180 - (39 + 35) = 106$. From the EAC it follows that $y = 39 + 35 = 74$.

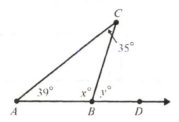

Other easy corollaries of the AST follow:

COROLLARY T50(d) In an isosceles right triangle, the measure of each base angle is 45.

COROLLARY T50(e) The measure of each angle of an equilateral triangle is 60.

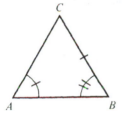 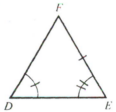

The accompanying diagrams illustrate what might be referred to as a side-angle-angle correspondence. The Third Angle Corollary implies that $\angle C \cong \angle F$, transforming the correspondence into an ASA correspondence, from which it follows that the triangles are congruent. We state this result as a theorem.

THEOREM 51 *The Side-Angle-Angle Theorem* (SAA): Given a correspondence between two triangles. If two angles and a side of one triangle are congruent to the corresponding parts of the other, then the triangles are congruent.

Note that the ASA theorem can be viewed as a special case of the SAA theorem.

The SAA Theorem provides an easy route to a familiar statement that is often taken for the definition of parallel lines.

THEOREM 52 Parallel lines are everywhere equidistant.

Hypothesis: $\overleftrightarrow{AB} \parallel \overleftrightarrow{CD}$

$\overline{AC} \perp \overleftrightarrow{CD}, \overline{BD} \perp \overleftrightarrow{CD}$

Prove: $AC = BD$

Proof:

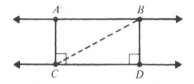

STATEMENTS	REASONS
1. Introduce auxiliary segment $\overline{CB}$	1. Line Postulate
2. $\overleftrightarrow{AB} \parallel \overleftrightarrow{CD}$; $\overline{AC} \perp \overleftrightarrow{CD}$; $\overline{BD} \perp \overleftrightarrow{CD}$	2. Given
3. $\overline{AC} \perp \overleftrightarrow{AB}$	3. T48
4. $\angle CAB$ and $\angle CDB$ are rt. $\angle s$	4. Definition of $\perp$
5. $\angle CAB \cong \angle CDB$	5. Rt. $\angle s \cong$
6. $\angle CBA \cong \angle BCD$	6. PAI
7. $\overline{CB} \cong \overline{CB}$	7. Reflexive Theorem (T6)
8. $\triangle ABC \cong \triangle DCB$	8. SAA
9. $\overline{AC} \cong \overline{BD}$	9. CPCTC

PROBLEM SET 5.4

1. Find the measure of the third angle of a triangle if the measures of the other two angles are as follows:

(a) 30, 60   (b) 23, 27   (c) 120, 59
(d) x, y (e) $30 + x, 30 - x$
(f) $x, 90$

2. In the figure, find x, y, w, and z.

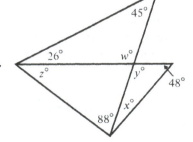

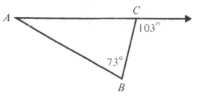

3. For the figure, find:

   (a) $m\angle A$
   (b) $m\angle ACB$

4. In the figure, $m\angle CBD = 122$, and $\overline{AB} \cong \overline{BC}$
   Find $m\angle A$.

5. In the figure, $m\angle A = 55$, $m\angle C = 50$, and
   $\overrightarrow{BE}$ bisects $\angle CBD$. Find $m\angle EBD$.

6. Given: $\overrightarrow{BE}$ bisects $\angle CBD$, and $\overline{AC} \parallel \overrightarrow{BE}$.
   Prove: $\triangle ABC$ is isosceles

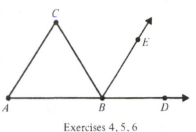

Exercises 4, 5, 6

7. The measure of one angle of a triangle is 1 more than twice the measure of the second, and 13 less than the measure of the third. Find each measure.

8. Given: $\triangle ABC$, A-B-D. If $m\angle A = 47$, and $m\angle CBD = 150$, find $m\angle C$.

9. For the given figure, find $m\angle A$.

10. How was the Parallel Postulate used in the proof of the AST? (*Hint:* It was *not* used as a justification for introducing $\overleftrightarrow{l} \parallel \overline{AB}$.)

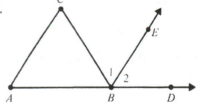

11. Give the missing reasons in the proof of the EAC.

12. Prove (indirectly): No triangle has more than one obtuse angle.

13. Complete this alternative proof of the AST.

    Given: $\triangle ABC$ with exterior angle $\angle CBD$
    Prove: $m\angle A + m\angle C = m\angle CBD$

    Proof:

	STATEMENTS	REASONS
1.	$\triangle ABC$ with exterior angle $\angle CBD$	1. Given
2.	Introduce $\overrightarrow{BE}$; $\overrightarrow{BE} \parallel \overline{AC}$	2. T?

14. Prove Corollary T50(b)

15. If the measure of a base angle of an isosceles triangle is twice the measure of a second angle, find the measure of the base angle.

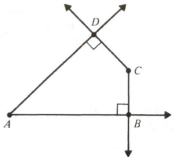

16. Given: The figure with $\angle C$ and $\angle A$; $\overrightarrow{CD} \perp \overrightarrow{AD}$; and $\overrightarrow{AB} \perp \overrightarrow{CB}$
    Prove: $\angle A$ and $\angle C$ are supplementary (*Hint:* Use Corollary T50(b).)

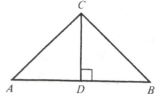

17. Given: $\triangle ABC$; $\angle ACB$ is a right angle; and $\overline{CD} \perp \overline{AB}$
    Prove: $\angle A \cong \angle BCD$

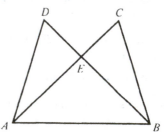

18. Given: The figure with $\angle D \cong \angle C$, and $\overline{AE} \cong \overline{BE}$
    Prove: $\overline{AD} \cong \overline{BC}$

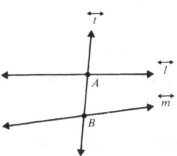

5.5 MANY PARALLEL LINES

As a final set of theorems about parallel lines, we will now consider what can occur when more than two parallel lines are cut by two transversals.

DEFINITION 5.4 Suppose $\overleftrightarrow{t}$ is a transversal of $\overleftrightarrow{l}$ and $\overleftrightarrow{m}$, intersecting them at A and B, respectively. Then $\overleftrightarrow{l}$ and $\overleftrightarrow{m}$ *intercept* the segment $\overline{AB}$ on the transversal $\overleftrightarrow{t}$.

Example $\overleftrightarrow{l}$ and $\overleftrightarrow{m}$ intercept $\overline{AB}$ on $\overleftrightarrow{t}$.

THEOREM 53 If three parallel lines intercept congruent segments on one transversal, then they intercept congruent segments on any transversal parallel to the given one.

Hypothesis: $\overleftrightarrow{l}$, $\overleftrightarrow{m}$, $\overleftrightarrow{n}$ are parallel lines with transversals $\overleftrightarrow{s}$ and $\overleftrightarrow{t}$; $\overleftrightarrow{s} \parallel \overleftrightarrow{t}$; $\overline{AB} \cong \overline{BC}$

Conclusion: $\overline{DE} \cong \overline{EF}$

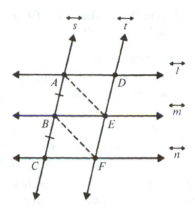

Proof: Introduce auxiliary segments $\overline{AE}$ and $\overline{BF}$. From *PAI* we get the following congruences: $\angle BAE \cong \angle DEA$ and $\angle DAE \cong \angle BEA$. Also, $\angle CBF \cong \angle EFB$ and $\angle EBF \cong \angle CFB$. The first two congruences, together with the ASA theorem, lead us to the conclusion that $\triangle BAE \cong \triangle DEA$. From this it follows that $\overline{AB} \cong \overline{DE}$. The second pair of congruences, by the same reasoning, implies that $\overline{BC} \cong \overline{EF}$. But by hypothesis, $\overline{AB} \cong \overline{BC}$. Therefore, $\overline{DE} \cong \overline{EF}$.

Our next theorem is a further generalization of this idea.

THEOREM 54 If three parallel lines intercept congruent segments on one transversal, then they intercept congruent segments on any transversal.

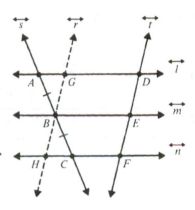

Hypothesis: $\overleftrightarrow{l}$, $\overleftrightarrow{m}$, $\overleftrightarrow{n}$ are parallel lines with transversals $\overleftrightarrow{s}$ and $\overleftrightarrow{t}$; $\overline{AB} \cong \overline{BC}$

Conclusion: $\overline{DE} \cong \overline{EF}$

Proof: There are two possibilities. Either $\overleftrightarrow{s} \parallel \overleftrightarrow{t}$ or $\overleftrightarrow{s} \not\parallel \overleftrightarrow{t}$. If $\overleftrightarrow{s} \parallel \overleftrightarrow{t}$, then the conclusion follows from T53.

On the other hand, suppose $\overleftrightarrow{s} \not\parallel \overleftrightarrow{t}$. Then through B we can introduce $\overleftrightarrow{r}$, $\overleftrightarrow{r} \parallel \overleftrightarrow{t}$, intersecting $\overleftrightarrow{l}$ and $\overleftrightarrow{n}$ in G and H, respectively. We continue the proof in two-column form.

STATEMENTS	REASONS
1. $\overline{AB} \cong \overline{BC}$	1. Hypothesis
2. $\angle BAG \cong \angle BCH$; $\angle BGA \cong \angle BHC$	2. PAI (Section 5.3)
3. $\triangle BAG \cong \triangle BCH$	3. SAA
4. $\overline{GB} \cong \overline{HB}$	4. CPCTC
5. $\overline{DE} \cong \overline{EF}$	5. T53

By repeated applications of T54, we obtain the same conclusion for any number of parallel lines.

COROLLARY T54(a) *The Egg-Slicer Corollary:* If three or more parallel lines intercept congruent segments on one transversal, then they intercept congruent segments on any transversal.

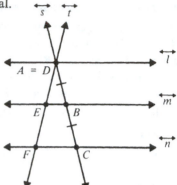

Theorem 54 does not preclude the possibility that $A = D$, as in the diagram shown. In such a case, $\overleftrightarrow{m}$ bisects one side of $\triangle FDC$. Theorem 54 then leads us to the conclusion that $\overline{DE} \cong \overline{EF}$. We restate these ideas as a corollary.

COROLLARY T54(b) If a line bisects one side of a triangle and is parallel to a second side, then it bisects the third side.

PROBLEM SET 5.5

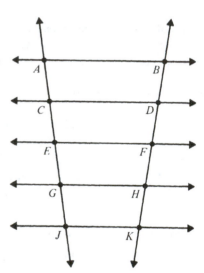

1. Given $\overleftrightarrow{AB} \parallel \overleftrightarrow{CD} \parallel \overleftrightarrow{EF} \parallel \overleftrightarrow{GH} \parallel \overleftrightarrow{JK}$; $AC = CE = EG = GJ$; and $BK = 80$. Find DF and HK.

2. Given $\overleftrightarrow{AB} \parallel \overleftrightarrow{CD} \parallel \overleftrightarrow{EF} \parallel \overleftrightarrow{GH} \parallel \overleftrightarrow{JK}$; $AC = CE = EG = GJ = 6$; and $BF = 10$. Find BD and BK.

Exercises 1, 2

3. Given: D is midpoint of $\overline{AC}$; $\overleftrightarrow{DE} \parallel \overleftrightarrow{AB}$; and $\overleftrightarrow{EF} \parallel \overleftrightarrow{AC}$
   Prove: F is midpoint of $\overline{AB}$

4. Given: D is midpoint of $\overline{AC}$; $\overleftrightarrow{DE} \parallel \overleftrightarrow{AB}$; and $\overleftrightarrow{EF} \parallel \overleftrightarrow{AC}$
   Prove: $\triangle DEC \cong \triangle FBE$

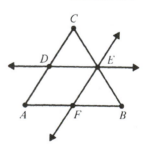

Exercises 3, 4

5. Complete the following proof of the converse of Corollary T54(b):

The segment joining the midpoints of two sides of a triangle is parallel to the third side.

Given: $\triangle ABC$; and D and E are midpoints of $\overline{AC}$ and $\overline{BC}$

Prove: $\overline{DE} \parallel \overline{AB}$

Proof (indirect): Suppose $\overline{DE} \not\parallel \overline{AB}$. Introduce $\overleftrightarrow{DF}$, $\overleftrightarrow{DF} \parallel \overline{AB}$, and $F \in \overleftrightarrow{BC}$.

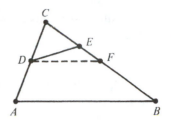

6. Using the theorem proven in problem 5, prove the following theorem:

The midpoints of two sides of a triangle are equidistant from the third side.

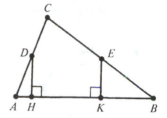

7. Given: $\overline{AB} \cong \overline{BC}$; $\overline{AH} \parallel \overline{BG} \parallel \overline{CF}$; and $\overline{LF} \parallel \overline{KG} \parallel \overline{JH}$

Prove: $\overline{JK} \cong \overline{KL}$

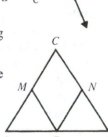

8. The following problem was suggested in Problem Set 4.4, problem 8. Now that we have the Parallel Postulate and some theorems following it, you should be able to complete the proof by using the converse of Corollary T54(b). See Problem 5 above.

Given: $\triangle ABC$ is isosceles with vertex C; and D, M, and N are the midpoints of $\overline{AB}$, $\overline{AC}$, and $\overline{BC}$, respectively

Prove: $\overline{MD} \cong \overline{CN}$

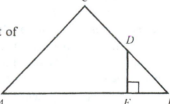

9. Given: $\triangle ABC$, $AC = BC$; D the midpoint of $\overline{BC}$; and $\overline{DE} \perp \overline{AB}$ at E

Prove: $EB = \frac{1}{4} AB$

10. Add the auxiliary segments shown in the diagram, and prove T54
using T52.

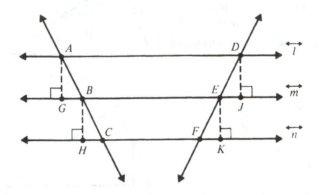

CHAPTER 5: REVIEW EXERCISES

1. For each of the following statements indicate whether it is *always*
true, *sometimes* but not always true, or *never* true.

(a) If two lines are cut by a transversal, then alternate interior
angles are congruent.

(b) Let l be a line. Then $\overleftrightarrow{l} \parallel \overleftrightarrow{l}$.

(c) Given $\triangle ABC$. Then $\overleftrightarrow{AB}$ is a transversal of $\overleftrightarrow{BC}$ and $\overleftrightarrow{AC}$.

(d) Given $\triangle ABC$ with A-B-D. If $\angle ABC \cong \angle CBD$, then $\angle BCA$
is acute.

(e) If a line is parallel to one side of a triangle, it is parallel to
some other side.

(f) If two lines do not intersect, then they are parallel.

(g) If two parallel lines are cut by a transversal, then the bisectors
of any two corresponding angles are parallel.

(h) If two lines are perpendicular to the same line, they are
parallel.

(i) Two medians of a triangle bisect each other.

(j) Given $\triangle ABC$. If $\angle A$ and $\angle B$ are complementary, then
$\triangle ABC$ is a right triangle.

(k) If the bisector of an exterior angle of a triangle is parallel to a
side of the triangle, the triangle is isosceles.

(l) If two lines are not coplanar, then they are skew.

(m) If three lines intercept congruent segments on one transversal,
then they intercept congruent segments on any transversal.

2. What value of x will guarantee that $\overset{\leftrightarrow}{l} \parallel \overset{\leftrightarrow}{m}$?

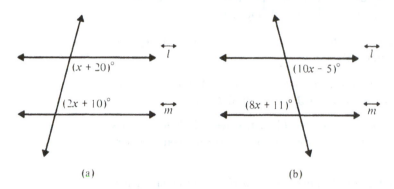

(a)   (b)

3. In Problem Set 4.8, problem 8, we derived the well-known law of reflection—that is, the angle of incidence is congruent to the angle of reflection. Using this result, calculate the measures of the angles of reflection as the light ray depicted in the figure is reflected from the surfaces represented by $\overset{\leftrightarrow}{AB}$, $\overset{\leftrightarrow}{BC}$, and $\overset{\leftrightarrow}{AD}$ (second reflection from $\overset{\leftrightarrow}{AD}$).

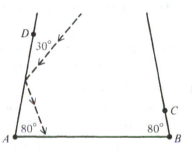

4. Given $\triangle ABC$, and A-B-D.

(a) If $m\angle CBD = 120$ and $m\angle C = 60$, what can you say about $\triangle ABC$?

(b) If $m\angle A = 45$ and $AC = BC$, find $m\angle CBD$.

(c) If $m\angle CBD = 135$ and $m\angle A = 3x + 5$, $m\angle C = 4x - 10$, find $m\angle A$, $m\angle C$.

5. If the vertex angle of an isosceles triangle has measure 40, find the measures of the base angles.

6. In problem 7 of Problem Set 5.5, must $\overset{\leftrightarrow}{AB}$ and $\overset{\leftrightarrow}{JK}$ be coplanar?

7. Given: $\overset{\rightarrow}{DB}$ bisects $\angle ADC$, and $\overline{DA} \cong \overline{AB}$
Prove: $\overset{\leftrightarrow}{AB} \parallel \overset{\leftrightarrow}{DC}$

8. Prove: In a plane, if a line is perpendicular to one of two intersecting lines, it is not perpendicular to the other. (Give an indirect proof.)

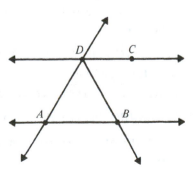

9. Prove: If the hypotenuse and one acute angle of one right triangle are congruent to the corresponding parts of a second right triangle, then the triangles are congruent.

10. List the methods you have at your disposal for proving two triangles are congruent.

11. Suppose $\triangle ABC$ with A–D–C and B–E–C such that $\overline{DE} \parallel \overline{AB}$ are given. If $AD = DC = 6$ and $BC = 10$, find BE and EC.

12. Prove: If two parallel lines are cut by a transversal, then bisectors of interior angles on the same side of the transversal are perpendicular.

13. Using the definition of *alternate interior angles* as a model, write a definition of *alternate exterior angles*. Provide a diagram.

14. Prove that a line parallel to the base of an isosceles triangle and intersecting the other two sides of the triangle at different points forms another isosceles triangle.

15. Prove: In a plane, if the sides of an angle are parallel to the sides of another angle, the two angles are either congruent or supplementary. (*Hint:* See problem 12 of Problem Set 5.3.)

16. The measure of one of the angles of a triangle is four times the measure of a second angle. The measure of an exterior angle at the third vertex is 80. Find the measure of each angle of the triangle.

17. Prove: If three parallel lines intercept congruent segments on one transversal, the parallel lines are equidistant from each other.

18. Let $\triangle ABC$ be given. If a line through A is perpendicular to the bisector of $\angle B$ at P, and another line through P is parallel to $\overline{BC}$ and intersects $\overline{AB}$ at M, prove that M is the midpoint of $\overline{AB}$.

19. Prove: In a triangle, if a median is half as long as the side which it bisects, then the triangle is a right triangle and the side is its hypotenuse.

20. Given: $\overline{AC}$ and $\overline{DB}$ intersect at E, with A–E–C and D–E–B such that $AD = BC$ and $\overline{AD} \parallel \overline{BC}$.
    Prove: $\overline{AC}$ and $\overline{DB}$ bisect each other at E.

21. Let $\triangle ABC$ be isosceles with vertex C. Let B–D–C. Draw $\overleftrightarrow{DE} \perp \overline{AB}$ at E. Let $\overleftrightarrow{DE}$ intersect $\overleftrightarrow{AC}$ at F. Prove $\triangle DCF$ is isosceles.

REFERENCES

Coxeter, H. M. S. "Non-Euclidean Geometry." In *The Mathematical Sciences*, edited by COSRIMS, pp. 52–59. Cambridge, Mass.: The MIT Press, 1969.
A brief but clear glimpse at some of the alternatives to Euclidean geometry.

Kline, Morris. *Mathematics: A Cultural Approach*. Reading, Mass.: Addison-Wesley Publishing Co., 1962. pp. 553–559, 572–577.
First, Kline gives a brief history of attempts to determine whether the Parallel Postulate is provable and the resolution of this problem. He then considers the mathematical and cultural implications of the emergence of non-Euclidean geometry. Very readable.

Geometry will draw the soul toward truth.

PLATO

CHAPTER SIX

SIMILARITY

6.1 INTRODUCTION

In the sixth century B.C., the illustrious mathematician Pythagoras founded a school which adopted the philosophical premise that the examination of everything in nature, geometric forms included, would yield properties that could be expressed in terms of whole numbers. According to legend, the Pythagoreans were onboard ship when one of them demonstrated to the others that, no matter what unit of measure was employed, it was impossible to express the lengths of the legs and hypotenuse of an isosceles right triangle as whole numbers. Since this fact violated Pythagorean doctrine, it was viewed as a serious threat by the Pythagoreans, who responded by throwing its heretical author overboard—or so the story goes.

Thus was generated a "crisis" in mathematics which lasted for nearly two hundred years, until Eudoxus invented his elegant and ingenious theory of ratio and proportion. Eudoxus' work is generally believed to have furnished the entire content of Book V of Euclid's *Elements*.

In previous chapters, we examined the congruence relation between geometric sets of points. Specifically, we defined congruence for segments, angles, and triangles. When we say that two such figures are congruent, we are saying that they are the same shape and the same size.

In this chapter we will begin to consider relations between figures that are the same shape but not necessarily the same size. We have a name for such figures: We say they are *similar*.

Eudoxus' theory of proportion provides the tools we will use to explore and refine the similarity relation. The notion of similarity will then lead us to a proof of what is known as the *Pythagorean Theorem*. How ironic that the discovery that was most disturbing to Pythagoras should lead to a deft verification of the theorem for which he is most honored!

Applications of the ideas developed in this chapter are incorporated into the scale models used by industrial designers, the process of photographic enlargement, and a host of other activities.

6.2   RATIO AND PROPORTION

Before proceeding with our development of the geometric notion of similarity, we pause for a brief review of some of the facts about ratio and proportion that you studied in arithmetic and algebra.

DEFINITION 6.1   Let x and y be real numbers, $y \neq 0$. The *ratio of x to y* is the number

$$\frac{x}{y}$$

Example 1   Let $x = 3$ and $y = 4$. Then the ratio of x to y is $\frac{3}{4}$ and the ratio of y to x is $\frac{4}{3}$.

DEFINITION 6.2   Any list of numbers such that it can be determined which number is first, second, third, and so on, is called a *sequence*.

Example 2   The elements of the set of natural numbers: 1, 2, 3, 4, . . . may be considered as a sequence.

Example 3   Given $\triangle ABC$ and $\triangle DEF$, the lists

$$AB, \ BC, \ AC$$

and

$$DE, \ EF, \ DF$$

are both sequences.

You should notice that a sequence may be infinite, as in Example 2, or it may be finite as in Example 3.

Example 4   Suppose we are given the two sequences

$$3, \ 4, \ 5$$

and

$$12, \ 16, \ 20$$

Each member of the second sequence is four times the corresponding entry in the first sequence, or equivalently,

$$\tfrac{12}{3} = \tfrac{16}{4} = \tfrac{20}{5} = 4$$

We also might write

$$\tfrac{3}{12} = \tfrac{4}{16} = \tfrac{5}{20} = \tfrac{1}{4}$$

Such sequences are said to be *proportional*.

DEFINITION 6.3   Let $a_1, a_2, a_3, \ldots$ and $b_1, b_2, b_3, \ldots$ be sequences of positive numbers. If

$$\frac{a_1}{b_1} = \frac{a_2}{b_2} = \frac{a_3}{b_3} = \cdots$$

then the sequences are *proportional*. Any such statement of equality is called a *proportion*.

Thus a proportion is a statement that two or more ratios (of positive numbers) are equal.

Example 5 Given $\triangle ABC$ and $\triangle DEF$ with lengths of sides as indicated. Consider the obvious correspondence, $ABC \leftrightarrow DEF$. Then the lengths of the corresponding sides are *proportional* since

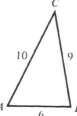
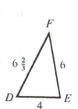

$$\frac{AB}{DE} = \frac{6}{4} = \frac{3}{2}$$

$$\frac{BC}{EF} = \frac{9}{6} = \frac{3}{2}$$

and

$$\frac{AC}{DF} = \frac{10}{6\frac{2}{3}} = \frac{3}{2}$$

We can write the following proportion:

$$\frac{AB}{DE} = \frac{BC}{EF} = \frac{AC}{DF}$$

(*Note:* In the future, we will adopt the widely used and simpler "the corresponding sides are proportional" when we *really* mean that their lengths are proportional.)

Example 6 Some proportions exhibit interesting properties. Consider the following proportions:

$$\frac{16}{4} = \frac{4}{1} \qquad \frac{3}{6} = \frac{6}{12} \qquad \frac{a}{b} = \frac{b}{c}$$

In each case, the denominator of the first ratio is the numerator of the second. Multiplying both sides of the third (and more general) proportion by bc, we obtain

$$ac = b^2$$

and since a, b, and c are all positive,

$$b = \sqrt{ac}$$

In the other two cases we see that

$$4 = \sqrt{(16)(1)} \quad \text{and} \quad 6 = \sqrt{(3)(12)}$$

We will find it useful to have a name for such a situation.

DEFINITION 6.4   Let

$$\frac{a}{b} = \frac{b}{c}$$

be a proportion. Then b is called the *geometric mean of a and c*. In the previous example we might have written

> 4 is the geometric mean of 16 and 1.
> 6 is the geometric mean of 3 and 12.

The following useful lemma summarizes many of the algebraic properties of proportions.

LEMMA 6.1   *Algebraic Properties of Proportions:* Let

$$\frac{a}{b} = \frac{c}{d}$$

be a proportion. Then

(a) $ad = bc$

(b) $\dfrac{d}{b} = \dfrac{c}{a}$

(c) $\dfrac{b}{a} = \dfrac{d}{c}$ Note that we have, in effect, taken the reciprocal of both sides of the expression $\dfrac{a}{b} = \dfrac{c}{d}$

(d) $\dfrac{a}{c} = \dfrac{b}{d}$

(e) $\dfrac{a + b}{b} = \dfrac{c + d}{d}$

(f) $\dfrac{a}{b} = \dfrac{c}{d} = \dfrac{a + c}{b + d}$

PROBLEM SET 6.2

1. Supply the numbers necessary to make each of the following sentences a proportion:

(a) $\dfrac{}{10} = \dfrac{3}{5}$ (b) $\dfrac{2}{3} = \dfrac{}{6} = \dfrac{6}{}$ (c) $\dfrac{15}{} = \dfrac{6}{36} = \dfrac{}{60}$

(d) $\dfrac{84}{} = \dfrac{7}{24} = \dfrac{}{72} = \dfrac{}{12}$

2. Show that the following pairs of sequences are proportional:

(a) 3, 6, 8   and   9, 18, 24

(b) 7, 1, 11, 10   and   21, 3, 33, 30

(c) 1, 2, 3, . . .   and   2, 4, 6, . . .

(d) $\dfrac{1}{5}, \dfrac{1}{4}, \dfrac{1}{3}, \dfrac{1}{2}$ and $\dfrac{1}{15}, \dfrac{1}{12}, \dfrac{1}{9}, \dfrac{1}{6}$

(e) $3a, 7a, 9a$ and $9b, 21b, 27b$

(f) $\dfrac{2}{3}, \dfrac{3}{4}, \dfrac{4}{5}$ and $\dfrac{4}{9}, \dfrac{1}{2}, \dfrac{8}{15}$

3. Determine the ratio of x to y if

(a) $3x = 2y$ (b) $ax = by$ (c) $2x - 4y = 0$

(d) $\dfrac{x + y}{x - y} = \dfrac{3}{2}$

4. Indicate which part of Lemma 6.1 justifies each statement.

(a) If $\dfrac{3}{x} = \dfrac{y}{7}$,   then $\dfrac{3}{y} = \dfrac{x}{7}$

(b) If $\dfrac{2}{3} = \dfrac{4}{x}$,   then $2x = 3.4$

(c) If $\dfrac{x}{2} = \dfrac{y}{3}$,   then $\dfrac{2}{x} = \dfrac{3}{y}$

(d) If $\dfrac{5}{4} = \dfrac{15}{12}$,   then $\dfrac{5 + 4}{4} = \dfrac{15 + 12}{12}$

(e) If $\dfrac{a}{2} = \dfrac{3}{5}$,   then $\dfrac{5}{2} = \dfrac{3}{a}$

5. From the following list of sequences, find which pairs of sequences are proportional.

(a) 2, 4, 6   (b) 2, 6, 4   (c) 7, 3, 5

(d) 1, 3, 2   (e) $2, \frac{6}{7}, \frac{10}{7}$ (f) 5, 10, 15

(g) $\frac{1}{3}, 1, \frac{2}{3}$

6. Solve each of the proportions for x.

(a) $\dfrac{x}{10} = \dfrac{3}{5}$ (b) $\dfrac{2}{x} = \dfrac{4}{7}$ (c) $\dfrac{2}{3} = \dfrac{x}{5}$ (d) $\dfrac{2}{7} = \dfrac{4}{x-2}$

(e) $\dfrac{2}{x+1} = \dfrac{2x+2}{9}$

7. Find the geometric mean of each pair.

(a) 4, 16   (b) 4, 9   (c) 9, 16   (d) 3, 12

(e) 2, 4   (f) 28, 63   (g) $\frac{1}{4}, \frac{1}{9}$ (h) 2, 2

8. The proportion

$$\frac{a}{b} = \frac{c}{d}$$

is sometimes read "a is to b as c is to d."

(a) If x is to 5 as 50 is to 10, find x.
(b) If x is to 5 as 5 is to x, find x.
(c) If 6 is to x as 10 is to 3, find x.

9. (a) Prove Lemma 6.1(a) using the multiplication law (see Appendix A).
(b) Use Lemma 6.1(a) to prove Lemma 6.1(b).
(c) Use Lemma 6.1(a) to prove Lemma 6.1(c).
(d) Use Lemma 6.1(a) to prove Lemma 6.1(d).
(e) Prove Lemma 6.1(e) using the addition law. (*Hint*: $b/b = d/d = 1$)
(f) Prove Lemma 6.1(f), using two of the other five parts of the lemma. One part may be used twice.

10. Suppose $\triangle ABC$ and $\triangle DEF$ are given and

$$\frac{AB}{DE} = \frac{BC}{EF} = \frac{AC}{DF}$$

What can you say about

$$\frac{AB + BC + AC}{DE + EF + DF} ?$$

Explain

11. The hypothesis of Lemma 6.1 asserts that $a/b = c/d$ is a proportion. Why is the hypothesis "Let $a/b = c/d$" insufficient?

12. (a) If

$$\frac{5}{p} = \frac{6}{q} = \frac{r}{10} = \frac{q}{12}$$

solve for $p, q,$ and r.

(b) If

$$\frac{3}{x} = \frac{y}{10} = \frac{5\sqrt{2}}{z} = \frac{6}{y}$$

solve for $x, y,$ and z.

13. Given:

$$\triangle ABC, \text{ with } \frac{CD}{DA} = \frac{CE}{EB}$$

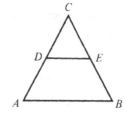

(a) Prove: $\dfrac{CD}{EC} = \dfrac{DA}{EB}$

(b) Prove: $\dfrac{DA}{CA} = \dfrac{EB}{CB}$

(c) Prove: $\dfrac{CD}{CA} = \dfrac{CE}{CB}$

14. We have already seen that if a and b are positive numbers, the *geometric mean* of a and b is $\sqrt{ab}$. The *arithmetic mean* of these numbers is defined to be $(a + b)/2$. Find the geometric mean and the arithmetic mean of each of the following pairs:

(a) 3, 27   (b) 8, 8   (c) 2, 4
(d) 7, 28   (e) 1, 16

Do you see enough of a pattern to guess at a general relationship between the geometric and arithmetic mean of two positive numbers? If so, state that relationship.

6.3   SIMILAR TRIANGLES

The notion of proportion is given geometric meaning in the following definition.

DEFINITION 6.5 Given a one-to-one correspondence between the vertices of two triangles, or a triangle and itself. The triangles are said to be *similar* if the corresponding angles are congruent and the (lengths of) the corresponding sides are proportional. The ratio of the (lengths of) corresponding sides of similar triangles is called their *ratio of similitude*.

NOTATION RULE 6.1   If $\triangle ABC$ is similar to $\triangle DEF$, we write

$$\triangle ABC \sim \triangle DEF$$

and we mean that $\angle A \cong \angle D$, $\angle B \cong \angle E$, $\angle C \cong \angle F$, and

$$\frac{AB}{DE} = \frac{BC}{EF} = \frac{AC}{DF} = k$$

the ratio of similitude.

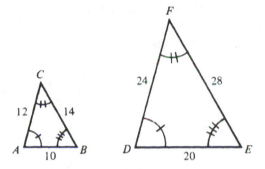

Example 1   Given the lengths and congruences as marked in the figure,

$$\frac{AB}{DE} = \frac{AC}{DF} = \frac{BC}{EF} = \frac{1}{2}$$

Thus $\triangle ABC \sim \triangle DEF$.

Our first similarity theorem is very easy to prove.

THEOREM 55   If two triangles are congruent, they are similar.

The definition of similar triangles specifies a large set of congruence and equals relations. As we shall see, certain subsets of this set are sufficient to guarantee that triangles are similar. The next theorems will find their first applications in the discovery of such subsets.

THEOREM 56 *The Basic Proportionality Theorem* (BPT): A line parallel to one side of a triangle and intersecting the interiors of the other two sides separates these sides proportionally.

Hypothesis: $\triangle ABC$; A-D-C; B-E-C; and $\overset{\leftrightarrow}{DE} \parallel \overline{AB}$
Conclusion:

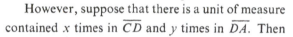

$$\frac{CD}{DA} = \frac{CE}{EB}$$

Proof: Our proof will rest on the assumption that there is a unit of measure contained a whole number of times in both $\overline{CD}$ and $\overline{DA}$. As the Pythagoreans knew, this is not always the case. At this stage in our development, the more general proof would require techniques beyond the scope of this course. We will return to this question in an exercise in Chapter 7.

However, suppose that there is a unit of measure contained x times in $\overline{CD}$ and y times in $\overline{DA}$. Then

$$\frac{CD}{DA} = \frac{x}{y}$$

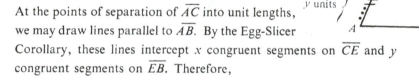

At the points of separation of $\overline{AC}$ into unit lengths, we may draw lines parallel to $\overline{AB}$. By the Egg-Slicer Corollary, these lines intercept x congruent segments on $\overline{CE}$ and y congruent segments on $\overline{EB}$. Therefore,

$$\frac{CE}{EB} = \frac{x}{y}$$

so by transitivity we have

$$\frac{CD}{DA} = \frac{CE}{EB}$$

In Problem Set 6.2, problems 13(b) and (c), you were asked, in effect, to prove the following easy and useful corollary of this theorem. If you have not done this, do it now. (*Hint:* Use Lemma 6.1.)

COROLLARY T56 *The Basic Proportionality Corollary* (BPC): A line parallel to one side of a triangle and intersecting the interiors of the

other two sides separates them into segments which are proportional
to these sides.

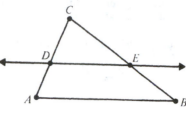

In terms of the diagram, we have the following:

Given: $\triangle ABC$; $A\text{-}D\text{-}C$; $B\text{-}E\text{-}C$; and $\overset{\leftrightarrow}{DE} \parallel \overline{AB}$
Prove:

(a) $\dfrac{CD}{CA} = \dfrac{CE}{CB}$ (b) $\dfrac{DA}{CA} = \dfrac{EB}{CB}$

Example 2 $CD = 3$, $AC = 7$, and $CE = 4$.
Find EB.

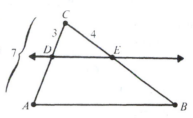

Method (a): By the definition of *between*, we
may conclude that

$$AD + 3 = 7 \quad \text{or} \quad AD = 4$$

Thus by the BPT,

$$\frac{3}{4} = \frac{4}{EB}$$

Solving for EB, we obtain

$$EB = \frac{16}{3}$$

Method (b): From the BPC and the definition of *between*, we
conclude that

$$\frac{4}{4 + EB} = \frac{3}{7}$$

Solving for EB, we obtain

$$28 = 3(4 + EB) \quad \text{or} \quad 28 = 12 + 3(EB)$$

Therefore,

$$EB = \frac{28 - 12}{3} = \frac{16}{3}$$

The following converse of the BPC is also well worth verifying.

THEOREM 57 If a line intersects two sides of a triangle, separating them into segments which are proportional to these sides, then it is parallel to the third side.

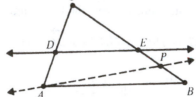

Given: $\triangle ABC$; A-D-C; B-E-C; and

$$\frac{CD}{CA} = \frac{CE}{CB}$$

Prove: $\overleftrightarrow{DE} \parallel \overline{AB}$

Proof (indirect): Suppose $\overleftrightarrow{DE} \not\parallel \overline{AB}$. Then through A we may introduce $\overleftrightarrow{AP}$, $\overleftrightarrow{AP} \parallel \overleftrightarrow{DE}$, $P \in \overleftrightarrow{BC}$, $P \neq B$. Considering $\triangle APC$, the BPC implies

$$\frac{CD}{CA} = \frac{CE}{CP}$$

We already have

$$\frac{CD}{CA} = \frac{CE}{CB}$$

and a slight bit of algebraic maneuvering will lead to $CP = CB$ from which it follows (by the Point-Plotting Postulate) that $P = B$. →|←

Theorems 56 and 57 are extremely useful in many branches of mathematics. You will find applications of the ideas contained in them later in this course (in Chapter 10 for example) as well as in the trigonometry and calculus courses you may take in the future.

PROBLEM SET 6.3

1. $\triangle ABC \sim \triangle DEF$, and the lengths of the corresponding sides are as marked. Find DE and EF. What is the ratio of similitude for these two triangles?

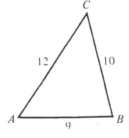
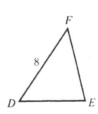

2. What is the ratio of similitude for congruent triangles?

3. The sides of a triangle are 2, 3, and 4 in., respectively. The longest side of a similar triangle is 6 in. Find the lengths of the other two sides.

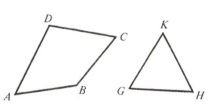

4. Can figure $ABCD$ be similar to $\triangle GHK$? Explain.

5. Given $\triangle ABC \sim \triangle DEF$.

    (a) If ratio of similitude is $\frac{4}{3}$ and $AB = 6$, $BC = 10$, and $AC = 12$, find DE, EF, and DF.

    (b) If $AB = 6$ and $EF = 5$, solve for BC in terms of DE.

    (c) If the ratio of similitude is $\frac{1}{3}$ and $m \angle A = 30$, find $m \angle D$.

    (d) $\triangle ABC$ is equilateral. $AB = 5$, and $DE = 7$. Find EF and DF.

6. Prove T55

7. (a) If $CD = 6$, $DA = 4$, and $CE = 8$, find EB

    (b) If $CD = 4$, $DA = 5$, and $EB = 7$, find CE

    (c) If $CD = 4$, $CA = 7$, and $CB = 12$, find CE

    (d) If $CD = 4$, $CA = 7$, and $EB = 3$, find CB

    (e) $CD = x$, $DA = x + 2$, $CE = x - 1$, and $EB = x$; solve for x

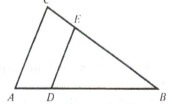

$\overline{DE} \parallel \overline{AB}$

8. Which of the following sets of lengths imply that $\overline{AC} \parallel \overline{DE}$?

    (a) $AD = 4$, $DB = 4$, $CE = 3$, and $EB = 3$

    (b) $AD = 4$, $DB = 8$, $CE = 3$, and $CB = 9$

    (c) $AB = 10$, $DB = 7$, $BC = 15$, and $EB = 10$

    (d) $AB = 7\frac{1}{2}$, $AD = 5$, $CB = 9$, and $EB = 6$

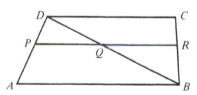

9. In the figure, $\overline{AB} \parallel \overline{PR} \parallel \overline{DC}$,

$$\frac{DQ}{QB} = \frac{3}{4}$$

$AD = 14$, and $BC = 12$. Find AP, PD, RC, and BR.

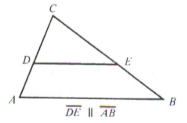

10. In the figure, find AC and BC.

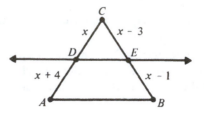

11. In the figure, $\triangle ABC \sim \triangle EDC$, $CE = 6$, $BE = 4$, and $AC = AD + 3$.
Find AC.

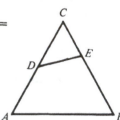

12. Given: $\triangle ABC$; and D, E, F are midpoints
Prove: $\triangle ABC \sim \triangle EFD$

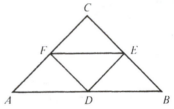

13. Given: $\triangle AFD \sim \triangle DEC$
Prove: $\angle B \cong \angle FDE$

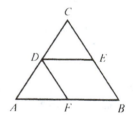

14. Use the accompanying figure to prove: If three parallel lines are cut by two transversals, the intercepted segments on the two transversals, are proportional.

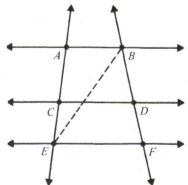

6.4 CONDITIONS GUARANTEEING SIMILARITY

We are now ready to reduce the number of conditions which must be met before we can state that two triangles are similar, much as we did in the case of triangle congruence.

THEOREM 58 *The Angle-Angle-Angle Similarity Theorem* (AAA∼): Given a one-to-one correspondence between the vertices of two triangles, or a triangle and itself. If the three angles of one of the triangles are congruent to the corresponding angles of the other, then the triangles are similar.

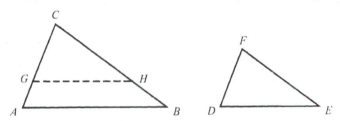

In terms of the figure, we have

Hypothesis: $\angle A \cong \angle D$; $\angle B \cong \angle E$; and $\angle C \cong \angle F$
Conclusion: $\triangle ABC \sim \triangle DEF$

Proof: Since the corresponding angles are congruent, we need only show that the corresponding sides are proportional. On $\overrightarrow{CA}$ and $\overrightarrow{CB}$, respectively, there exist points G and H such that $\overline{CG} \cong \overline{FD}$ and $\overline{CH} \cong \overline{FE}$. Since $\angle C \cong \angle F$, we may conclude that $\triangle GCH \cong \triangle DFE$. By CPCTC we have $\angle CGH \cong \angle D$. But $\angle D \cong \angle A$. Therefore, $\angle CGH \cong \angle A$. Then it follows from T41 that $\overline{GH} \parallel \overline{AB}$. The Basic Proportionality Corollary can now be used to conclude that

$$\frac{CG}{AC} = \frac{CH}{BC}$$

However, $CG = FD$ and $CH = FE$. Thus we may write

$$\frac{FD}{AC} = \frac{FE}{BC}$$

Note that these are ratios of corresponding sides. To show that

$$\frac{FD}{AC} = \frac{DE}{AB}$$

we proceed analogously. Two easy corollaries follow.

177

COROLLARY T58(a) *The Angle-Angle Corollary* (AA~): If two angles of one triangle are congruent to two angles of a second triangle, then the triangles are similar.

COROLLARY T58(b) A line parallel to one side of a triangle and intersecting the interiors of the other two sides determines a triangle similar to the given one.

Example 1 In the figure, $\overline{DE} \parallel \overline{AB}$, $AD = 2$, $DC = 3$, and $AB = 4$.
Find DE.

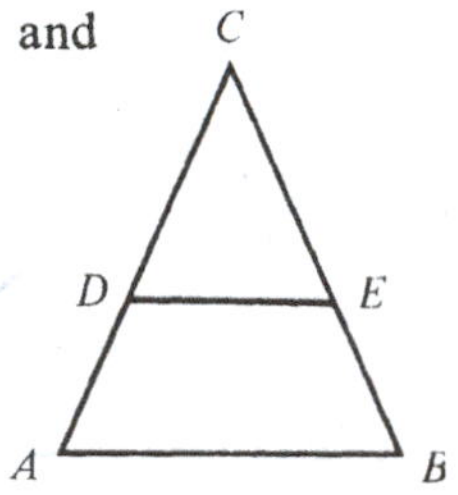

$$\frac{DE}{AB} = \frac{CD}{AC}$$

implies

$$\frac{DE}{4} = \frac{3}{5}$$

Therefore,

$$DE = 4 \times \frac{3}{5} = \frac{12}{5} = 2\frac{2}{5}$$

From the properties of congruence for angles and Lemma 6.1, we can derive a helpful lemma.

LEMMA 6.2 Triangle similarity is reflexive, symmetric, and transitive.

The proof of this lemma is left to the reader. It finds its application in the proofs of our next two theorems.

THEOREM 59 *The Side-Angle-Side Similarity Theorem* (SAS~): Given a one-to-one correspondence between the vertices of two triangles, or a triangle and itself. If two sides of one triangle are proportional to the corresponding sides of the other and the included angles are congruent, then the triangles are similar.

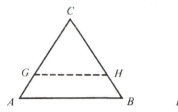

Hypothesis: $\triangle ABC$; $\triangle DEF$; $\angle C \cong \angle F$; and

$$\frac{AC}{DF} = \frac{BC}{EF}$$

Conclusion: $\triangle ABC \sim \triangle DEF$

Proof: (an outline): Assuming $AC > DF$, we introduce $\overline{GH} \parallel \overline{AB}$ such that $GC = DF$ and $HC = EF$. Then $\triangle GHC \cong \triangle DEF$ and $\triangle ABC \sim \triangle GHC$, which implies that $\triangle ABC \sim \triangle DEF$.

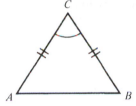
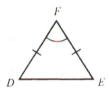

Example 2   Suppose two isosceles triangles have congruent vertex angles, as in the figure. Obviously,

$$\frac{AC}{DF} = \frac{BC}{EF}$$

Therefore, the SAS$\sim$ Theorem applies, and we may write $\triangle ABC \sim \triangle DEF$.

THEOREM 60   *The Side-Side-Side Similarity Theorem* (SSS$\sim$): Given a one-to-one correspondence between the vertices of two triangles, or a triangle and itself. If the corresponding sides are proportional, then the triangles are similar.

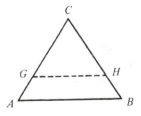
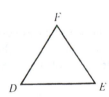

Hypothesis: $\triangle ABC$; $\triangle DEF$ with

$$\frac{AB}{DE} = \frac{BC}{EF} = \frac{AC}{DF}$$

Prove: $\triangle ABC \sim \triangle DEF$

Proof:

STATEMENTS	REASONS
1. $\dfrac{AB}{DE} = \dfrac{BC}{EF} = \dfrac{AC}{DF}$	1. Given
2. On $\overrightarrow{CA}$ and $\overrightarrow{CB}$ there exist G and H, respectively, such that $GC = DF$ and $HC = EF$	2. Point-Plotting postulate
3. $\angle C \cong \angle C$	3. Reflexive law
4. $\dfrac{BC}{HC} = \dfrac{AC}{GC}$	4. S1, S2, and substitution
5. $\triangle GHC \sim \triangle ABC$	5. Why?
6. Therefore, $\dfrac{GH}{AB} = \dfrac{HC}{BC}$	6. Why?
7. $GH = AB\,\dfrac{HC}{BC}$	7. Multiplication law
8. Therefore, $GH = AB\,\dfrac{EF}{BC}$	8. S2, S7, and substitution
9. But $DE = AB\,\dfrac{EF}{BC}$	9. S1 and multiplication law
10. Therefore, $GH = DE$	10. Transitivity
11. $\triangle GHC \cong \triangle DEF$	11. S2, S10, and ?
12. $\triangle GHC \sim \triangle DEF$	12. Why?
13. Therefore, $\triangle ABC \sim \triangle DEF$	13. Similarity is transitive

The student who does not feel comfortable manipulating expressions that involve square roots (radicals) should review that material in Appendix A.

PROBLEM SET 6.4

1. Indicate which of the following figures contain enough information to prove a pair of triangles to be similar. State the similarity relationship and indicate which of AA$\sim$, SAS$\sim$, and SSS$\sim$ you would use to prove it.

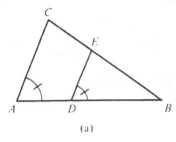

(a)

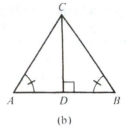

(b)

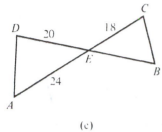

(c)

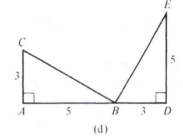

(d)

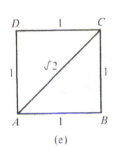

(e)

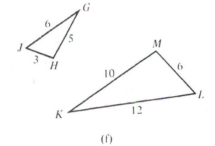

(f)

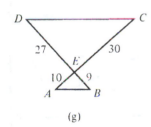

(g)

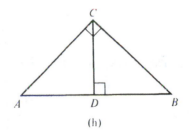

(h)

2. In the figure, $\overline{DE} \parallel \overline{BC}$.

   (a) If $AD = 6$ and $DC = 4$, what is $\dfrac{AE}{AB}$?

   (b) If $\dfrac{AE}{EB} = 2$, what is $\dfrac{AD}{DC}$?

   (c) If $AE = 5$ and $EB = 4$, what is $\dfrac{DE}{BC}$?

   (d) If $AD = 6$, $DC = 4$, and $BC = 9$, find DE

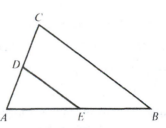

3. A line from the top of a flagpole to the ground just passes over the top of a smaller pole 6 ft high and meets the ground at a point 4 ft from the base of the smaller. How high is the flagpole if its base is 10 ft. from the base of the smaller pole?

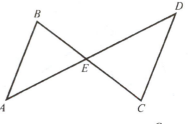

4. Given: $\overline{AB} \parallel \overline{DC}$

Prove: $\triangle ABE \sim \triangle DCE$

5. (a) If $\overline{AB} \parallel \overline{DE}$ and $\overline{AC} \parallel \overline{BD}$, which triangles in the figure are similar?

(b) Prove your answer to part (a).

6. Given $\triangle DEF$, $DF = 20$, $DE = 16$, and $EF = 24$. If $D\text{-}H\text{-}F$, $E\text{-}K\text{-}F$, and $FH = 15$, and $FK = 18$, find HK.

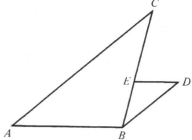

7. Given the information indicated in the figure.

(a) State the similarity relationship between the two triangles.

(b) What theorem justifies this?

(c) Solve for BD.

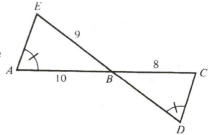

8. Given: $\overline{DA} \perp \overline{AB}$; $\overline{CE} \perp \overline{DB}$; and $\overline{CB} \perp \overline{AB}$

Prove: $\triangle ABD \sim \triangle ECB$

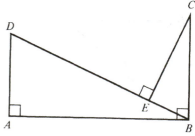

9. *The Ironing Board Theorem, II*

Given: $AE = k \cdot EC$; $BE = k \cdot ED$

Prove: $\overleftrightarrow{DC} \parallel \overleftrightarrow{AB}$.

10. Use the SAS$\sim$ theorem to prove that corresponding medians of similar triangles have the same ratio as any two corresponding sides.

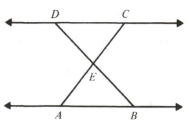

11. Use the AA~ corollary to prove that the corresponding altitudes of similar triangles have the same ratio as any two corresponding sides.

12. State the converse of the SSS~ theorem. Is is true? Explain.

13. How are Corollary T58(b), T55, and Lemma 6.2 each used in the proof of the SAS~ theorem?

14. Use Corollary T58(a) to prove: The segment joining the midpoints of two sides of a triangle is parallel to the third side.

·15. Given: $\triangle ABC$; $\angle A = \angle B$; A-D-C; B-E-C; and $\overline{AD} \cong \overline{BE}$
     Prove: $\triangle ABC \sim \triangle DEC$

16. Indicate with a sketch why there is no SSA~ theorem.

17. Supply the missing reasons in the proof of T60.

6.5 RIGHT TRIANGLE SIMILARITIES AND THE PYTHAGOREAN THEOREM

In problem 1(h) of Problem Set 6.4 you were led to infer our next theorem, which readily follows from the AA~ corollary.

THEOREM 61   The altitude to the hypotenuse of a right triangle forms two triangles, each of which is similar to the other and to the original right triangle.

Hypothesis: $\triangle ABC$ is a right triangle with right
             angle $\angle C$, and $\overline{CD}$ is the altitude to
             the hypotenuse
Conclusion: $\triangle ABC \sim \triangle ACD \sim \triangle CBD$

Proof: Consider $\triangle ABC$ and $\triangle ACD$. Each triangle has a right angle and acute angle $\angle A$. Therefore, by the AA~ corollary,

$$\triangle ABC \sim \triangle ACD$$

Next, consider $\triangle ABC$ and $\triangle CBD$. Each triangle has a right angle and acute angle $\angle B$. It follows then that

$$\triangle ABC \sim \triangle CBD$$

Since similarity is symmetric and transitive, we may write

$$\triangle ABC \sim \triangle ACD \sim \triangle CBD$$

Consider the accompanying figure. Since $\triangle ABC \sim \triangle ACD$, we may write

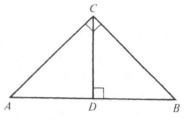

$$\frac{AD}{AC} = \frac{AC}{AB}$$

That is, AC is the geometric mean between AD and AB. Likewise, $\triangle ABC \sim \triangle CBD$ implies that

$$\frac{BD}{BC} = \frac{BC}{AB}$$

or, BC is the geometric mean between BD and AB.

Finally, $\triangle ACD \sim \triangle CBD$ implies that

$$\frac{AD}{CD} = \frac{CD}{BD}$$

or, CD is the geometric mean between AD and BD.

We can summarize these results as a theorem.

THEOREM 62 Given a right triangle and an altitude to the hypotenuse:

(a) The altitude to the hypotenuse is the geometric mean between the segments into which it separates the hypotenuse.

(b) Each leg is the geometric mean between the hypotenuse and the segment determined by the altitude to the hypotenuse which is adjacent to that leg.

Example 1 Given the right triangle and lengths indicated, T62 allows us to write the following:

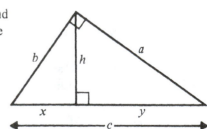

(a) $h = \sqrt{xy}$, or $h^2 = xy$
(b) $a = \sqrt{cy}$, or $a^2 = cy$
(c) $b = \sqrt{cx}$, or $b^2 = cx$

Example 2 In the figure, $CD = 3$, $BD = 4$.

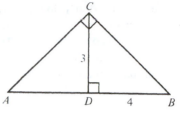

(a) Then T62(a) implies

$$3 = \sqrt{4 \cdot AD}$$

$$9 = 4 \cdot AD$$

$$AD = \tfrac{9}{4} = 2\tfrac{1}{4}$$

Thus

$$AB = 4 + 2\tfrac{1}{4} = 6\tfrac{1}{4}$$

(b) And T62(b) can then be used to write

$$(BC)^2 = 4 \cdot (6\tfrac{1}{4}) = 4 \cdot \tfrac{25}{4} = 25$$

Thus

$$BC = 5$$

We are now in a position to prove the most famous theorem of all mathematics. It is probably the most important as well.

THEOREM 63 *The Pythagorean Theorem* (PT): The square of the length of the hypotenuse of a right triangle is equal to the sum of the squares of the lengths of the legs.

Proof: Consider the right triangle in the figure, where $\overline{CD}$ is the altitude to the hypotenuse. As we have seen (Example 1), T62 implies that

$$a^2 = cy \quad \text{and} \quad b^2 = cx$$

Therefore,

$$a^2 + b^2 = cy + cx$$

$$a^2 + b^2 = c(y + x) \quad \text{(distributive law, Appendix A)}$$

But $y + x = c$. (Why?) Therefore,

$$a^2 + b^2 = c^2$$

Although hundreds of proofs of this theorem exist, few are as direct as this one. It is not known how, or even whether, Pythagoras himself actually proved it. That a proof was developed by the

Pythagoreans seems fairly likely, however. They probably used a method of proof similar to one you will use in an exercise in Chapter 7, after we discuss the concept of area. Tradition tells us Pythagoras was so overjoyed with the discovery that he celebrated the occasion by sacrificing an ox to the gods.

Example 3   If the lengths of the legs of a right triangle are 8 and 15, we can find the length of the hypotenuse as follows:
The PT implies that $8^2 + 15^2 = c^2$, c the length of the hypotenuse, so that

$$64 + 225 = c^2$$

$$c^2 = 289$$

$$c = \sqrt{289} = 17$$

Example 4   In the figure, we find x by noting that the PT implies

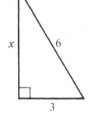

$$x^2 + 3^2 = 6^2$$

$$x^2 = 36 - 9 = 27$$

$$x = \sqrt{27} = \sqrt{9 \cdot 3} = 3\sqrt{3}$$

THEOREM 64   *Converse of the Pythagorean Theorem:* If a triangle has sides of lengths a, b, and c, and $a^2 + b^2 = c^2$, then the triangle is a right triangle and the side of length c is the hypotenuse.

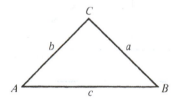
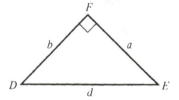

Proof: Suppose we are given $\triangle ABC$ with $a^2 + b^2 = c^2$ and *right* triangle $\triangle DEF$ with lengths of legs a and b and hypotenuse d. Then, by the PT, $a^2 + b^2 = d^2$. But $a^2 + b^2 = c^2$. Therefore, $c = d$ and $\triangle ABC \cong \triangle DEF$ by the SSS theorem. Therefore, $\angle C \cong \angle F$ and since F is a right angle, so is $\angle C$.

Applications of the PT to problems outside of mathematics abound. Many, such as the following example, involve measurement.

Example 5 Two ships leave the same West Coast port P at
12 noon. Ship A travels due west at 20 miles per hour. Ship B
travels due south at 25 miles per hour. After two hours, their
positions are shown by the adjacent figure, where $PA = 40$ miles
and $PB = 50$ miles. Therefore, the square of their distance apart
is given by

$$(AB)^2 = (PA)^2 + (PB)^2$$

$$= 1600 + 2500 = 4100$$

Thus their distance apart is

$$AB = \sqrt{4100} = 10\sqrt{41}$$

$$\approx 64.03 \text{ miles (to the nearest hundredth)}$$

PROBLEM SET 6.5

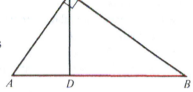

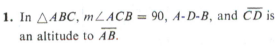

1. In $\triangle ABC$, $m\angle ACB = 90$, A-D-B, and $\overline{CD}$ is
an altitude to $\overline{AB}$.

    (a) Find AB if $CD = 6$ and $DB = 3$
    (b) Find BD if $AD = 4$ and $CD = 8$
    (c) Find CD if $AB = 20$ and $DB = 4$
    (d) Find AC if $BD = 5$ and $AD = 3$
    (e) Find BD if $CB = 6$ and $AD = 5$

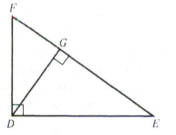

2. Given right $\triangle DEF$ and altitude to the hypotenuse $\overline{DG}$. If $FG = 9$ and $DG = 12$, find FD,
DE, and EF.

3. Which of the following triples of numbers could be lengths of the
sides of right triangles?

    (a) 3, 4, 5   (e) 5, 12, 13
    (b) 7, 24, 25   (f) 9, 13, 15
    (c) 1, 1, $\sqrt{2}$ (g) 2, 4, $2\sqrt{3}$
    (d) 2, 3, 4   (h) $2\frac{1}{2}$, $4\frac{1}{4}$, $5\frac{1}{4}$

4. If the length of a leg of an isosceles right triangle is 6, find the
altitude to the hypotenuse.

5. Given the right triangle $\triangle ABC$.

   (a) If $b = 6$ and $c = 8$, find a
   (b) If $a = 12$ and $b = 6$, find c
   (c) If $b = 4$ and $c = 4$, find a
   (d) If $a = 2$ and $c = \sqrt{3}$, find b

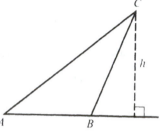

6. A man walks 400 yards (yd) due south, and then 300 yd due east. How far is he from his starting point?

7. A plane flies 80 miles north, 60 miles east, and 20 miles south. How far is the plane from its starting point?

8. If a and b are lengths of the legs and c the length of the hypotenuse of a right triangle, show that for any positive number k, ka, kb, and kc are also lengths of the sides of a right triangle.

9. In the figure, $AB = 3$, $BC = 7$, and $AC = 9$. Find h, the altitude to $\overline{AB}$.

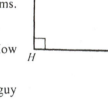

10. On the baseball diamond depicted in the figure, consecutive base paths are perpendicular. Let $H =$ home plate, $F =$ first base, $S =$ second base, $T =$ third base, $P =$ pitcher's mound. The bases are 90 ft apart and $PH = 60\frac{1}{2}$ ft. Assume that the pitcher's mound and the bases are points, and solve the following problems.

   (a) Find the distance from home plate to second base.
   (b) On many infield plays, the pitcher must run to first base. How far is it from the pitcher's mound to first base?

11. A TV antenna is being tied to a flat horizontal roof using three guy wires. Each wire is fastened to the pole at a point 20 ft above the roof and fastened to the roof at a point 12 ft from the foot of the pole. Approximately how many feet of wire will be necessary?

12. Show that the numbers

$$n, \quad \frac{n^2 - 4}{4}, \quad \frac{n^2 + 4}{4}$$

may represent the lengths of sides of a right triangle. What additional condition will allow us to change the word "may" to "must"?

13. Find the length of each side of a right triangle if the second longest side is 4 in. longer than twice the length of the shortest side, and the longest side is 2 in. longer than the second longest side.

14. Find the length of an altitude for an equilateral triangle with side of length 2.

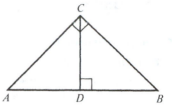

15. Given right triangle $\triangle ABC$, hypotenuse $\overline{AB}$, attitude to the hypotenuse $\overline{CD}$. If $CD = 12$ and $AB = 25$, find AC and BC.

16. Given: $\overline{BG} \parallel \overline{CD}$; $\overline{GF} \parallel \overline{DE}$

Prove: $\dfrac{AF}{AB} = \dfrac{EF}{BC}$

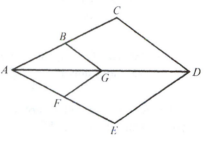

17. If, in the PT, a, b, and c, the lengths of the legs and the hypotenuse, respectively, are integers, then they are called a *Pythagorean triple*. Examples are 3, 4, 5 and 5, 12, 13 (see problem 3 above). The Greeks of Pythagoras's time knew that if u and v are relatively prime (no common whole number divisors greater than 1) and if $u > v$, then $a = 2uv$, $b = u^2 - v^2$, and $c = u^2 + v^2$ constitute a Pythagorean triple. For example, let $u = 2$, $v = 1$ and then $a = 2 \cdot 2 \cdot 1 = 4$, $b = 2^2 - 1^2 = 3$, and $c = 2^2 + 1^2 = 5$.

(a) Generate the Pythagorean triples needed to complete the following chart:

u	v	a	b	c
3	1			
4	2			
5	3			
4	3			
5	4			
7	2			

(b) Notice that in the first three rows, u and v have the same parity (i.e., they are either both odd or both even). Does it appear that triples produced from a u and v with the same parity differ in some significant way from those generated by a u and v with different parity? Make a conjecture.

6.6 SOME APPLICATIONS OF THE PYTHAGOREAN THEOREM

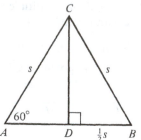

Consider equilateral triangle $\triangle ABC$, which has side of length s, as in the figure, with altitude $\overline{CD}$. $\overline{CD}$ is also a median. Hence $DB = \frac{1}{2}s$. $m\angle B = 60$, which implies that $m\angle DCB = 30$. Since $\triangle DBC$ is similar to every other 30-60-90 right triangle, we can generalize this result.

THEOREM 65 One leg of a right triangle is half as long as the hypotenuse if and only if the angle opposite that leg has measure 30.

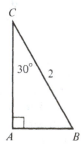

The following corollaries of T65 can be swiftly obtained using the Pythagorean Theorem (PT).

COROLLARY T65(a) Given an equilateral triangle with side of length s. The length of each altitude is $\dfrac{s\sqrt{3}}{2}$.

The proof is left as an exercise.

COROLLARY T65(b) Given a 30-60-90 right triangle. The length of the side opposite the 60° angle is $\sqrt{3}/2$ times the length of the hypotenuse, and $\sqrt{3}$ times the length of the side opposite the 30° angle.

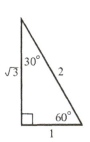

Again the proof is easy and is left as an exercise.

The triangle shown here might be considered the prototype of all 30-60-90 triangles. It is similar to all other such triangles and can be a useful tool while performing certain calculations.

Example 1 Given $\triangle ABC$, $AC = 6$. $m\angle C = 30$, $m\angle B = 60$. Since $\triangle ABC$ is similar to the prototype, we may write

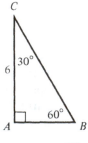

$$\frac{6}{\sqrt{3}} = \frac{AB}{1}$$

$$AB = \frac{6}{\sqrt{3}} = \frac{6\sqrt{3}}{3} = 2\sqrt{3}$$

Also

$$\frac{6}{\sqrt{3}} = \frac{BC}{2}$$

$$BC = \frac{12}{\sqrt{3}} = \frac{12\sqrt{3}}{3} = 4\sqrt{3}$$

This second result could also be obtained directly from Corollary T65(b).

Consider isosceles right triangle $\triangle ABC$ with sides of length a and hypotenuse b. The PT implies that

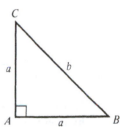

$$a^2 + a^2 = b^2$$

$$2a^2 = b^2$$

a and b are positive real numbers. Therefore,

$$\sqrt{2a^2} = \sqrt{b^2}$$

$$\sqrt{2}\, a = b$$

We state this result as a theorem.

THEOREM 66 The hypotenuse of an isosceles right triangle is $\sqrt{2}$ times as long as a leg.

Again, we may consider the triangle shown as the prototype of all isosceles right triangles, since it is similar to all other isosceles right triangles.

Example 2   Given isosceles right triangle $\triangle ABC$, right angle $\angle A$.

(a) Suppose one leg has length 13. Then by T66, the hypotenuse has length $13\sqrt{2}$.

(b) Suppose the hypotenuse has length 3. Then, since $\triangle ABC$ is similar to the prototype, we may find x, the length of a leg as follows:

$$\frac{x}{1} = \frac{3}{\sqrt{2}}, \qquad x = \frac{3}{\sqrt{2}} = \frac{3\sqrt{2}}{2}$$

Example 3 *Measuring the height of a monument:* A tourist with an angle measuring device (a protractor usually works) can pace off the distance AB in the figure, and then easily calculate CD as

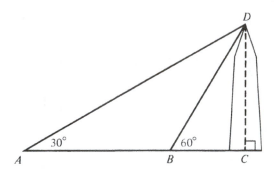

follows. Let $AB = x$. Then $BD = x$. (Why?) But BD is the hypotenuse of $\triangle BDC$. Therefore, by Corollary T65(b),

$$DC = \frac{\sqrt{3}}{2}(BD) = \frac{\sqrt{3}}{2}x$$

PROBLEM SET 6.6

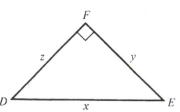

1. $\triangle ABC$ is a 30-60-90 triangle, and $m\angle B = 60$. Find the length of the other sides, given the following:

(a) $z = 4$ (d) $x = 2\sqrt{3}$
(b) $x = 3$ (e) $y = 4$
(c) $y = 3\sqrt{3}$ (f) $z = 2\sqrt{3}$

2. $\triangle DEF$ is an isosceles right triangle, $m\angle F = 90$. Find the lengths of the other sides, given the following:

(a) $x = 2\sqrt{2}$ (c) $x = 3$
(b) $y = 4$ (d) $z = 2\sqrt{2}$

3. Find the altitude of an equilateral triangle with side of length 6.

4. If $\triangle ABC$ is a 30-60-90 triangle with right angle $\angle C$, and $AC + BA = 6$, find BC.

5. $\triangle DEF$ is an isosceles right triangle. $\angle F$ is a right triangle.

(a) If $DF + FE = 24$, find DE
(b) If $DE + EF = 10$, find DF (*Hint:* let $DF = x$ and express DE and EF in terms of x.)

6. Given the figure, $m\angle A = 45$, $m\angle B = 30$, $\overline{CD} \perp \overline{AB}$, and $CD = 6$. Calculate:

(a) AC　　(b) BC　　(c) DB　　(d) AB

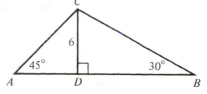

7. The measure of a base angle of an isosceles triangle is 30. The base has length 6.

(a) Find the lengths of the other sides.
(b) Calculate the altitude to the base.

8. Given $\triangle ABC$, $AB = 5\sqrt{3}$, $BC = 10\sqrt{3}$, and $AC = 15$. Find:

(a) $m\angle A$　　　　　(b) $m\angle B$

9. *Measuring the distance across a river:* Starting at A, directly across the river from C, a surveyor walks to point B, where he measures the angle between his path and his line of sight to C. If $AB = 100$ yd, find AC, the distance across the river if

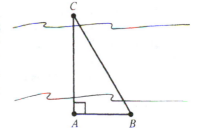

(a) $m\angle B = 45$　　　　(b) $m\angle B = 60$

10. Find the altitude of an equilateral triangle if a side has length 10.

11. The measure of a base angle of an isosceles triangle is 45. The base has length 6.

(a) Find the lengths of the other sides.
(b) Calculate the altitude to the base.

12. Given $\triangle DEF$, $DE = 2\sqrt{2}$, $EF = 2\sqrt{2}$, and $DF = 4$. Find:

(a) $m\angle D$　　　　　(b) $m\angle E$

13. If two of the altitudes of an isosceles right triangle are 6 in., find the length of the third altitude.

14. In the figure, $m\angle D = 45$, $DE = 5$, and $DF = 14$. Find EF.

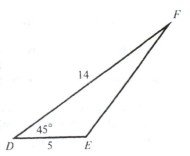

CHAPTER 6: REVIEW EXERCISES

1. For each of the following statements, indicate whether it is *always* true, *sometimes* but not always true, or *never* true.

   (a) A statement that two or more ratios are equal is a proportion.
   (b) If $ac = b^2$, then b is the geometric mean of a and c.
   (c) If two triangles are similar, they are congruent.
   (d) If two triangles are congruent, they are similar.
   (e) Every pair of equilateral triangles are similar.
   (f) Every pair of isosceles right triangles are similar.
   (g) Let $a > 0$, $b > 0$. Then the geometric mean between a and b is greater than $(a + b)/2$.
   (h) If the altitude to the hypotenuse of a right triangle is one half the length of a leg of that triangle, then the triangle is isosceles.
   (i) If the altitude to the hypotenuse of a right triangle is one half the length of the hypotenuse, then the triangle is isosceles.
   (j) Given $\triangle ABC$ and $\triangle DEF$. If $\angle A \cong \angle E$, and $\angle B \cong \angle F$, then $\triangle ABC \sim \triangle DEF$.
   (k) Given $\triangle ABC$ and exterior angle $\angle CBD$. If $\triangle DBC \sim \triangle ABC$, then $\angle CBD$ is obtuse.

2. Complete the following statements:

   (a) If $3x = 4y$, then $\dfrac{x}{y} = $ _____

   (b) If $\dfrac{5}{x} = \dfrac{2}{3}$, then $x = $ _____

   (c) If $\dfrac{x + y}{y} = \dfrac{24}{20}$, then $\dfrac{x}{y} = $ _____

   (d) $\dfrac{2}{3} = \dfrac{-}{9} = \dfrac{8}{-} = \dfrac{3}{-}$

3. Find the geometric mean between the following pairs:

   (a) $(18, 32)$　　　　(b) $(3\sqrt{3}, 9\sqrt{3})$　　　(c) $(3\frac{2}{3}, 2\frac{5}{11})$

4. If 3 is the geometric mean between 2 and p, find p.

5. Sketch two 4-sided figures whose corresponding sides are proportional, but which are not similar—that is, which are not the same size and the same shape.

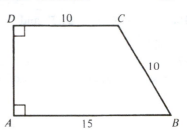

6. Sketch two 4-sided figures whose corresponding angles are congruent, but which are not similar.

7. Given the figure, find AD.

8. In the figure, $AB = 2$, $BE = 3$, and $DC = 4$. Find:

    (a) DE (b) EC (c) AC

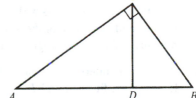

9. If the altitude to the hypotenuse of a right triangle is half the length of one of the legs, what is the measure of the angle opposite that leg?

10. The shadow of a telephone pole is 20 ft long. The shadow of a 6-ft tall man is 5 ft long. How high is the telephone pole?

11. The lengths of the sides of a triangle are 10, 15, and 20 in. A 6-in. segment parallel to the shortest side has its endpoints in the other two sides. Find the lengths of the segments into which the other two sides are separated.

12. Given: $\triangle ABC$ with altitudes $\overline{BD}$ and $\overline{AE}$
    Prove: (a) $\triangle AEC \sim \triangle BDC$
           (b) In $\triangle ABC$, the product of an altitude and its corresponding base is a constant

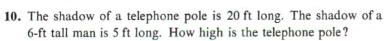

13. Given: $\triangle ABC$ is a right triangle with right angle C, and

$$\frac{AD}{AC} = \frac{AC}{AB}$$

    Prove: $CD \perp AB$

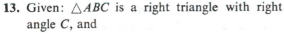

14. Given: $\triangle EAF \sim \triangle CBF$; $\overline{AE} \cong \overline{BC}$; and F is the midpoint of $\overline{EC}$
    Prove: $\triangle ECD$ is isosceles

15. The altitude to the hypotenuse of a right triangle separates the hypotenuse into segments of length 5 and 20. Find the length of that altitude and the legs.

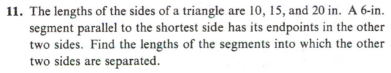
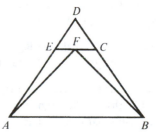

16. If $\triangle ABC \sim \triangle DEF$ and $\triangle ABC \sim \triangle EDF$, what can you conclude?

17. If a side of one equilateral triangle is congruent to an altitude of a second, what is their ratio of similitude?

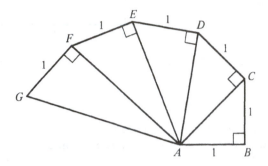

18. Calculate AC, AD, AE, AF, and AG. Two patterns are emerging here. State them as well as you can.

19. Can the altitude to the hypotenuse of a right triangle be equal to the length of one of the legs of that triangle?

20. Suppose the geometric mean of a and b is equal to their arithmetic mean (see problem 14 in Problem Set 6.2). What is $a - b$? Explain.

21. The sides of a triangle have lengths 7, 14, and 19. The sum of the lengths of the sides (perimeter) of a similar triangle is 160. What is the length of its longest side?

22. The altitude to the hypotenuse of a right triangle separates the hypotenuse into segments with lengths 3 and 9. Find the lengths of the legs and the length of the altitude of the triangle. What are the measures of the acute angles of the triangle?

23. $\overline{AC}$ and $\overline{BC}$ intersect at E such that $\overline{AB} \parallel \overline{CD}$ and $AB = 2 \cdot CD$. If $AC = 21$, find AE and EC.

24. If the lengths of the legs of a right triangle are a and b, find the length, h, of the altitude to the hypotenuse in terms of a and b.

25. An isosceles triangle has legs of length 13 and a base of length 10. Find length of an altitude to the base.

26. The lengths of the sides of a right triangle are 3 in. and 4 in. Find the lengths of the segments into which the altitude to the hypotenuse separates the base.

27. Draw a 30-60 right triangle. Draw the bisector of the 60° angle. Find the ratio of the lengths of segment into which it separates the opposite side.

28. $\triangle ABC$ is a right triangle with its right angle at C. The bisector of $\angle B$ intersects $\overline{AC}$ at D, and the bisector of the exterior angle at B intersects $\overleftrightarrow{AC}$ at E. If $BD = 9$ and $BE = 12$, what are the lengths of the sides of $\triangle ABC$?

29. A triangle has sides of length 3, 6, and 8. The bisectors of the largest interior angle and the smallest exterior angle intersect the line containing the opposite side at A and B, respectively. If X is the vertex of the smallest angle of the triangle, find AX and BX.

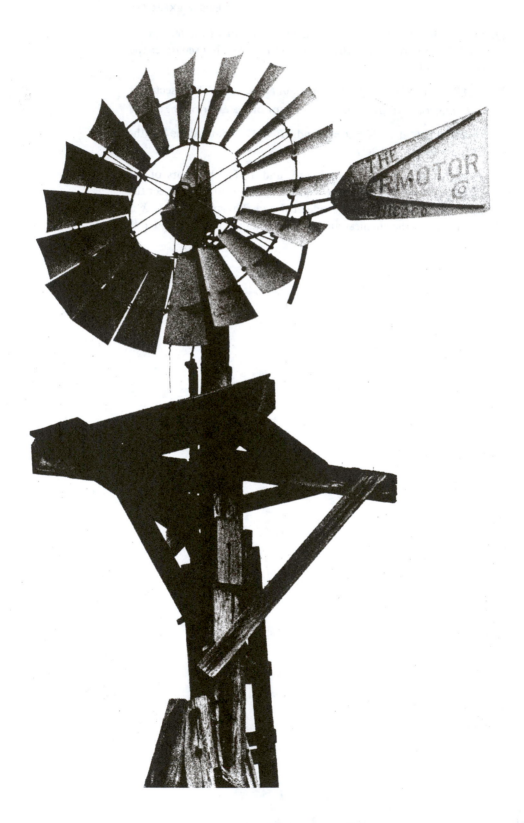

There is something fascinating about science. One gets such wholesale returns of conjecture out of such a trifling investment of fact.

MARK TWAIN (1835–1910)

CHAPTER SEVEN

POLYGONS AND AREA

7.1 INTRODUCTION

In Chapter 4 we defined a *triangle* as the union of the segments joining three noncollinear points. It is fairly obvious that given three noncollinear points, only one trio of such segments is possible. On the other hand, suppose four distinct coplanar points are given, no three of which are collinear. Using these four points as endpoints of four segments, many different-looking figures are possible. Some are pictured in Fig. 7.1. Triangles and sets of points like those depicted in parts (d) and (e) belong to a large class of sets called *polygons* (from Greek, meaning "many angles"). One might say that we form a polygon by fitting three or more segments together end-to-end in such a way that they do not intersect other than at their endpoints.

Notice that parts (d) and (e) differ from each other in that the region bounded by part (e) is convex, whereas the region bounded by part (d) is not. Polygons that are the boundaries of convex regions will receive most of our attention in this chapter.

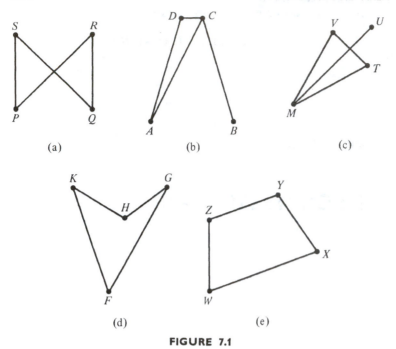

(a)   (b)   (c)

(d)   (e)

FIGURE 7.1

From past experience we know that associated with every polygon is something called its *area*. You probably recall some formulas for computing areas. This chapter provides derivations of many of these area formulas.

7.2 POLYGONS

We can summarize the informal discussion in the introduction with a lengthy definition.

DEFINITION 7.1   Let $P_1, P_2, P_3, \ldots, P_n$, $n \geq 3$, be a sequence of n distinct coplanar points. The union of the n segments $\overline{P_1P_2}, \overline{P_2P_3}, \ldots \overline{P_{n-1}P_n}, \overline{P_nP_1}$ is a *polygon* if both the following conditions are satisfied:

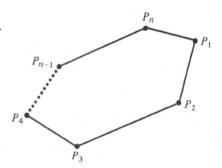

1. No two segments with a common endpoint are collinear.
2. No two segments intersect at an interior point of either segment.

The *n* distinct points are the *vertices* of the polygon, and the *n* segments $\overline{P_1P_2}, \overline{P_2P_3}, \ldots, \overline{P_{n-1}P_n}, \overline{P_nP_1}$ are its *sides*. The sum of the lengths of the sides is called the *perimeter* of the polygon. The end-points of a side are called *consecutive vertices*. Intersecting sides are called *consecutive sides*. An angle determined by a pair of con-secutive sides is called an *angle of the polygon*. If the vertices of the angles of a polygon are consecutive vertices, the angles are called *consecutive angles*. A segment whose endpoints are nonconsecutive vertices of a polygon is called a *diagonal* of that polygon. The following table completes the definition:

NUMBER OF SIDES	CLASSIFICATION OF POLYGON
3	triangle
4	quadrilateral
5	pentagon
6	hexagon
7	heptagon
8	octagon
9	nonagon
10	decagon

Example Consider the pentagon shown. Its five sides are $\overline{AB}$, $\overline{BC}, \overline{CD}, \overline{DE},$ and $\overline{EA}$. We can also name its five angles: $\angle A, \angle B, \angle C, \angle D,$ and $\angle E$. $\overline{AB}$ and $\overline{BC}$ are consecutive sides. $\overline{AB}$ and $\overline{CD}$ are nonconsecutive sides. $\angle C$ and $\angle D$ are con-secutive angles. $\angle C$ and $\angle E$ are nonconsecu-tive angles. The perimeter (a number!) is $AB + BC + CD + DE + EA$, and $\overline{AD}$ is a diagonal of the pentagon.

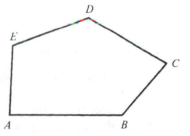

NOTATION RULE 7.1 Suppose $P_1, P_2, P_3, \ldots, P_n$ are the vertices of a polygon as in Definition 7.1, $n \geq 4$. We will use the symbol $P_1P_2P_3 \ldots P_n$ to name such a polygon. Thus, in the preceding example, we have the polygon *ABCDE*.

DEFINITION 7.2 A polygon is *convex* if no two points of it lie on opposite sides of any line containing a side of the polygon. A convex polygon is a *regular polygon* if all of its sides are congruent (equi-lateral) and all of its angles are congruent (equiangular).

From the sketches of polygons shown in Fig. 7.2 it can be seen that to say a polygon is convex is to say that it is the boundary of a convex set.

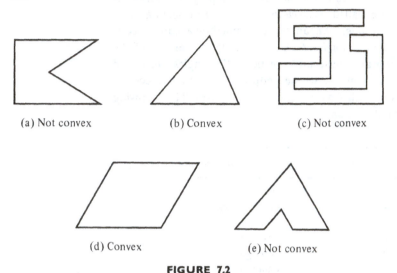

(a) Not convex   (b) Convex   (c) Not convex

(d) Convex   (e) Not convex

FIGURE 7.2

In the following pages we will give most of our attention to polygons that are convex.

PROBLEM SET 7.2

1. Given the polygon shown.

   (a) Classify and name the polygon.
   (b) Name three pairs of consecutive angles.
   (c) Name three pairs of nonconsecutive sides.
   (d) Name all the diagonals which contain A.
   (e) What is the perimeter of this polygon?

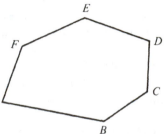

2. Indicate which of the figures (see top of page 203), are polygons. Of these, which are convex? For the figures which are not polygons, indicate which part of Definition 7.1 they fail to satisfy.

3. Is a convex polygon a convex set? Explain.

4. Sketch an equilateral, equiangular polygon that is *not* regular.

5. What are the two other names for a regular triangle?

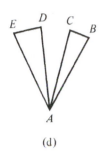
(a)

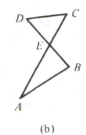
(b)

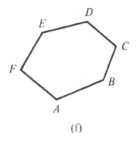
(c)

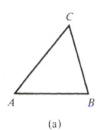
(d)

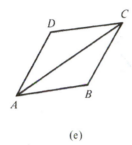
(e)

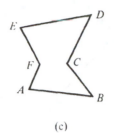
(f)

6. Complete the following chart for *convex* polygons.

POLYGON	NUMBER OF DIAGONALS FROM A SINGLE VERTEX	NUMBER OF VERTICES	NUMBER OF DIAGONALS
triangle	0		
quadrilateral	1	4	2
pentagon			
hexagon	3	6	9
octogon			
n-gon (an n-sided polygon)		n	

7. Using the Angle Sum Theorem and the following figures, calculate the sum of the measures of the angles of a convex

   (a) Quadrilateral.   (b) Pentagon.   (c) Hexagon.

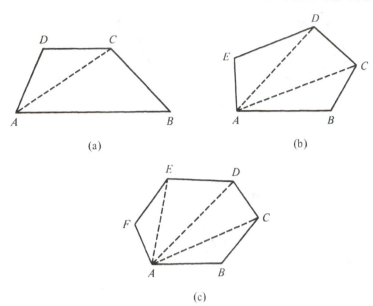

(a)   (b)

(c)

8. The figures for problem 7 should make it easy for you to accept the following generalization:

> The sum of the measures of the angles of
> a convex polygon of n sides is $(n - 2) \cdot 180$.

(a) Use this result to calculate the sum of the measures of the angles of a triangle, a convex hexagon, a convex octagon, and a convex decagon.

(b) Calculate the measure of each angle of a regular quadrilateral, pentagon, and hexagon.

(c) How may sides has a convex polygon if the sum of the measures of its angles is equal to 1800? 2160?

(d) The measure of an angle of a regular polygon is 135. How many sides has the polygon?

9. (a) What is the sum of the measures of the exterior angles (one at each vertex) of a triangle? (*Hint:* Add the sums of measures of the exterior and corresponding interior angles, and then subtract the sum of the measures of the angles of the triangle.)

(b) Calculate the sum of the measures of the exterior angles (one at each vertex) of a convex quadrilateral.

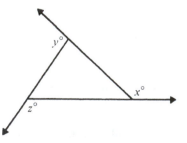

(c) Perform the same task for a convex pentagon.

(d) Do you see a pattern? Make a conjecture.

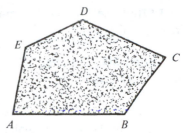

10. In the figure, the shaded region represents the
interior of convex polygon *ABCDE*. Try to
construct a definition for the interior of a convex polygon.

11. Calculate the perimeters of the following figures:

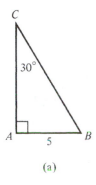

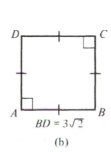

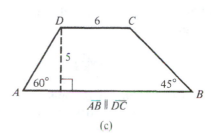

(a)   (b)   (c)

12. The perimeter of △*ABC* is 47. If *A*, *B*, and *C* are midpoints of
$\overline{DE}$, $\overline{EF}$, and $\overline{DF}$, respectively, what is the perimeter of △*DEF*?

13. We have already defined the congruence and similarity relations
for triangles. Generalize these ideas further by composing a good
mathematical definition for each of the following:

(a) Congruent polygons.   (b) Similar polygons.

7.3 QUADRILATERALS

Some types of quadrilaterals are important enough that we give
them names of their own.

DEFINITION 7.3 A quadrilateral having exactly one pair of parallel
sides is called a *trapezoid* (from the Greek meaning
table-shaped). The parallel sides are called *bases*
of the trapezoid. The segment joining the midpoints
of the nonparallel sides is called the *median*.

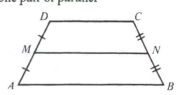

In the figure, $\overline{DC} \parallel \overline{AB}$. $\overline{MN}$ is a median.

DEFINITION 7.4 A quadrilateral having two pairs of parallel sides is called a *parallelogram*.

NOTATION RULE 7.2 Parallelogram *ABCD* will be denoted $\square ABCD$.

You should notice that each side of a parallelogram is contained in a transversal of the lines containing two other sides. In the previous figure, for example, $\overleftrightarrow{AD}$ is a transversal of $\overleftrightarrow{AB}$ and $\overleftrightarrow{DC}$. This fact makes the following theorem obvious.

THEOREM 67 The consecutive angles of a parallelogram are supplementary.

Thus, in the accompanying figure, if $m\angle A = x$, then $m\angle B$ and $m\angle D = (180 - x)$. And if $m\angle B = 180 - x$, then T67 implies that $m\angle C = 180 - (180 - x) = x$. And so we have Theorem 68.

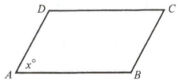

THEOREM 68 The nonconsecutive angles of a parallelogram are congruent.

Example In the figure, *ABCD* is a parallelogram, and $m\angle A = 125$. Therefore

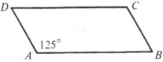

$$m\angle B = 180 - 125 = 55$$

$$m\angle C = 125$$

$$m\angle D = 55$$

Theorem 67 makes it possible for us to define a well-known object economically, as follows.

DEFINITION 7.5 If a parallelogram has a right angle it is called a *rectangle*. Consecutive sides may be referred to as a *base* and an *altitude* of the rectangle.

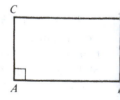

All of us know that a rectangle has *more* than one right angle. In fact,

COROLLARY T67 A rectangle has four right angles.

The reason that we state this as a theorem rather than as part of the definition is basically a matter of aesthetics and good taste. Certainly it is easier to verify that a parallelogram is a rectangle if we need to discover the existence of only *one* right angle as opposed to four.

THEOREM 69 Each diagonal separates a parallelogram into two congruent triangles.

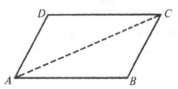

Given: $\square ABCD$; diagonal $\overline{AC}$
Prove: $\triangle ACD \cong \triangle CAB$

The proof employs PAI (Section 5.3) and is left as an exercise.
From T69, many mathematical blessings flow. For example, using CPCTC we obtain:

THEOREM 70 The nonconsecutive sides of a parallelogram are congruent.

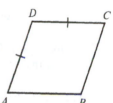

DEFINITION 7.6 If two consecutive sides of a parallelogram are congruent, the parallelogram is a *rhombus*.

From the previous theorem we obtain:

COROLLARY T70(a) All the sides of a rhombus are congruent.
Using the Perpendicular Bisector Corollary and Corollary T70a the next corollary easily follows.

COROLLARY T70(b) The diagonals of a rhombus perpendicularly bisect each other.

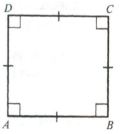

DEFINITION 7.7 If a parallelogram is a rectangle and a rhombus, then it is a *square*.

Corollaries T67 and T70(a) justify the markings on the square in the diagram.

THEOREM 71 The diagonals of a parallelogram bisect each other.

The proof involves PAI and T70 and is left as an exercise.

Theorems 67–71 give us some of the properties of parallelograms. Next we consider properties of a quadrilateral that are sufficient to conclude that the quadrilateral is a parallelogram. We begin by stating a lemma that you verified in the previous problem set.

LEMMA 7.1 The sum of the measures of the angles of a convex quadrilateral is 360.

THEOREM 72 If both pairs of nonconsecutive angles of a quadrilateral are congruent, the quadrilateral is a parallelogram.

Proof: Using the measures indicated in the figure, Lemma 7.1 implies that $2x + 2y = 360$. Therefore,

$$x + y = 180$$

Hence

$$\angle A \text{ and } \angle D \text{ are supplementary}$$

$$\angle D \text{ and } \angle C \text{ are supplementary}$$

Therefore, $\overline{AB} \parallel \overline{DC}$ and $\overline{AD} \parallel \overline{BC}$, from which it follows (by definition) that $ABCD$ is a parallelogram.

The next three theorems are easy to prove if diagonals are introduced.

THEOREM 73 If both pairs of nonconsecutive sides of a quadrilateral are congruent, the quadrilateral is a parallelogram.

THEOREM 74 If two sides of a quadrilateral are parallel and congruent, the quadrilateral is a parallelogram.

THEOREM 75 If the diagonals of a quadrilateral bisect each other, the quadrilateral is a parallelogram.

PROBLEM SET 7.3

1. Given $\square ABCD$. If $m\angle A = 50$, find the measures of the other angles.

2. In the figure $\overline{AC} \parallel \overline{DE}$, $m\angle A = 50$, and $m\angle B = 30$. Find

    (a) $m\angle ADE$ (b) $m\angle CED$
    (c) $m\angle C$

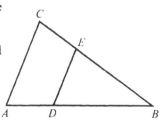

3. Given: Trapezoid $ABCD$; M is the midpoint of $\overline{AD}$; and $\overline{MN} \parallel \overline{AB}$

    Prove: $\overline{MN}$ is a median

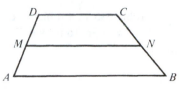

4. Given: Trapezoid $ABCD$; $m\angle A = m\angle B = 45$; $DC = 6$; and $AD = 4$

    (a) Calculate AB.   (b) Find the perimeter.

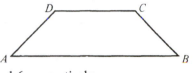

5. Prove T69 in two-column form.

6. If the base and altitude of a rectangle are 10 and 6, respectively, find the perimeter.

7. Given $\square ABCD$

    (a) If $m\angle A = (4x - 5)$ and $m\angle C = (3x + 12)$, find the measures of all four angles.

    (b) If $m\angle A = (3x + 5)$ and $m\angle B = (9x - 5)$, find the measures of all four angles

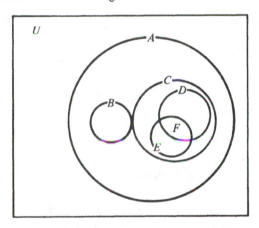

8. The Venn diagram shown illustrates relationships between the following classes of geometric figures: polygons, parallelograms, quadrilaterals, rectangles, squares, rhombuses, and trapezoids. Match these classes with the letters A, B, C, D, E, F, and U, used to name them in the diagram.

9. Prove T71

10. (a) Give an example of a quadrilateral with perpendicular diagonals that is not a square.

    (b Prove: If the diagonals of a parallelogram are perpendicular, the parallelogram is a rhombus. (Do *not* use congruent △).

11. Use problem 5 in Problem Set 5.5 together with the suggestive figure shown here to prove: The median of a trapezoid is parallel to the bases.

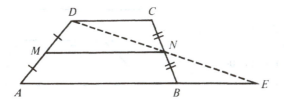

12. The angle bisectors of the angles of a parallelogram that is not a rhombus determine a rectangle. Why is a rhombus excluded from the hypothesis?

13. Prove T73

14. A trapezoid is said to be *isosceles* if its nonparallel sides are congruent.

    (a) Given: $ABCD$ is an isosceles trapezoid

        Prove: $\angle A \cong \angle B$

    (b) Use 14(a) to prove that the diagonals of an isosceles trapezoid are congruent.

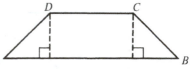

15. Which of the following are always true for a parallelogram? A rectangle? A rhombus? A square?

    (a) The diagonals bisect the angles of the quadrilateral.

    (b) The diagonals are perpendicular.

    (c) The consecutive angles are congruent.

    (d) The nonconsecutive angles are supplementary.

    (e) The diagonals are congruent.

    (f) The diagonals are $\sqrt{2}$ times as long as the length of each side.

    (g) The diagonals are contained in lines of symmetry for the figure.

16. Prove T74

17. $ABCD$ is a rectangle. If $AB = 8$ and $BC = 6$, find the perimeter of the quadrilateral formed by joining the midpoints of its sides.

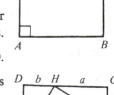

18. Calculate the length of a diagonal of a square with perimeter 20.

19. Given: $ABCD$ is a square; and E, F, G, and H separate the sides into segments of length a and b as in the figure.

Prove: (a) $EFGH$ is a rhombus.

        (b) $EFGH$ is a square.

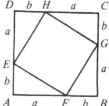

7.4 POLYGONAL REGIONS AND AREA

As we have seen in Problem Set 7.2, a convex polygon of n sides can be "decomposed" into $(n - 2)$ triangles by drawing all the diagonals from one of its vertices. The pentagon shown has been "decomposed" into three triangles. The fact that we can do this motivates the following development.

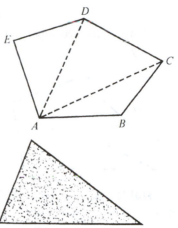

DEFINITION 7.8 The union of a triangle and its interior is called a *triangular region*.

DEFINITION 7.9 A *polygonal region* is the union of a finite number of coplanar triangular regions, no two of which have any interior points in common.

Example 1 Each of the shaded figures that follow is a polygonal region (as is their union). The dashed lines indicate how the figure may be viewed as a union of triangular regions.

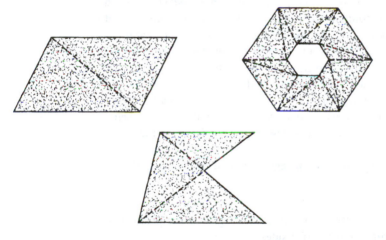

In this section we will begin to compute areas of polygonal regions. To compute an area is to assign a number to a polygonal region. We go about this in a manner analogous to that of assigning measures to segments and angles. [See P13 (Section 3.3).]

POSTULATE 19 *The Area Postulate:* To each polygonal region there corresponds a unique positive real number.

DEFINITION 7.10 The number assigned to a polygonal region by the Area Postulate is called its *area*.

A glance at the accompanying two figures will make Postulate 20 intuitively clear.

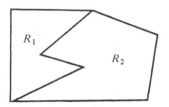
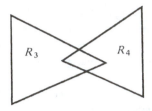

POSTULATE 20 *The Area Addition Postulate:* If two polygonal regions do not intersect other than in a finite number of segments and points, then the area of their union is the sum of their areas.

POSTULATE 21 If two triangles are congruent, then the triangular regions determined by them have the same area.

Two remarks seem in order here. First, areas are *numbers* assigned to *regions* (sets of points). Do not make the common mistake of referring to regions as areas. In the same vein, avoid the temptation to confuse a polygon (a set of points) with its perimeter (a number).

Our second remark has to do with the units of measure we use when we compute areas. We have seen that the choice is arbitrary for units of distance. Although the same is true for area, it is a good idea to be consistent and to use square yards when distance is being measured in yards, square meters when distance is being measured in meters, and so on.

Before actual computations can begin, we need another postulate— a familiar formula.

POSTULATE 22 The area of a rectangular region is the product of the lengths of a pair of consecutive sides.

A certain economy of language is effected if we adopt the familiar practice of referring to the area of a polygonal region more simply as the area of the *polygon* bounding that region. Further economy is introduced below.

NOTATION RULE 7.3 Given polygon $P_1P_2P_3 \ldots P_n$. We will denote the area of polygon $P_1P_2P_3 \ldots P_n$ by the symbol $a(P_1P_2P_3 \ldots P_n)$. In the case of a triangle, $\triangle ABC$, we write $a\triangle ABC$.

Example 2   Consider rectangle $ABCD$ in the figure.

$$a(ABCD) = 3 \cdot 5 = 15$$

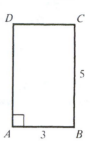

COROLLARY P22   Let s be the length of a side of a square. Then the area of that square (region) is s^2.

LEMMA 7.2   The area of a right triangle is half the product of the lengths of its legs.

Given: $\triangle ABC$ with right angle $\angle A$
Prove: $a\triangle ABC = \frac{1}{2}AB \cdot AC$

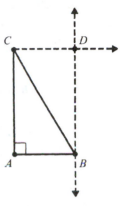

Proof: Through C and B there exist lines that are parallel to $\overline{AB}$ and $\overline{AC}$, respectively, intersecting at some point D. $ABDC$ is a parallelogram, and since $\angle A$ is a right angle, it is also a rectangle. Therefore, $a(ABDC) = AB \cdot AC$. From the Area Addition Postulate we know that

$$a(ABDC) = a\triangle ABC + a\triangle DCB$$

But $\triangle ABC \cong \triangle DCB$ (T69). Therefore, $a\triangle ABC = a\triangle DCB$. So we may write

$$AB \cdot AC = a(ABDC) = a\triangle ABC + a\triangle ABC = 2(a\triangle ABC)$$

from which it follows that $a\triangle ABC = \frac{1}{2}AB \cdot AC$.

A more general result follows.

THEOREM 76   The area of a triangle is half the product of any altitude and the corresponding base.

Proof: Let $\triangle ABC$ be any triangle. Let $\overline{CD}$ be the altitude from vertex C. Then there are three cases to consider.

Case 1: $D = A$

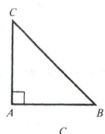

Case 2: $A\text{-}D\text{-}B$

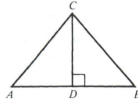

Case 3: $D \notin \overline{AB}$

The theorem has already been shown to be true for case 1 (Lemma 7.2). The other cases involve Lemma 7.2 and the Area Addition Postulate, and are left as exercises.

Example 3 To compute the area of $ABCD$, we add $\overline{CE} \perp \overline{AB}$, A-E-B, to the figure. Then $a(ABCD) = a(AECD) + a \triangle CEB$. Since $CE = 6$ and $CE = \sqrt{3} \cdot EB$, we obtain

$$EB = \frac{CE}{\sqrt{3}} = \frac{6}{\sqrt{3}} = \frac{6\sqrt{3}}{3} = 2\sqrt{3}$$

Thus $a \triangle CEB = \frac{1}{2} CE \cdot EB = \frac{1}{2} \cdot 6 \cdot 2\sqrt{3} = 6\sqrt{3}$. Also, $a(AECD) = 6 \cdot 10 = 60$. Therefore, $a(ABCD) = 60 + 6\sqrt{3}$.

Consider the equilateral triangle shown. Corollary T65(a) (Section 6.6) tells us that

$$h = \frac{s\sqrt{3}}{2}$$

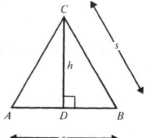

Since $a \triangle ABC = \frac{1}{2} hs$, we have the following result.

COROLLARY T76 The area of an equilateral triangle with side of length s is given by

$$\frac{s^2 \sqrt{3}}{4}$$

PROBLEM SET 7.4

1. Decompose each of the following polygonal regions (see top of facing page), into the smallest number of triangular regions that you can.

2. Find the perimeter and the area of a rectangle if a pair of consecutive sides measure 8 and 10.5 ft.

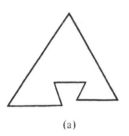

(a)

(b)

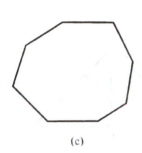

(c)

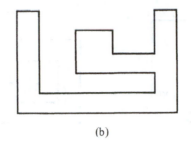

(d)

3. Find the perimeter and the area of the square shown if $AC = 3\sqrt{2}$.

4. The area of a rectangle is 208 sq in. If one side has length 13 in., find the lengths of the other three sides.

5. Find the area of a rectangle with the given base and altitude:

(a) 6 ft and 8 ft

(b) 10 ft and 3.5 ft

(c) 3 ft and 2 yd

(d) 5 cm and 6 cm

(e) ⅓ miles and ⅖ miles

(f) 3.2 in. and 1.4 in.

6. The cost of a certain brand of carpeting is $14 per sq yd. How much will such carpeting cost for a rectangular room that measures 12 by 15 ft?

7. Compute the areas of the following polygonal regions. All angles not otherwise marked are right angles.

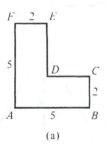

(a)

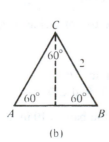

(b)

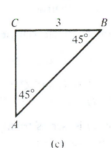

(c)

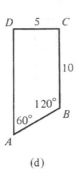

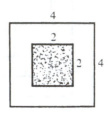

(d)   (e) (The unshaded region)

8. (a) $CD = 6$, and $AB = 9$. Find $a\triangle ABC$.
  (b) $a\triangle ADC = 48$, and $a\triangle ABC = 90$. Find $a\triangle CDB$.
  (c) $m\angle A = m\angle B = 45$, and $CD = 4$. Find $a\triangle ABC$.
  (d) $AC = 4.5$, $BC = 6$, and $AB = 7.5$. Find $a\triangle ABC$.

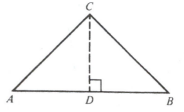

9. (a) How is the area of a rectangle changed if the length of its base is doubled but the altitude remains the same?
  (b) How is the area of a rectangle changed if the length of each of its sides is doubled? Tripled?
  (c) How is the area of a triangle changed if the length of each side is doubled? Tripled?

10. How many square tiles, 9 in. on a side, are necessary to completely tile a floor measuring 9 by 12 ft?

11. How many square tiles, 9 in. on a side, are necessary to tile the floor of the L-shaped room depicted?

12. A rectangle and a triangle have equal areas and equal bases. How do their altitudes compare?

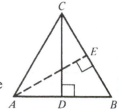

13. Find the areas of the given triangles (see top of facing page):

14. Given: $\triangle ABC$; and $\overline{AE}$ and $\overline{CD}$ as altitudes

  (a) If $AE = 6$, $BC = 4$, and $CD = 3$, find AB
  (b) If $AB = 22$, $AE = 10$, and $BC \doteq 30$, find CD

15. Find the area of an isosceles triangle if the base is 10 in. and the congruent sides measure 13 in.

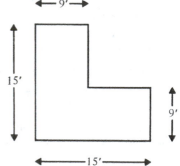

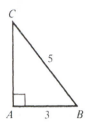

(a)

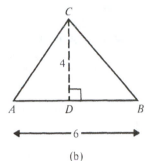

(b)

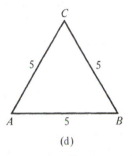

(c)

(d)

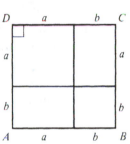

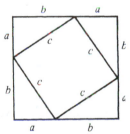

16. If the area of an equilateral triangle is equal to $4\sqrt{3}$, find the perimeter.

17. What well-known algebraic identity is illustrated by the accompanying figure? [*Hint:* What is $a(ABCD)$?]

18. The Pythagoreans are thought to have used a figure like the one shown here together with the result of problem 17 to prove the Pythagorean Theorem. Using this figure, prove that $a^2 + b^2 = c^2$.

19. Write out the proof for case 2 of T76.

20. In problem 14 of Problem Set 6.2, you were led to discover the relationship between the geometric and arithmetic means of two positive numbers, to wit:

$$\sqrt{xy} \le \frac{x+y}{2}$$

Show that this is equivalent to saying that a square has the largest area of all rectangles of given perimeter. (*Hint:* Square both sides of the above inequality.)

7.5 MORE ABOUT AREA

DEFINITION 7.11 A segment having its endpoints in the lines containing parallel sides of a quadrilateral, and perpendicular to one of them (and hence to both), is called an *altitude* to each of those sides.

As before, it is convenient to refer to the length of the altitude as the altitude.

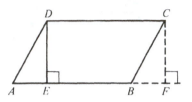

Example 1   In $\square ABCD$, $\overline{DE}$ is an altitude, as is $\overline{CF}$.

In trapezoid $GHJK$, $\overline{KL}$, $\overline{NM}$, and $\overline{PH}$ are all altitudes.

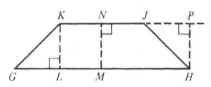

Consider trapezoid $WXYZ$. If we introduce diagonal $\overline{WY}$, two triangles are formed: $\triangle WYZ$ and $\triangle WXY$. If $\overline{YZ}$ is the base of $\triangle WYZ$, $\overline{WT}$ is an altitude. Therefore, $a\triangle WYZ = \frac{1}{2}WT \cdot YZ$. If $\overline{WX}$ is the base of $\triangle WXY$, $\overline{YS}$ is an altitude, and $a\triangle WXY = \frac{1}{2}YS \cdot WX$. But YS and WT are also names for the altitude of $WXYZ$. If we call this altitude h, we may write

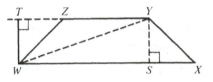

$$a(WXYZ) = a\triangle WYZ + a\triangle WXY$$
$$= \tfrac{1}{2}WT \cdot YZ + \tfrac{1}{2}YS \cdot WX$$
$$= \tfrac{1}{2}h \cdot YZ + \tfrac{1}{2}h \cdot WX$$
$$= \tfrac{1}{2}h(YZ + WX)$$

Summarizing, we have Theorem 77.

THEOREM 77   The area of a trapezoid is half the product of the altitude and the sum of the bases.

In a like manner, we can prove Theorem 78.

THEOREM 78   The area of a parallelogram is the product of the length of a side and the altitude to that side.

Example 2 Given $\square ABCD$, $m\angle A = 45$, $AB = 10$, and $BC = 15$. If we let $\overline{DE}$ be an altitude to $\overline{AB}$, $\triangle AED$ is an isosceles right triangle which means that

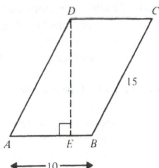

$$DE = \frac{15}{\sqrt{2}} = \frac{15\sqrt{2}}{2}$$

Therefore,

$$a(ABCD) = 10 \cdot \frac{15\sqrt{2}}{2} = 75\sqrt{2}$$

You should notice that the formula for the area of a rectangle is a special case of T78.

Suppose $\triangle ABC \sim \triangle DEF$, and their ratio of similitude is k. Let $\overline{CH}$ and $\overline{FK}$ be the altitudes from C and F, respectively. Since $\triangle AHC \sim \triangle DKF$ (by AA~), if we let $CH = h$, then $FK = kh$. Therefore,

$$a\triangle ABC = \tfrac{1}{2}hc \quad \text{and} \quad a\triangle DEF = \tfrac{1}{2}kh \cdot kc$$

It then follows that

$$\frac{a\,\triangle DEF}{a\,\triangle ABC} = \frac{\frac{1}{2}kh \cdot kc}{\frac{1}{2}h \cdot c} = k^2$$

We state this result as a theorem.

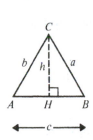
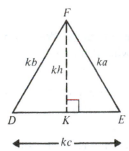

THEOREM 79 If two triangles are similar, then the ratio of their areas is the square of the ratios of any two corresponding sides.

Theorem 79 gives us a clear view of the solution to problem 9(c) in Problem Set 7.4 which asked: "How is the area of a triangle changed

if the length of each side is doubled? Tripled?" If the length of each
side is doubled, the ratio of similitude between the new triangle and the
old is 2. Therefore, the ratio of their areas is 2^2 or 4. If the length of
each side is tripled, T79 tells us that the area becomes 9 times as large.

DEFINITION 7.12   Given a one-to-one correspondence between the
vertices of two polygons, or a polygon and itself, such that the
corresponding sides and the corresponding angles are congruent, then
the *polygons are congruent.* If the corresponding angles are congruent
and the lengths of the corresponding sides are proportional, then
the *polygons are similar.*

In Section 6.2 we introduced Lemma 6.1(f), which states that if

$$\frac{a}{b} = \frac{c}{d}$$

then

$$\frac{a}{b} = \frac{a+c}{b+d}$$

It is easy to extend this as follows: If

$$\frac{a}{b} = \frac{c}{d} = \frac{e}{f} = k$$

then

$$\frac{a+c+e}{b+d+f} = k$$

and similarly for any number of ratios. We can give this geometric
meaning by stating the following theorem.

THEOREM 80   If two polygons are similar, then the ratio of their
perimeters is the ratio of any pair of corresponding sides.

Example 3   Suppose the lengths of the sides of a triangle are
2, 3, and 4, and a similar triangle has perimeter 180. If it has sides
of length a, b, and c, respectively, we may write

$$\frac{a}{2} = \frac{b}{3} = \frac{c}{4} = \frac{a+b+c}{9} = \frac{180}{9} = 20$$

Thus

$$\frac{a}{2} = 20 \quad \text{which implies } a = 40$$

$$\frac{b}{3} = 20 \quad \text{which implies } b = 60$$

$$\frac{c}{4} = 20 \quad \text{which implies } c = 80$$

Decomposing similar polygons into similar triangles gives justification for the following extension of T79.

THEOREM 81 If two polygons are similar, then the ratio of their areas is the square of the ratios of any pair of corresponding sides.

The next chapter is about circles. If you do not already own a drawing compass, now is the time to obtain one. The traditional "dime-store" compass should suffice.

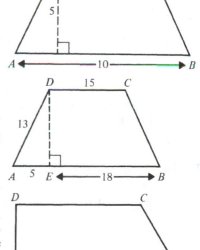

PROBLEM SET 7.5

1. *ABCD* is a trapezoid, and $\overline{AB} \parallel \overline{CD}$. If the lengths of the segments are as marked, find *a(ABCD)*.

2. *ABCD* is a trapezoid, and $\overline{AB} \parallel \overline{CD}$. If the lengths of the segments are as marked, find the area of the trapezoid.

3. *ABCD* is a trapezoid, $\overline{AB} \parallel \overline{CD}$, $\overline{DA} \perp \overline{AB}$, $AB = 27$, $DC = 15$, and $BC = 13$. Compute *a(ABCD)*.

4. *ABCD* is an isosceles trapezoid (see problem 14, Problem Set 7.3), and $m\angle A = m\angle B = 45$. If $AB = 18$ and $DC = 10$, find *a(ABCD)*.

5. Given: $\square DEFG$, $DE = 20$, and the altitude to $GF = 10$. Find *a(DEFG)*.

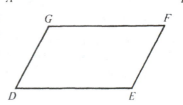

6. The area of a given parallelogram is 280 sq in. If the altitude to a base is 16 in., how long is that base?

7. $ABCD$ is a rhombus, $m\angle A = 150$, and $AB = 10$. Find $a(ABCD)$.

8. Given: $\square ABCD$; $m\angle A = 120$; $AB = 10$; and $BC = 6$. Find $a(ABCD)$.

9. One pair of corresponding sides of two similar triangles have lengths of 16 and 20, respectively.

   (a) The perimeter of the smaller triangle is 40. Find the perimeter of the larger.

   (b) If the area of the smaller triangle is 64, find the area of the larger.

10. The perimeter of a parallelogram is 34. Its area is 36. Find the perimeter of a similar parallelogram with area 64.

11. Find the ratios of the areas of the triangles shown. $m\angle C = m\angle F = 30$

12. Find the area of the trapezoid shown.

13. The lengths of the sides of a polygon are 2, 5, 6, 9, and 10. The perimeter of a similar polygon is 24. Find the lengths of its sides.

14. In the parallelogram shown, $AB = 10$, $BC = 4$, and $GH = 6$. Find:

   (a) $a(ABCD)$ (b) EF

15. The altitude of one equilateral triangle equals the length of one side of another equilateral triangle. What is the ratio of their areas?

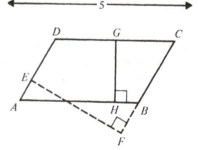

CHAPTER 7: REVIEW EXERCISES

1. For each of the following statements indicate whether it is *always* true, *sometimes* but not always true, or *never* true.

   (a) The consecutive angles of a polygon are supplementary.

   (b) The areas of similar polygons are proportional to their perimeters.

   (c) Rectangles are similar polygons.

(d) If $a\triangle ABC = 9(a\triangle DEF)$, then $AB = 3DE$

(e) A triangular region is a polygonal region.

(f) If two polygons have the same area, then they are similar.

(g) If a triangle and a parallelogram have equal areas and equal bases, then their altitudes are equal.

(h) The diagonals of a square are perpendicular.

(i) If the diagonals of a quadrilateral are perpendicular, the quadrilateral is a rhombus.

(j) The nonconsecutive angles of a quadrilateral are congruent.

(k) If a rhombus is not a square, then it is not a rectangle.

(l) The area of the union of two triangular regions is the sum of their areas.

(m) The nonparallel sides of a trapezoid have different lengths.

(n) A rhombus is symmetric with respect to each of its diagonals.

(o) A trapezoid is symmetric with respect to one of its diagonals.

2. Find the area of each of the following polygons.

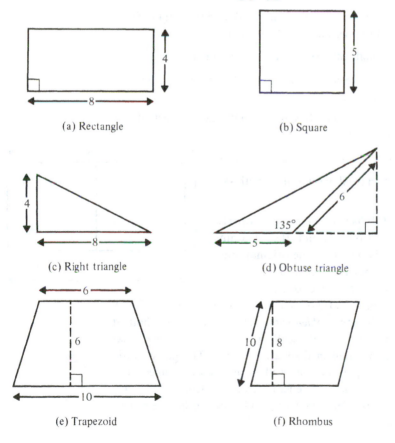

(a) Rectangle

(b) Square

(c) Right triangle

(d) Obtuse triangle

(e) Trapezoid

(f) Rhombus

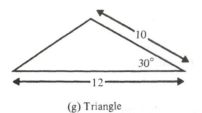

(g) Triangle

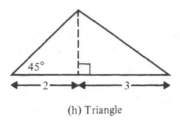

(h) Triangle

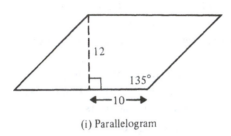

(i) Parallelogram

3. The measure of one angle of a rhombus is 60, and the length of one side is 6 in. Find the perimeter and the area.

4. Prove: The diagonals of a rectangle are congruent.

5. Let $ABCDEF$ be a regular hexagon. If $AC = 10\sqrt{3}$, find AB. Calculate the perimeter.

6. Prove: The area of a rhombus is one half the product of the lengths of its diagonals.

7. Prove: If $\overline{AB} \parallel \overline{CD}$, then $a\triangle ABC = a\triangle ABD$

8. Find the area of an equilateral triangle with side of length 5.

9. Rectangle $ABCD$ has an interesting property. If a square is cut from one end, the remaining rectangle $BCFE$ is similar to the original one.

   Set up a proportion between lengths of corresponding sides and solve for x.

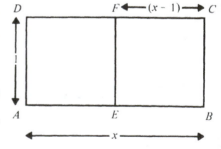

This number x, the ratio between the base and altitude of such a rectangle, is called the *golden ratio*. The early Greeks thought that this was the most aesthetically pleasing of all rectangular shapes and used it often in their architecture. The golden ratio has mathematical significance as well, being closely connected to Pascal's triangle (which we encountered in Chapter 1), the geometry of the regular pentagon, and many other seemingly disjoint topics.

10. Show that the area of a right triangle is one half the product of the geometric mean of the lengths of the segments into which the altitude of the hypotenuse separates the hypotenuse and the arithmetic mean of those lengths. A sketch should make this one easy.

11. Prove T75

12. Given: $\triangle ABC$ and median $\overline{CD}$
    Prove: $a\triangle ADC = a\triangle DBC$

13. $ABCD$ is a quadrilateral. What about $CDAB$? What about $ACBD$? Explain.

14. The diagonal of a square is 14. Calculate its perimeter and its area.

15. Prove case 3 of T76

16. Given rectangle $ABCD$; $AB = 5$; $AC = 10$. Find $a(ABCD)$.

17. Is a square a convex polygon? A convex set? Explain.

18. Find the perimeter of a rhombus with area 20 if one of its diagonals is half the length of the other.

REFERENCES

Beck, Anatole; Bleicher, M. N.; and Crowe, D. W. *Excursions into Mathematics*. New York: Worth Publishers, 1969. pp. 147–153.
   A brief history of the problem of area is given here. The Greeks had no satisfactory solution and none was obtained until the last century.

Haldane, J. B. S. "On Being the Right Size." In *The World of Mathematics*, edited by J. R. Newman, Vol. 2, pp. 952–57. New York: Simon and Schuster, 1956.
   We have learned that the area of a polygon increases as the square of the increase of the lengths of the sides. For solid figures, the ratio of the volumes of "similar" figures is the *cube* of the ratio of corresponding edges. This leads to surprising results in the plant and animal kingdoms.

Stein, Sherman K. *Mathematics: The Man-Made Universe*. San Francisco: W. H. Freeman and Co., Publishers, 1969. Chap. 7.
   If you did the tiling problems in Problem Set 7.4, you will be interested in what kinds of tiling patterns are possible.

The eye is the first circle.

RALPH WALDO EMERSON (1803–1882)

CHAPTER EIGHT

CIRCLES

8.1 INTRODUCTION

The county seat of Pickaway County, Ohio, is a small city named Circleville. The city was laid out in a circular plan in 1810. The fascination of Circleville's founders with the circular motif is nothing new in human history.

The Old Testament description of the building of Solomon's temple indicates that it was circular; the floor plan of the Pantheon in Rome is circular with a circular hole in the center of the dome to admit the Sun God.

In the latter part of the nineteenth century, American Indians were forced from their ancestral homes and relocated on reservations often more than a thousand miles away. Not the least of the trauma faced by these people in their new surroundings was the fact that they were compelled to live in rooms that were rectangular. They believed that the circle was a holy form and had traditionally constructed their own dwellings accordingly. As a result, they suffered mental and spiritual anguish in dwellings that were not circular.

Mathematicians, too, have been moved to attach special significance to this figure. Proclus, in his commentary on Euclid's Elements wrote, "The circle is the first, the most simple, and the most perfect figure."

8.2 SOME DEFINITIONS

Each of us by now has a fairly clear notion of what a circle is. Without reading further, try to write a definition for the word circle. Compare your definition with what follows.

DEFINITION 8.1 The set of all points in a plane which are a given distance from a given point in that plane is called a *circle*. The given point is the *center* of the circle and the given distance is called the *radius* of the circle.

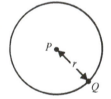

A circle is named by naming its center. In the figure, *P* is the center of circle *P* and *r* is its radius. We also refer to any segment joining a circle to its center as a radius (*plural:* radii) of that circle. Thus, in the figure, $\overline{PQ}$ is a radius of circle *P*.

DEFINITION 8.2 Let circle *P* have radius *r*. The set of all points in the plane of circle *P* that are less than *r* units from *P* is called the *interior* of the circle. The set of all points that are more than *r* units from *P* is called the *exterior* of the circle. The union of a circle and its interior is called a *circular region*.

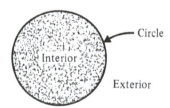

As the definition implies, a circle may be viewed as the boundary of a circular region.

DEFINITION 8.3 Given circle *P*. A segment whose endpoints are points of the circle is called a *chord* of the circle (from the Latin word for "string" [cord] or "small rope"). A chord which contains the center *P* is called a *diameter* of the circle. (The word diameter is also used to refer to the *length* of such a chord.) A line which contains a chord is called a *secant* (from the Latin, meaning cutting). A line which lies in the plane of the circle and which intersects the circle in exactly one point is called a *tangent* (from the Latin, meaning touching); the point of intersection is called the *point of tangency*, or the *point of contact*.

Example Consider circle P in the figure. $\overline{AB}$, $\overline{CD}$, and $\overline{EF}$ are all chords of circle P. $\overline{AB}$ is a diameter. AB is the diameter. $\overleftrightarrow{EF}$ is a secant; $\overleftrightarrow{TS}$ is a tangent, T being the point of tangency.

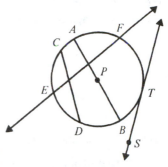

In the figure, $\overleftrightarrow{l}$ is tangent to circle Q at T. With a protractor, measure the angle that $\overleftrightarrow{QT}$ makes with l. Make a few sketches of tangents to circles and measure the angle determined by the tangent and the radius drawn to the point of contact. What do you conclude? Hopefully, something like the next theorem.

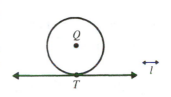

THEOREM 82 Each tangent to a circle is perpendicular to the radius drawn to the point of tangency.

Proof (indirect): Suppose $\overline{PT}$ is not perpendicular to l. Then there exists some point S, $S \in l$ such that $\overline{PS} \perp l$. Also there exists a point R on the ray opposite $\overrightarrow{ST}$ such that $SR = ST$. Therefore, $\overline{PS}$ is the perpendicular bisector of $\overline{TR}$, and it follows that $PT = PR$. However, if $PT = PR$, R lies on (is an element of) circle P, and we have *two* points of circle P on tangent l. →|←

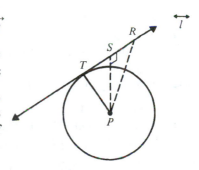

The converse of this theorem is also true.

THEOREM 83 If a line is perpendicular to a radius at its outer end, it is tangent to the circle.

Proof: The idea of the proof is to show that no *other* point of the given line can be a point of the circle. The proof is left as an exercise.

PROBLEM SET 8.2

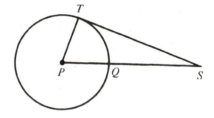

1. In the figure shown, give names introduced in this section for each of the following:

   (a) $\overline{AB}$ (d) $\overleftrightarrow{BC}$ (g) PA

   (b) P (e) $\overleftrightarrow{TS}$ (h) $\overline{PA}$

   (c) $\overline{CB}$ (f) T

2. Sketch circle R with radius $\overline{CR}$, tangent $\overleftrightarrow{CP}$, secant $\overleftrightarrow{CT}$, and chord $\overline{CS}$ (use compass and ruler).

3. Sketch circle P with chords $\overline{AB}$, $\overline{BC}$, and $\overline{AC}$.

4. Sketch circle Q with radius $\overline{QR}$. If $P \in \overleftrightarrow{QR}$ with P-Q-S and $PQ < QR$, and $S \in \overleftrightarrow{QR}$ with Q-R-S and $QR < QS$, sketch P and S.

5. Sketch circle Q with chords $\overline{AB}$, $\overline{BC}$, $\overline{CD}$, and $\overline{AD}$ if A and D are on the same side of $\overleftrightarrow{BC}$, and C and D are on the same side of $\overleftrightarrow{AB}$.

6. Sketch circle P with radius 1 in. Locate A and B so that $PA = 0.5$ in. and $PB = 1.5$ in.

   (a) Which point lies in the interior of circle P? Why?
   (b) Which point lies in the exterior of circle P? Why?
   (c) Find the maximum and minimum values for AB.

7. Given $\overleftrightarrow{ST}$ tangent to circle P at T.

   (a) If $TP = 9$ and $SP = 41$, find TS
   (b) If $TS = 12$ and $PQ = 6$, find PS
   (c) If $TS = 10$ and $QS = 6$, find PT

8. The radius of circle Q is 5 in. $QR = 13$ in. A line containing R is tangent to circle Q at S. Find QS.

9. Which law for inequalities assures us that a circle, its interior, and its exterior are mutually disjoint sets?

10. Given circle P with radius r. Define the interior of circle P using set-builder notation.

11. Which theorem implies that every circle has a tangent at each of its points?

12. Given: Circle P; $\overline{AB}$ is a diameter; chords $\overline{BC}$ and $\overline{BD}$; and $\overline{BC} \cong \overline{BD}$
Prove: $\angle CBP \cong \angle DBP$

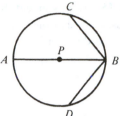

13. Given: Circle P; $\overline{AB}$ a diameter; $\overleftrightarrow{BT}$ is tangent to circle P at B; and $\overleftrightarrow{AS}$ is tangent to circle P at A
Prove: $\overleftrightarrow{AS} \parallel \overleftrightarrow{BT}$

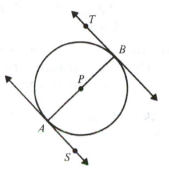

14. Given: $\overleftrightarrow{PT}$ and $\overleftrightarrow{PS}$ are tangent to circle Q
Prove: (a) $\overline{PT} \cong \overline{PS}$
(b) $\angle TPQ \cong \angle SPQ$

15. Suppose you are given circle P and line l in the plane of circle P. If a point of $\overleftrightarrow{l}$ lies in the interior of circle P, what would you call line $\overleftrightarrow{l}$?

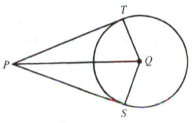

16. Follow the hint given in the text and prove T83.

8.3   MORE ON CHORDS, TANGENTS, AND SECANTS

The two figures that accompany T84 and T85 should give you a clear view of the proofs for these theorems. The proofs themselves are left as exercises.

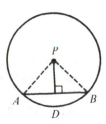

THEOREM 84 The perpendicular segment from the center of a circle to a chord of that circle bisects the chord.

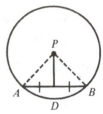

THEOREM 85 If a segment bisecting a chord which is not a diameter passes through the center of the circle, then that segment is perpendicular to the chord.

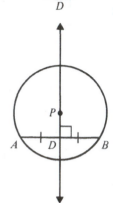

In the accompanying figure, it *appears* that $\overleftrightarrow{PD}$, the perpendicular bisector of $\overline{AB}$, passes through the center of circle P. Indeed it does. The Perpendicular Bisector Theorem asserts that $\overleftrightarrow{PD}$ must contain *all* of the points equidistant from A and B. The center of the circle is surely one of these. We summarize this result as a theorem.

THEOREM 86 In the plane of a given circle, the perpendicular bisector of a chord passes through the center of the circle.

DEFINITION 8.4 Circles are *concentric* if they are coplanar and have the same center. Distinct circles are *tangent* if they are coplanar and are tangent to the same line at the same point. If the centers of tangent circles are on the same side of their common tangent, they are *internally tangent*. Otherwise, they are *externally tangent*.

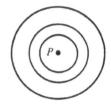

Concentric circles

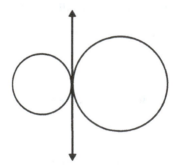

Internally tangent   Externally tangent

Congruent figures, as we have defined them, are the same size and the same shape. All circles are the same shape.

DEFINITION 8.5 Circles *are congruent* if their radii are congruent.

We will use the same congruence symbol ($\cong$) as in previous settings.

The next theorem is an obvious consequence of the definition of "circle" and "congruent circles."

THEOREM 87   Every circle is congruent to itself.

The proof of the next theorem should be divided into two parts and is left as an exercise.

THEOREM 88   Chords of congruent circles are congruent iff they are equidistant from their centers.

In problem 15 of Problem Set 8.2 you were asked to consider a coplanar line and circle, such that the line intersected the interior of the circle. You probably concluded that the line must be a secant. To show that this is the case, we need to consider two possibilities. First, if the given point of intersection is the center of the circle, the Point-Plotting Postulate (PPP) guarantees two points at distances from the center equal to the radius that are on the given line, and thus the line *is* a secant. The second possibility is depicted in the accompanying figure. If we let

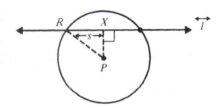

$\overleftrightarrow{PX} \perp \overleftrightarrow{l}$, $X \in \overleftrightarrow{l}$, we can conclude that X is in the interior of circle P. (Why?) Therefore, if r is the radius of the circle, $PX < r$ and $(PX)^2 < r^2$, from which it follows that

$$r^2 - (PX)^2 > 0$$

Letting $s = \sqrt{r^2 - (PX)^2}$, the **PPP** again guarantees the existence of two points on $\overleftrightarrow{l}$, each s units from X. Naming one of these points R, we may write

$$s = XR = \sqrt{r^2 - (PX)^2}$$

and squaring both sides of the latter equality, obtain

$$(XR)^2 = r^2 - (PX)^2$$

or

$$(PX)^2 + (XR)^2 = r^2$$

and by the converse of the Pythagorean Theorem, we can conclude that R is r units from P; that is, R is a point of circle P. Theorem 89 formalizes this result.

THEOREM 89 If a line and a circle are coplanar and if the line intersects the interior of the circle, then the line contains two points of the circle.

PROBLEM SET 8.3

1. A chord 10 in. long is 12 in. from the center of a circle. What is the radius of the circle?

2. A circle has radius 6 in. and a chord of length 8 in. How far is the chord from the center of the circle?

3. A chord is 4 in. from the center of a circle of radius 7 in. How long is the chord?

4. A chord intersects a radius at its outer end determining a 30° angle. If the radius is 8 ft, how long is the chord?

5. Prove T84

6. Given two concentric circles with center P. If $\overline{AB}$, a chord of one, is tangent to the other, and their radii are 4 in. and 6 in., find the length of $\overline{AB}$.

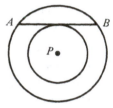

7. Two circles P and Q are tangent at T. Circle P is congruent to circle Q. Are they internally or externally tangent?

8. What property does T87 attach to circle congruence?

9. Given: Two concentric circles with center P; and $\overline{AB}$ and $\overline{CD}$ are chords of the larger circle and tangent to the smaller circle
   Prove: $\overline{AB} \cong \overline{CD}$

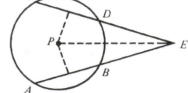

10. Prove T85

11. Find the radius of a circle if a 4-in. chord is twice as far from the center as an 8-in. chord.

12. Given: Circle P; nonparallel chords $\overline{AB}$, $\overline{CD}$; and $\overline{AB} \cong \overline{CD}$
    Prove: $\overline{CE} \cong \overline{AE}$

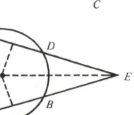

13. Prove that no circle contains three collinear points. This result adds the word "exactly" to T89.

14. Given: Circles P and Q are tangent at T
Prove: P, Q, and T are collinear

15. (a) Restate T88 as two theorems.
    (b) Prove whichever of the two theorems of (a) asserts that chords are congruent if they are equidistant from the center. (*Hint:* Use T84.)
    (c) Prove the other half of T88.

16. Suppose circles P and Q intersect as in the figure. Prove that $\overline{PQ}$ is the perpendicular bisector of $\overline{AB}$.

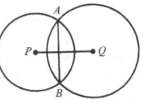

17. Given: Congruent circles P and Q; $P \in$ circle Q; $Q \in$ circle P; and circles P and Q intersect at A and B
Prove: $\triangle PQA$ is equilateral

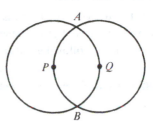

8.4   CIRCULAR ARCS AND ANGLES

We begin this section with some definitions.

DEFINITION 8.6   An angle is called a *central angle* if its vertex is the center of a circle.

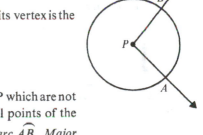

In the figure, $\angle BPA$ is a central angle.

DEFINITION 8.7   If A and B are two points of circle P which are not endpoints of a diameter, then the union of A, B, and all points of the circle which are in the interior of $\angle BPA$ is called *minor arc* $\overarc{AB}$. *Major arc* $\overarc{AB}$ is the union of A, B, and all points of the circle which are in the

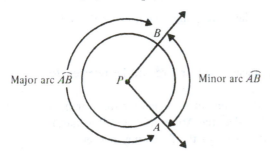

Major arc $\overarc{AB}$ Minor arc $\overarc{AB}$

exterior of $\angle BPA$. If A and B are endpoints of a diameter of circle P, then the union of A and B with all of the points which lie in either one of the half planes determined by $\overleftrightarrow{AB}$ is called a *semicircle*. P is called the *center* of any of these arcs. In each case, A and B are referred to as *endpoints* of the arc, and any other point of the arc is called an *interior point*.

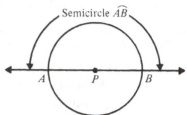

Consider the circle shown. Since there are two arcs, a minor arc and a major arc, with endpoints A and B, to refer to $\overparen{AB}$ might be confusing. When such a situation arises, we generally include an interior point in the name of the arc. Thus, in this case, we may write $\overparen{AXB}$ to refer to the minor arc and $\overparen{AYB}$ to refer to the major arc. If the context makes it clear which arc is referred to, we simply write $\overparen{AB}$.

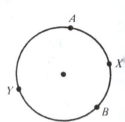

Since measures have been assigned to other sets of points, we assign them to arcs in the following natural way.

DEFINITION 8.8 The *degree measure* of any arc $\overparen{AXB}$, denoted $m\,\overparen{AXB}$, is equal to

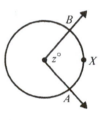

1. The measure of the corresponding central angle if $\overparen{AXB}$ is a minor arc.

$$m\,\overparen{AXB} = z$$

2. 180 if $\overparen{AXB}$ is a semicircle.

$$m\,\overparen{AXB} = 180$$

3. 360 minus the measure of the corresponding minor arc if $\overparen{AXB}$ is a major arc.

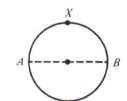

From now on, the degree measure of an arc will be referred to simply as its measure.

We have already introduced notions of betweenness for segments (the definition of *between*) and angles (the Angle Addition Postulate). The latter of these is used in the proof of the following theorem.

$$m\,\overparen{AXB} = (360 - z)$$

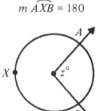

THEOREM 90 *The Arc Addition Theorem:* If X is an interior point of $\overset{\frown}{AB}$, then $m\,\overset{\frown}{AXB} = m\,\overset{\frown}{AX} + m\,\overset{\frown}{XB}$.

The verification of this theorem requires the consideration of five cases. In Problem Set 8.4 you will be asked to supply a proof for one of the cases. The others have similar though more wearisome proofs.

Example 1 In the figure, $\overline{AB}$ is a diameter of circle P, and $m\angle BPC = 58$. Thus $m\,\overset{\frown}{BC} = y = 58$. Since $\overset{\frown}{BCA}$ is a semicircle, we can write $m\,\overset{\frown}{BCA} = 180$. The Arc Addition Theorem implies that $m\,\overset{\frown}{BCA} = m\,\overset{\frown}{BC} + m\,\overset{\frown}{CA}$. Substituting, we obtain $180 = 58 + x$, or $x = 180 - 58 = 122$.

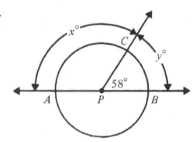

Example 2 In the figure are two concentric circles with center Q. Since $m\angle BQA = 42$, it follows that the degree measure of both minor arcs $\overset{\frown}{AB}$ and $\overset{\frown}{CD}$ is 42. Thus we see that the measure of an arc does not depend upon the size of the circle.

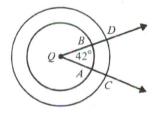

We have consistently defined congruence to mean that two objects are the same shape and the same size. We do the same for arcs.

DEFINITION 8.9 Two arcs are *congruent* if they are subsets of congruent circles and have equal measures.

In the accompanying figure, we say that $\angle AXB$ is *inscribed in* $\overset{\frown}{AXB}$.

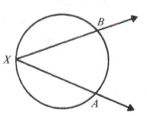

DEFINITION 8.10 An angle is *inscribed* in an arc if both of the following are true:

1. The vertex of the angle is an interior point of the arc.
2. Each side of the angle contains an endpoint of the arc.

In each of the figures shown, $\angle BAC$ is said to *intercept the arc(s)* named.

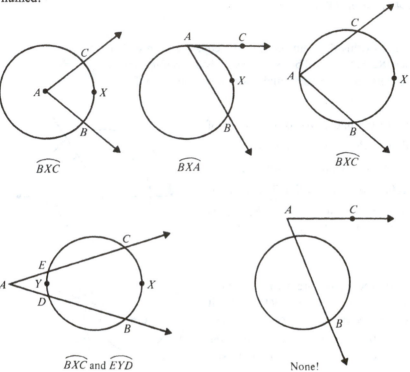

$\overset{\frown}{BXC}$ $\overset{\frown}{BXA}$ $\overset{\frown}{BXC}$

$\overset{\frown}{BXC}$ and $\overset{\frown}{EYD}$ None!

DEFINITION 8.11   An angle *intercepts* an arc if all the following conditions are met:

1. Each endpoint of the arc is a point of the angle.
2. Each ray of the angle contains at least one endpoint of the arc.
3. The interior of the arc lies in the interior of the angle.

PROBLEM SET 8.4

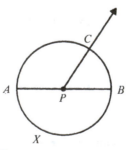

1. Given circle P at the right:

   (a) Name the central angles.
   (b) Name the minor arcs.
   (c) Name the semicircles.
   (d) Name the major arcs.
   (e) If $m\angle CPB = 50$, find $m\,\overset{\frown}{BC}$ (minor arc).

(f) What is $m\ \overset{\frown}{BCA}$?

(g) If $m\angle CPB = 50$, find $m\ \overset{\frown}{CA}$ (minor arc).

(h) Name two congruent arcs.

2. Given circle P; $\overrightarrow{PE}$ bisects $\angle BPD$. Find the measures of the following minor arcs if $\overline{AB} \perp \overline{DC}$.

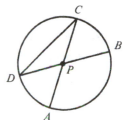

(a) $\overset{\frown}{DA}$ (b) $\overset{\frown}{BD}$ (c) $\overset{\frown}{BE}$ (d) $\overset{\frown}{EA}$

3. Given circle P. If $m\angle CPD = 120$, find:

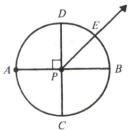

(a) $m\ \overset{\frown}{BC}$ (minor arc)   (b) $m\ \overset{\frown}{BA}$ (minor arc)

(c) $m\ \overset{\frown}{DAC}$ (d) $m\ \overset{\frown}{CDA}$

4. Given the figure.

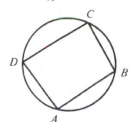

(a) Name the arc intercepted by $\angle DCB$.

(b) Name the angle that intercepts $\overset{\frown}{DCB}$.

(c) Name the angle inscribed in $\overset{\frown}{ABC}$.

(d) Name the arc in which $\angle ADC$ is inscribed.

(e) If $m\ \overset{\frown}{BCD} = 170$, what is $m\ \overset{\frown}{BAD}$?

5. Given the figure and $\overrightarrow{BT}$ tangent at B. Name the arc(s) intercepted by

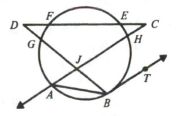

(a) $\angle D$ (b) $\angle DJC$ (c) $\angle JBT$

(d) $\angle JAB$ (e) $\angle CAB$ (f) $\angle C$

6. Prove the Arc Addition Theorem for the case in which $\overset{\frown}{AXB}$ is a minor arc.

7. Given circle P, $A \in$ circle P, and $B \in$ circle P. The measure of major arc $\overset{\frown}{AB}$ is 30 less than twice the measure of minor arc $\overset{\frown}{AB}$. Find the measure of both arcs.

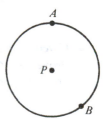

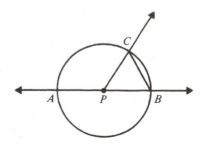

8. In the figure, $\overline{AB}$ is a diameter of circle P. If $m\angle PCB = 60$, what is the measure of minor arc $\overset{\frown}{BC}$?

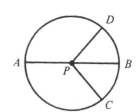

9. Given: Circle P and $\angle DPB \cong \angle CPB$
   Prove: $\overset{\frown}{DB} \cong \overset{\frown}{CD}$

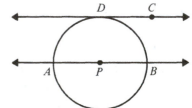

10. Given circle P. If $\overset{\leftrightarrow}{DC} \parallel \overset{\leftrightarrow}{AB}$ and $\overset{\leftrightarrow}{DC}$ is tangent to circle P at D, find $m\,\overset{\frown}{AD}$ and $m\,\overset{\frown}{DB}$.

11. Using a compass, draw circle P with inscribed acute angle $\angle ABC$ such that $P \in \overset{\rightarrow}{BA}$. Draw $\overset{\rightarrow}{PC}$. How do $\angle ABC$ and $\angle APC$ seem to be related? (Use a protractor here if necessary.) Repeat the process for some other angle $\angle DEF$, $P \in \overset{\rightarrow}{ED}$, and such that $\angle DEF \not\cong \angle ABC$.

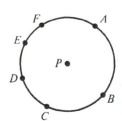

12. Given circle P and points A, B, C, D, E, and F. Draw and measure (with a protractor) $\angle AFB$, $\angle AEB$, $\angle ADB$, and $\angle ACB$. Make a conjecture.

13. Draw semicircle $\overset{\frown}{AXB}$ and inscribed angle $\angle AXB$. Using a protractor, estimate $m\angle AXB$. Repeat this process a few times and make a conjecture.

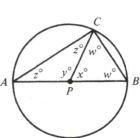

14. Given: Circle P and diameter $\overline{AB}$
    Prove: $\angle ACB$ is a right angle (*Hint:* Use ITT and EAT, and give an algebraic proof.)

15. Given: Circle P and $m \overgroup{AXB} = m \overgroup{BYC}$
      Prove: $\triangle ADP \cong \triangle CDP$

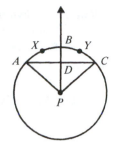

16. Given: Circle $P \cong$ circle Q and $\angle A \cong \angle C$
      Prove: $m \overgroup{AB} = m \overgroup{CD}$

8.5   MORE ARCS, ANGLES, AND CHORDS

Problems 11 and 13 of Problem Set 8.4 may have already led you to make a conjecture such as the following.

THEOREM 91   The measure of an inscribed angle is one-half the measure of its intercepted arc.

Hypothesis: Let inscribed angle $\angle ABD$ intercept $\overgroup{AD}$ on circle C
Conclusion: $m \angle ABD = \frac{1}{2} m \overgroup{AD}$

Proof: There are actually three cases to consider.

Case 1: $C \in \angle ABD$

Case 2: $C \in \mathcal{I} \angle ABD$

Case 3: $C \in \mathcal{E} \angle ABD$

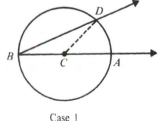

Case 1

The proof for case 1 follows. $m \overgroup{AD} = m \angle ACD$, and $m \angle ACD = m \angle ABD + m \angle CDB$ (Exterior Angle Theorem). Therefore, $m \overgroup{AD} = m \angle ABD + m \angle CDB$. But $\triangle BCD$ is isosceles. Thus $m \overgroup{AD} = 2m \angle ABD$. It follows that

$$m \angle ABD = \tfrac{1}{2} m \overgroup{AD}$$

Case 1 is used to prove cases 2 and 3. These proofs are left as exercises.

Two fairly obvious corollaries follow. The first you were asked to prove in a more complicated fashion in problem 14 of Problem Set 8.4. It is one of the theorems attributed to that earliest deductive mathematician, Thales.

COROLLARY T91(a)   An angle inscribed in a semicircle is a right angle.

COROLLARY T91(b)   Angles inscribed in the same arc or in congruent arcs are congruent.

Example 1   Given the figure, $m \stackrel{\frown}{AC} = 100$, and $m \stackrel{\frown}{BC} = 130$. Therefore, $m \stackrel{\frown}{AB} = 360 - (100 + 130) = 130$. By T91 we have:

$$m \angle A = \tfrac{1}{2} m \stackrel{\frown}{BC} = 65$$

$$m \angle B = \tfrac{1}{2} m \stackrel{\frown}{AC} = 50$$

$$m \angle C = \tfrac{1}{2} m \stackrel{\frown}{AB} = 65$$

The next definition is needed for the following theorem.

DEFINITION 8.12   Let $\overleftrightarrow{AB}$ be tangent to any circle at A. Then $\overrightarrow{AB}$ is called a *tangent ray* to that circle at A. Let $\overleftrightarrow{CD}$ be a secant of a circle, $\overline{CD}$ a chord. Then $\overrightarrow{CD}$ (and $\overrightarrow{DC}$) is a *secant ray* of that circle.

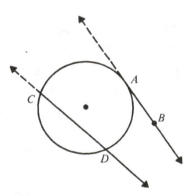

THEOREM 92   If a secant ray and tangent ray have a common endpoint, then the measure of the angle formed is one half the measure of the intercepted arc.

Proof: There are two cases to consider.

Case 1: The center of the circle is in the exterior of the angle, as in the figure. Adding the auxilary segments shown, we can write (letting lowercase letters represent measures of angles or arcs):

$$y + z = 90$$

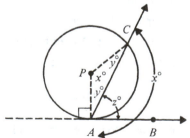

which implies that

$$2y + 2z = 180$$

From the Angle Sum Theorem we conclude that

$$2y + x = 180$$

Therefore, $x = 2z$, or $z = \frac{1}{2}x$.

Case 2: The center of the circle is not in the exterior of the angle. The proof for this case is left as an exercise.

THEOREM 93 In the same circle or congruent circles, if two chords that are not diameters are congruent, then the corresponding minor arcs are congruent.

Given: Circle $P \cong$ circle Q, and $\overline{AB} \cong \overline{CD}$
Prove: Minor arc $\overset{\frown}{AB} \cong$ minor arc $\overset{\frown}{CD}$

Proof: By the SSS theorem $\triangle APB \cong \triangle CQD$. Therefore, $\angle P \cong \angle Q$. Since $m\overset{\frown}{AB} = m\angle P$ and $m\overset{\frown}{CD} = m\angle Q$, it follows that $m\overset{\frown}{AB} = m\overset{\frown}{CD}$, or $\overset{\frown}{AB} \cong \overset{\frown}{CD}$.

THEOREM 94 In the same circle or congruent circles, if two arcs are congruent, then the corresponding chords are congruent.

Proof: There are three cases to consider.

Case 1: The congruent arcs are minor arcs. The circles shown suggest a simple proof using congruent triangles.

Case 2: The congruent arcs are semicircles. Diameters of congruent circles are congruent.

Case 3: The congruent arcs are major arcs. The proof here is left as an exercise.

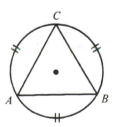

Example 2 In the figure, $\overset{\frown}{AB} \cong \overset{\frown}{BC} \cong \overset{\frown}{AC}$. Thus it follows that $\overline{AB} \cong \overline{BC} \cong \overline{AC}$, and $\triangle ABC$ is equilateral.

PROBLEM SET 8.5

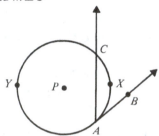

1. Given the figure.

(a) If $m\angle C = 70$, what is $m\ \widehat{AB}$?

(b) If $m\ \widehat{BC} = 100$ and $m\ \widehat{AC} = 120$, what is $m\angle B$? $m\angle C$? $m\angle A$?

(c) If $m\angle B = m\angle C = 70$, find $m\ \widehat{AB}$, $m\ \widehat{BC}$, $m\ \widehat{ACB}$

(d) If $\widehat{AB} \cong \widehat{AC}$ and $m\angle A = 40$, find $m\angle B$ and $m\angle C$

2. Given the figure. $\overrightarrow{AB}$ and $\overrightarrow{AC}$ are tangent and secant rays, respectively.

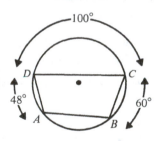

(a) If $m\ \widehat{AXC} = 160$, what is $m\angle BAC$?

(b) If $m\ \widehat{AYC} = 210$, what is $m\angle BAC$?

(c) If $m\angle BAC = 75$, what is $m\ \widehat{AYC}$?

3. Given the figure. Find $m\angle A$, $m\angle B$, $m\angle C$, and $m\angle D$.

4. Draw a quadrilateral with vertices on a given circle. Measure (with a protractor) pairs of nonconsecutive angles. Do this several times. Compare with the results of problem 3. Do you notice a pattern? If so, make a conjecture.

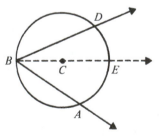

5. Use the figure to complete the proof for case 2 of T91.

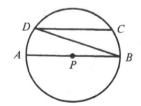

6. Given circle P in the figure with $\overline{AB} \parallel \overline{CD}$.

(a) If $m\angle ABD = 40$, find $m\ \widehat{AD}$ and $m\ \widehat{DC}$

(b) If $m\ \widehat{DC} = 110$, find $m\angle CDB$ and $m\ \widehat{AD}$

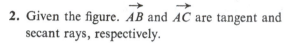

7. $ABCDE$ is a regular pentagon.

   (a) Find $m \stackrel{\frown}{DE}$ (b) Find $m \angle AFB$

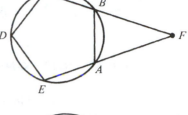

8. A quadrilateral is said to be inscribed in a circle if each of its vertices is on the circle.
Prove: The nonconsecutive angles of an inscribed quadrilateral are supplementary.

9. Prove case 2 of T92 (*Hint:* Introduce an auxiliary secant ray forming an acute angle with the tangent ray and employ the Angle Addition Postulate.)

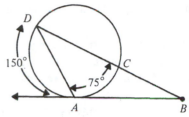

10. In the figure, $\overleftrightarrow{AB}$ is tangent to the circle at A. Find $m \angle B$.

11. Prove case 3 of T94

12. Given: Circle P with diameter $\overline{AB}$; $\overline{CD} \perp \overline{AB}$
Prove: CD is the geometric mean of AD and DB

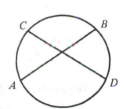

13. Given: $\overline{AB} \cong \overline{CD}$

    Prove: $\stackrel{\frown}{AC} \cong \stackrel{\frown}{BD}$

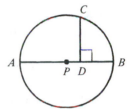

14. Use the diagram to prove: If two chords intersect in the interior of a circle, then the measure of each angle formed is one half the sum of the measures of the arcs intercepted by that angle and its vertical angle. (*Hint:* Use the Exterior Angle Theorem.)

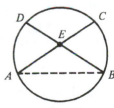

15. $\overrightarrow{PA}$ and $\overrightarrow{PB}$ are tangents to circle Q at A and B, respectively. If $m \angle APB = 60$, what are the measures of major arc $\stackrel{\frown}{AB}$ and minor arc $\stackrel{\frown}{AB}$?

16. Given: Chords $\overline{AB}$ and $\overline{DC}$ intersect at E
Prove: $DE \cdot EC = AE \cdot EB$ (*Hint:* Use similar triangles.)

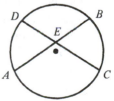

17. Given: Circles P and Q are internally tangent at T; $P \in$ circle Q;
$\overline{ST}$ and $\overline{RT}$ are chords of circles Q and P, respectively; and R-S-T
Prove: S is the midpoint of $\overline{RT}$

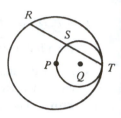

8.6 REGULAR POLYGONS AND CIRCLES

Suppose circle Q is divided into six congruent arcs, as in the figure.
The six congruent chords determined by these arcs form a regular
hexagon, since if we add to the figure the six radii $\overline{QA}, \overline{QB}, \dots, \overline{QF}$, it
is clear that we form six congruent isosceles triangles. Since the six
triangles are congruent, they all have the same area, so that using the
Area Addition Postulate we have

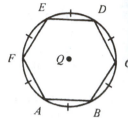

$$a(ABCDEF) = 6(a\triangle AQB)$$

Let $\overline{QH}$ be the distance from Q to $\overline{AB}$ (and thus the altitude to $\overline{AB}$ in
$\triangle AQB$). Then the area of the hexagon becomes

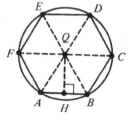

$$a(ABCDEF) = 6(\tfrac{1}{2}QH \cdot AB)$$

There is nothing magic about the number 6. To divide a circle into
n congruent arcs will always afford us the opportunity to draw a
regular n-gon and to calculate its area in the above fashion.

The definitions that follow allow us to calculate such areas more
systematically.

DEFINITION 8.13 If the vertices of a polygon belong to the same
circle, the polygon is *inscribed* in the circle. The circle is *circumscribed*
about the polygon. If the polygon is regular, the center of the circle is
the *center of the polygon* and the distance from the center to a side of
the polygon is called the *apothem*.

Many readers will notice that we have defined the center and
apothem of *inscribed* regular polygons. It can be shown that *every*
regular polygon can be inscribed in a circle.[1] Therefore, *every* regular
polygon has a center and an apothem. In the previous discussion of
regular hexagon $ABCDEF$, Q and QH were its center and apothem,
respectively.

[1] See Charles F. Brumfiel, Robert E. Eicholz, and Merrill E. Shanks, *Geometry*
(Reading, Mass.: Addison-Wesley Publishing Co., 1960), p. 194.

Suppose $ABCDE\ldots$ is a regular n-gon with center P and apothem h. Clearly, then, $a(ABCDE \ldots) = n(a\triangle PBC) = n(\frac{1}{2}h \cdot BC) = \frac{1}{2}h(n \cdot BC)$. But $n \cdot BC$ is the perimeter of the polygon, so we may write Theorem 95.

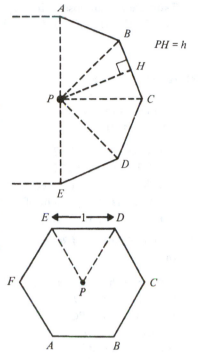

$PH = h$

THEOREM 95 The area of a regular polygon is the product of one half its apothem and its perimeter.

Example 1   Given a regular hexagon with side of length 1 in. We can calculate its area easily. First we note that, $m\angle EPD = 60$, which implies that $\triangle EPD$ is equilateral. Thus the apothem h becomes the altitude of an equilateral triangle with side of length 1, which Corollary T65(a) (Section 6.6) tells us is

$$\frac{1 \cdot \sqrt{3}}{2} \quad \text{or} \quad \frac{\sqrt{3}}{2}$$

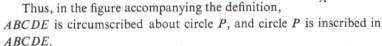

The perimeter, p, is obviously 6, so we have:

$$a(ABCDEF) = \tfrac{1}{2}hp = \frac{1}{2}\frac{\sqrt{3}}{2} \cdot 6 = \frac{3\sqrt{3}}{2}$$

DEFINITION 8.14   If each of the sides of a polygon is tangent to the same circle, the polygon is *circumscribed about the circle*. The circle is *inscribed* in the polygon.

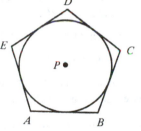

Thus, in the figure accompanying the definition, $ABCDE$ is circumscribed about circle P, and circle P is inscribed in $ABCDE$.

Example 2   Suppose a regular hexagon with perimeter 60 is circumscribed about a circle P. We can see from the figure that the radius r of the inscribed circle is the apothem of the hexagon. $\triangle PAB$ is equilateral with side of length 10, so that

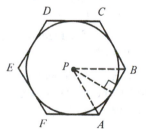

$$r = \frac{10\sqrt{3}}{2} = 5\sqrt{3}$$

Example 3 Suppose a circle Q of radius 2 is inscribed in a square

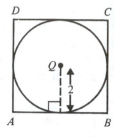

$ABCD$. Clearly, $a(ABCD) = 4^2 = 16$.

PROBLEM SET 8.6

1. In the figure, $ABCDEF$ is a regular hexagon. The diameter of circle P is 10.

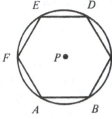

   (a) What is $m\angle APB$?
   (b) What is $m\angle ABC$?
   (c) What is the perimeter of $ABCDEF$?
   (d) What is the apothem of $ABCDEF$?
   (e) What is $a(ABCDEF)$?

2. In the figure, $\triangle ABC$ is an equilateral triangle. Circle P has radius 4.

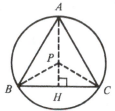

   (a) Prove that $\triangle APB \cong \triangle BPC$.
   (b) What is $m\angle CPB$?
   (c) What is $m\angle PCB$?
   (d) Calculate PH.
   (e) Calculate BC.
   (f) Calculate the perimeter of $\triangle ABC$.
   (g) Calculate $a\triangle ABC$.

3. The perimeter of a square is 24. What is its apothem?

4. Calculate the area and the perimeter of a square if its apothem is 6.

5. A circle is circumscribed about a rectangle with base 6 in. and altitude 4 in. What is the radius of the circle?

6. Find the perimeter and the area of a square inscribed in a circle of radius 2.

7. Given circle P with radius 4. Find the radius of the circumscribed square.

8. Find the perimeter of a regular triangle with apothem equal to 5 in.

9. Find the area of a regular hexagon if a side has length 6.

10. In the figure, circle P with radius 1 is inscribed in a 30-60-90 triangle $\triangle ABC$. Find the perimeter and the area of $\triangle ABC$.

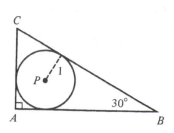

11. An isosceles right triangle is inscribed in a circle with diameter $\sqrt{2}$. What is the area of the triangle?

12. A 30-60 right triangle is inscribed in a circle with radius 4. What is the area of the triangle?

13. Given a regular polygon, its inscribed and circumscribed circles have the same center. What is the ratio of the radii of the inscribed and circumscribed circles for a square?

14. In problem 8 of Problem Set 8.5 you were asked to show that the nonconsecutive angles of an inscribed quadrilateral were supplementary. Do you think the same relationship holds for a circumscribed quadrilateral? If so, prove it. If not, give an example showing the relationship does not hold.

8.7 CIRCUMFERENCE AND AREA

The circumference of a circle is a number that represents what might be called that circle's perimeter. However, the word "perimeter" is not quite adequate since its earlier usage associated it with lengths of segments. A circle contains no segments. (Recall problem 13, Problem Set 8.3.) However, a glance at the accompanying figure should convince you that the value of the circumference of a circle can be approximated by calculating the perimeters of inscribed regular n-gons. Furthermore, it should be clear from the figure that the approximations get better as n gets larger. If we denote the perimeter of an inscribed regular n-gon and the circumference of the circle by P and C, respectively, we symbolize this notion by writing

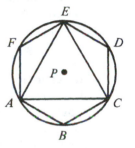

$$P_n \rightarrow C$$

and we say that the perimeter of the n-gons approaches the circumference as a limit. These ideas are summarized in the following definitions.

DEFINITION 8.15 The *circumference* of a circle is the limit of the perimeters of the inscribed regular polygons.

Definition 8.15 does not give enough information to calculate a circle's circumference. Theorem 96 moves us closer to that goal.

THEOREM 96 The ratio of the circumference to the diameter is the same for all circles.

We omit the proof since it involves some mathematical ideas that are quite sophisticated and beyond the scope of this book. The interested student may wish to consult Moise and Downs's *College Geometry*.[2]

The Greek letter π (pi) is generally used to represent this ratio, so we may write

$$\frac{C}{2r} = \pi$$

or more familiarly, $C = 2\pi r$.

As the formula $C/2r = \pi$ suggests, a circle with diameter equal to 1 has a circumference equal to π. Therefore, to calculate the perimeter of inscribed regular polygons for such a circle is to approximate the numerical value of π. Archimedes (287–212 B.C.), who many consider to be the greatest mathematician of antiquity, did essentially this. He computed the perimeters of inscribed and circumscribed regular 96-gons and found that π is between $3\frac{1}{7}$ and $3\frac{10}{71}$. Given the clumsy numerical notation of Archimedes' time, the above result was a remarkable feat. Archimedes' other mathematical achievements include the derivations of formulas for the computation of areas and volumes of geometric figures, formulas that you will meet in the following pages.

In addition to being a mathematician, Archimedes was also a brilliant scientist and engineer. *Archimedes' Principle*, for bodies floating or submerged in a liquid, is well known to science students. When, in 214 B.C., his native city of Syracuse, Sicily was besieged by Marcellus and his Roman legions, Archimedes designed machines that were used to defend the city. Although the Romans vastly outnumbered the defenders, the defenses held for two years.

Although Archimedes was not aware of the fact, the number π is irrational, that is, it cannot be written as the ratio of two integers. This fact was not demonstrated until the latter half of the eighteenth century and countless attempts had been made to find a rational value for π. Subsequently, the search has been for closer and closer rational decimal approximations to π. Today, electronic computers can continue to refine the approximation until the plug is pulled. For no earthly reason, certainly, π has been computed to over 100,000 decimal places. We will use whichever of the two approximations, 3.14 and $3\frac{1}{7}$, seems most appropriate.

---

[2] Edwin E. Moise and Floyd L. Downs, Jr., *College Geometry* (Reading, Mass.: Addison-Wesley Publishing Co., 1971), p. 423.

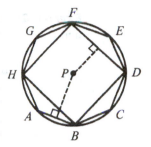

Example 1 The radius of a given circle is 28. Then the circumference is given by $C = 2\pi \cdot 28$, which, if we approximate π by $3\frac{1}{7}$, becomes $2 \cdot 3\frac{1}{7} \cdot 28 = 176$.

Intuitively, you should have little difficulty accepting the idea that $a(ABCDEFGH)$ more closely approximates the area of the circular region than does $a(BDFH)$. In fact, if we let A_n represent the area of a regular polygon of n sides and let A represent the area of the circular region, it seems reasonable to write

$$A_n \to A$$

Thus we make the following definition.

DEFINITION 8.16 The *area* of a circular region is the limit of the areas of the inscribed regular polygons.

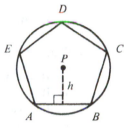

Hereafter, we shall adopt the common practice of referring to this area more simply as the *area of the circle*.

Before we can calculate such an area, we must first use the definition to derive a formula. Recall that the area of an inscribed regular n-gon with apothem h is given by

$$A_n = \tfrac{1}{2}hp_n \text{ where } p_n \text{ is the perimeter.}$$

As n gets larger, h approaches the radius r, and p_n approaches the circumference C. Symbolically, we write: Let $A_n \to A$. Then $h \to r$ and $p_n \to C$. Thus $A_n \to \tfrac{1}{2}rC$. Therefore, $A = \tfrac{1}{2}rC$. But $C = 2\pi r$. Hence

$$A = \tfrac{1}{2}r(2\pi r) = \pi r^2$$

THEOREM 97 The area of a circle with radius r is πr^2.

Example 2 Suppose a circle has diameter 4 in. Then the radius is 2 in. and the area

$$A = \pi(4^2) = 16\pi \text{ sq in.}$$

or approximately $16(3.14) = 50.24$ sq in.

Example 3   Suppose the circumference of a circle is 18π. We can find its area by first solving for r, and then using T97.

$$C = 18\pi = 2\pi r$$

implies

$$r = 9$$

Therefore,

$$A = \pi(9^2) = 81\pi$$

PROBLEM SET 8.7

1. Find the circumference and the area of a circle with radius

   (a) 2   (b) 3   (c) 4   (d) $\dfrac{1}{\pi}$

2. Find the radius of a circle if

   (a) Its diameter is 10.
   (b) Its circumference is 44 (let $\pi = 3\frac{1}{7}$).
   (c) Its area is 16π.
   (d) Its circumference is 62.8 in. (let $\pi = 3.14$).

3. A wheel has a 26-in. diameter. Use 3.14 for π and calculate:

   (a) The distance the wheel travels in one revolution.
   (b) The number of revolutions made by the wheel as it rolls for 1 mile (5,280 ft).

4. Two circles have radii for 4 and 8, respectively. What is the ratio of their areas?

5. Two circles have circumferences of 16 and 24, respectively.

   (a) What is the ratio of their radii?
   (b) What is the ratio of their areas?

6. Prove: The ratio of the areas of two circles is the square of the ratio of their radii.

7. A square with perimeter equal to 12 is inscribed in a circle. Find the circumference and the area of the circle.

8. In the figure, circle P has radius 4. Find the area of the shaded region if $\triangle ABC$ is equilateral.

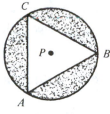

9. A side of a regular hexagon is 4 in. long.

   (a) What is the circumference of its inscribed circle?
   (b) What is the circumference of its circumscribed circle?

10. Find the radius of a circle if its circumference is numerically equal to its area.

11. A rectangular barn is 40 ft long and 16 ft wide. A horse is tied outside the barn at one corner with a rope that is 30 ft long. Over how many square feet of ground is the horse able to graze?

12. The centers of the two wheels in the diagram are 10 in. apart. If the radii of the wheels are 4 in., how long is the belt that is stretched around them?

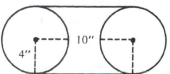

13. We can define ℓ, the *length of an arc* $\overset{\frown}{AB}$ by stating the following proportions:

$$\frac{\ell \overset{\frown}{AB}}{C} = \frac{m \overset{\frown}{AB}}{360}$$

where C is the circumference of the circle.

   (a) Find $\ell \overset{\frown}{AB}$ if $m \overset{\frown}{AB} = 45$ and the radius is 10 in.
   (b) Find $\ell \overset{\frown}{AB}$ if $m \overset{\frown}{AB} = 120$ and the radius is 6 in.
   (c) The length of a 90° arc is 6 in. Find the radius of the circle.
   (d) $AB = 2$ in. and $m \overset{\frown}{AB} = 60$. Find $\ell \overset{\frown}{AB}$.

14. A *sector* of a circle is a region bounded by two radii and an arc of the circle. As one might expect, the area of a sector is proportional to the measure of its intercepted arc. In terms of the figure,

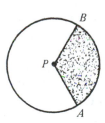

$$\frac{\text{area of sector}}{\pi r^2} = \frac{m \overset{\frown}{AB}}{360}$$

   (a) The radius of a circle is 2 in. What is the area of a sector having an arc of 60°? Of 135°? Of 180°?
   (b) Given circle P with $\overset{\frown}{AB}$, $m \overset{\frown}{AB} = 60$. If $AB = 6$ in., what is the area of the corresponding sector?

(c) If all the arcs (including the circle) have radius 2 in., what is the area of the shaded figure? (*Hint:* Consider half of one petal.)

8.8 CONSTRUCTIONS

We have been drawing geometric figures using the compass, the straightedge, and measuring devices such as the protractor and the ruler. Since we have introduced postulates involving measurement (sometimes called metric postulates), it has been legitimate to employ the latter two devices. The early Greeks, however, did not include metric notions in their treatment of geometry. Euclid's *Elements*, for example, contains no mention of either angular measure or distance. In drawing geometric figures, the Greeks used only the compass and the straightedge, according to the following two rules:

1. The straightedge may be used to draw a segment of indefinite length through any two given distinct points.
2. The compass may be used to draw a circle or an arc of a circle with any given point as center and any second given point on the circle or arc.

Mathematicians subsequently abridged the second of these rules, allowing use of the compass as dividers. That is, we permit the compass to be used to draw a circle (or arc), given its center and given a specified segment's length as its radius. It turns out that the abridged rules are equivalent to the original Euclidean rules.

Drawings carried out with the compass and straightedge, according to the preceding rules, are called *constructions*. Before doing some constructions, we need to state a postulate that declares that things are as they appear.

POSTULATE 23 *The Two-Circle Postulate:* Given circles P and Q with radii a and b, respectively. Let c be the distance between their centers. If each of a, b, and c is less than the sum of the other two, then the circles intersect in exactly two points, and the two points of intersection lie on opposite sides of the line containing their centers.

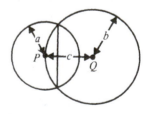

Euclid actually should have stated the Two-Circle Postulate before proving the very first theorem of *The Elements*. This oversight is the most famous of the flaws discovered in Euclid's monumental work. Actually, the Two-Circle Postulate is a logical consequence of our other postulates, but the proof is so long and involved[3] that it seems better here to adopt the statement as a postulate.

CONSTRUCTION 1   To copy a given segment on a given ray.

Given: $\overline{AB}$; $\overrightarrow{CD}$

Construct: A segment on $\overrightarrow{CD}$, congruent to $\overline{AB}$

Step 1: With A as center, draw an arc containing B.

Step 2: With C as center, draw an arc with radius AB, intersecting $\overrightarrow{CD}$ at X.

$\overline{CX}$ is the desired segment.

CONSTRUCTION 2   To copy a given angle, given a ray and one of the half planes determined by that ray.

Given: $\angle BAC$, $\overrightarrow{PQ}$ determining half plane $\mathscr{H}$
Construct: $\angle QPR$; $R \in \mathscr{H}$; $\angle QPR \cong \angle BAC$

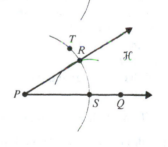

Step 1: With A as center, draw an arc intersecting $\overrightarrow{AB}$ at D and $\overrightarrow{AC}$ at E.

Step 2: With P as center, draw an arc $\overset{\frown}{ST}$, $T \in \mathscr{H}$, intersecting $\overrightarrow{PQ}$ at S, with radius AE.

Step 3: With S as center and radius ED, draw an arc intersecting $\overset{\frown}{ST}$ at R.

Step 4: Draw $\overrightarrow{PR}$.

$\angle QPR$ is the desired angle.

In the above construction, it should be easy to see that, although $\overline{DE}$ and $\overline{SR}$ are not drawn, $\triangle AED$ and $\triangle PRS$ are congruent triangles by the SSS theorem. This it what makes the construction succeed. $\angle A$ and $\angle P$ are corresponding parts of these triangles.

[3] E. E. Moise, *Elementary Geometry from an Advanced Standpoint* (Reading, Mass.: Addison-Wesley Publishing Co., 1963) pp. 198–201.

CONSTRUCTION 3　To copy a given triangle, given a ray and one of the half planes determined by that ray.

Given: $\triangle ABC$, $\overrightarrow{PQ}$ determining half plane $\mathcal{H}$

Construct: $\triangle PRS$; $R \in \overrightarrow{PQ}$; $S \in \mathcal{H}$; $\triangle PRS \cong$
$\qquad\qquad \triangle ABC$

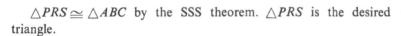

Step 1: Using construction 1, construct $\overline{PR}$ on $\overrightarrow{PQ}$,
$\qquad\quad \overline{PR} \cong \overline{AB}$.

Step 2: Draw circles P and R with radii AC and BC, respectively.

$\qquad$ Let S be the intersection of these circles in $\mathcal{H}$.

Step 3: Draw $\overline{RS}$ and $\overline{PS}$.

$\qquad \triangle PRS \cong \triangle ABC$ by the SSS theorem. $\triangle PRS$ is the desired triangle.

CONSTRUCTION 4　To construct the $\angle$ bisector of a given angle.

Given: $\angle A$

Construct: $\overrightarrow{AP}$, the bisector of $\angle A$

Step 1: Draw an arc of any convenient radius, using A as center and intersecting the sides of $\angle A$ at C and D.

Step 2: With D as center and AD as radius, draw circle D.

Step 3: With C as center and AD as radius, draw circle C.

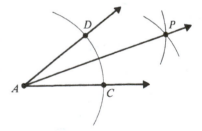

$\qquad$ By the Two-Circle Postulate, these circles intersect in two points on opposite sides of $\overleftrightarrow{DC}$. Let P be the point on the opposite side of $\overleftrightarrow{DC}$ from A.

Step 4: Draw $\overrightarrow{AP}$.

$\qquad \overrightarrow{AP}$ is the desired ray. Notice that $\triangle ADP \cong \triangle ACP$ so that $\angle DAP \cong \angle CAP$.

CONSTRUCTION 5　To bisect a given segment.

Given: Segment $\overline{CB}$

Construct: A bisector of $\overline{CB}$

Step 1: Draw a circle C with radius CB.

Step 2: Draw circle B with radius CB.

Step 3: Draw the line $\overleftrightarrow{l}$ connecting X and Y, the points of intersection of these two circles.

Note that, by the Two-Circle Postulate, this line intersects the line containing C and B. Call this point of intersection M. M is the desired point; that is, M bisects $\overline{CB}$.

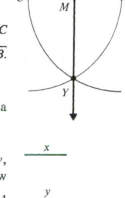

In the above construction, X and Y are each equidistant from C and B. Thus $\overleftrightarrow{XY}$ not only bisects $\overline{CB}$, it is perpendicular to $\overline{CB}$. (Why?) Thus we have also performed Construction 6.

CONSTRUCTION 6 To construct the perpendicular bisector of a segment.

Example 1 Suppose we are given two segments of lengths x and y, and we wish to construct a segment with length $x + y$. First, draw arbitrary ray $\overrightarrow{PQ}$. Then construct $\overline{PA}$ such that $PA = x$, and construct $\overline{AB}$ such that P-A-B, and $AB = y$. $\overline{PB}$ is the desired segment.

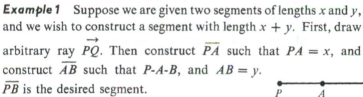

Example 2 Suppose we wish to construct a right triangle with legs of length x and y as in Example 1. We might begin by drawing an arbitrary ray $\overrightarrow{PQ}$ and constructing a segment $\overline{PS}$ of length $2y$ on

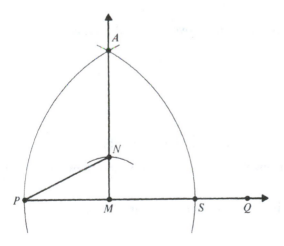

that ray. Next construct $\overleftrightarrow{AM}$ the $\perp$ bisector of $\overline{PS}$, where M is the midpoint of $\overline{PS}$. Then construct $\overline{MN}$ on $\overrightarrow{MA}$ such that $MN = x$. Thus either of the triangles $\triangle PMN$ and $\triangle SMN$ is the desired triangle.

Example 3 The first theorem of Euclid's *Elements* involved the construction of an equilateral triangle. We begin by drawing an arbitrary segment $\overline{AB}$. Draw circle A and circle B, each with radius AB. Let C be one of the points of intersection of these two circles. $\triangle ABC$ is equilateral. Do you see where the Two-Circle Postulate fits in?

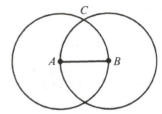

Example 4 Suppose we wish to construct a triangle with sides of lengths 1 in., $1\frac{1}{2}$ in., and 2 in. We begin by marking off (using a ruler) segments of each of these lengths. Then we draw an arbitrary ray $\overrightarrow{PQ}$ and use Construction 3, first constructing $\overline{PX}$, $PX = 2$ in. Then we draw circles P and X of radius $1\frac{1}{2}$ and 1 in., respectively.

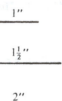

1''

$1\frac{1}{2}$''

2''

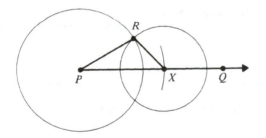

Let R be a point of intersection of these two circles. $\triangle PXR$ is the desired triangle.

PROBLEM SET 8.8

1. Some, but not all, drawings are constructions. Explain.

2. Construct a triangle, if possible, given sides with the following lengths.

   (a) $\frac{1}{2}$ in., 1 in., and $1\frac{1}{4}$ in.   (b) 1 in., 2 in., and $2\frac{1}{2}$ in.
   (c) 1 in., 2 in., and 3 in.

3. Given segment $\overline{AB}$. Construct an equilateral triangle with $\overline{AB}$ as one side.

4. Given $\overline{AB}$ and $\overline{CD}$ as shown.

(a) Construct a segment of length $AB + CD$.
(b) Construct a segment of length $CD–AB$.

5. (a) Draw acute $\triangle ABC$, and construct the three angle bisectors.
(b) Draw obtuse triangle $\triangle DEF$, and construct the three angle bisectors.
(c) Draw right triangle $\triangle GHJ$, and construct the three angle bisectors.
(d) Do you notice anything? If so, make a conjecture.

6. Construct a $60°$ angle. (*Hint:* See Example 3.)

7. Construct a $30°$ angle.

8. Construct an isosceles right triangle.

9. Construct an isosceles triangle if the base is 1 in. and the second side has length 2 in.

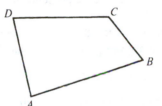

10. Construct a copy of the figure given.

11. Given segments of lengths 1 and $1\frac{1}{2}$ in., construct an isosceles trapezoid having a $1\frac{1}{2}$-in. lower base, 1-in. legs, and lower base angles with measure $30°$.

12. Given segment $\overline{AB}$. Construct the set of all points C such that $m\angle ACB = 90$.

13. Given circle P, point X, and $X \in$ circle P. Construct the tangent to circle P at X.

14. Construct $\triangle ABC$ given two angles and a side opposite one of them, as shown. (*Hint:* First construct $\angle C$ using the Supplement Postulate.)

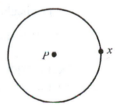

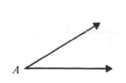

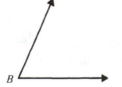

$$\overline{}\ x$$

$CB = x$

8.9 MORE CONSTRUCTIONS

CONSTRUCTION 7 To construct the line perpendicular to a given line at a given point on that line.

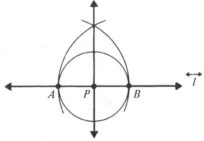

Given: Line $\overleftrightarrow{l}$; $P \in \overleftrightarrow{l}$

Construct: The perpendicular to $\overleftrightarrow{l}$ at P

Our strategy here will be to turn P into the midpoint of some segment of $\overleftrightarrow{l}$, so that by constructing the perpendicular bisector of that segment, we will have achieved the desired result.

Step 1: Draw a circle with P as its center. The points of intersection of $\overleftrightarrow{l}$ with circle P we will call A and B.

Step 2: Construct the perpendicular bisector of $\overline{AB}$.

CONSTRUCTION 8 To construct a perpendicular to a given line through a given point not on that line.

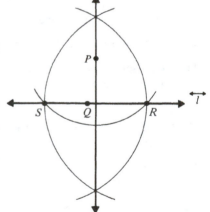

Given: Line $\overleftrightarrow{l}$ and P; $P \notin \overleftrightarrow{l}$

Construct: A line perpendicular to $\overleftrightarrow{l}$ and containing P

Step 1: Let Q be any point of $\overleftrightarrow{l}$.

Step 2: Draw circle P with radius r, $r > PQ$. Thus $\overleftrightarrow{l}$ must intersect circle P in two points. Call them S and R.

Step 3: Use Construction 6 to construct the perpendicular bisector of $\overline{SR}$.

This line is the desired result.

CONSTRUCTION 9 Through a given point, to construct a line parallel to a given line.

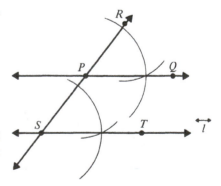

Given: Line $\overleftrightarrow{l}$ and P; $P \notin \overleftrightarrow{l}$

Construct: A line through P and parallel to $\overleftrightarrow{l}$

Step 1: Let $S \in \overleftrightarrow{l}$ and draw $\overleftrightarrow{PS}$. Let $T \in \overleftrightarrow{l}$.

Step 2: Let R-P-S. With $\overrightarrow{PR}$ as one side, construct (use Construction 2) $\angle RPQ$, such that $\angle RPQ \cong \angle PST$.

$\overleftrightarrow{PQ}$ is the desired line. The Corresponding Parallel Theorem assures this.

Construction 9 and the Egg-Slicer Corollary are used to divide a given segment into n congruent segments in the following construction.

CONSTRUCTION 10 To divide a given segment into n congruent segments.

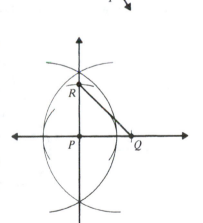

Given: Segment $\overline{AB}$

Construct: n congruent segments on $\overline{AB}$ intersecting only at endpoints (Here we let $n = 5$.)

Step 1: Draw ray $\overrightarrow{AP}$, $B \notin \overrightarrow{AP}$.

Step 2: On ray $\overrightarrow{AP}$ draw n (5) congruent consecutive segments $\overline{AP_1}$, $\overline{P_1P_2}$, $\overline{P_2P_3}$, ..., $\overline{P_{n-1}P_n}$ (here $\overline{P_{n-1}P_n}$ is $\overline{P_4P_5}$).

Step 3: Draw $\overline{P_nB}$ (here $\overline{P_5B}$).

Step 4: Use Construction 9 to construct lines through P_1, P_2, ..., P_{n-1} (here P_4), all of which are parallel to $\overline{P_nB}$.

Step 5: These rays intersect $\overline{AB}$ in B_1, B_2, ..., B_{n-1}.

Then by the Egg-Slicer Corollary, $\overline{AB_1}$, $\overline{B_1B_2}$, $\overline{B_2B_3}$, ..., $\overline{B_{n-1}B}$ are the desired segments.

Example 1 To construct an isosceles right triangle, we first draw a line $\overleftrightarrow{PQ}$ and then construct a perpendicular to $\overleftrightarrow{PQ}$ at P. On this perpendicular, construct $\overline{PR}$, $\overline{PR} \cong \overline{PQ}$. $\triangle QPR$ is isosceles.

Example 2 Given a segment of length x, we can construct a segment of length $x\sqrt{3}$ by first recalling that in a 30-60-90 triangle, the length of the side opposite the 60° angle has length $\sqrt{3}$ times the length of the side opposite the 30°

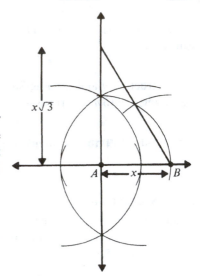

angle. Then we proceed, as in Example 1, by constructing a perpendicular to $\overline{AB}$ at A. Next, we construct a 60° angle at B, carrying out the first steps of the construction of an equilateral triangle.

Example 3 Consider three congruent circles, each of which is externally tangent to the other two. Notice that the segments joining their centers form an equilateral triangle. Thus if we wish to construct three such circles (given their radius AB), we proceed as follows.

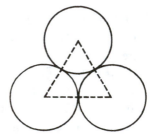

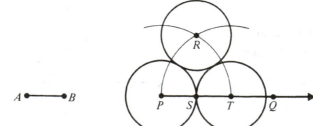

Draw a ray $\overrightarrow{PQ}$ and copy $\overline{AB}$ on it twice in succession, forming segments $\overline{PS}$ and $\overline{ST}$. Draw circle P with radius AB and circle T with radius AB. To draw the third tangent circle, we need only locate its center R. We do this by finding the third vertex of equilateral triangle $\triangle PTR$ with side $\overline{PT}$. Then we draw circle R with radius AB.

PROBLEM SET 8.9

1. Construct a 45° angle. (*Hint:* See Example 1.)

2. Construct a 75° angle. Note that 45 + 30 = 75.

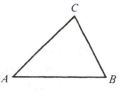

3. (a) Copy $\triangle ABC$ and construct the perpendicular bisectors of its sides.
    (b) Construct a right triangle and the perpendicular bisectors of its sides.
    (c) Draw an obtuse triangle and construct the perpendicular bisectors of its sides.
    (d) Do you notice anything? If so, make a conjecture.

4. Construct a square.

5. Given circle P, construct its circumscribed square.

6. Construct a rhombus given the lengths of its diagonals.

7. Given one angle and the length of one side of a rhombus, construct the rhombus.

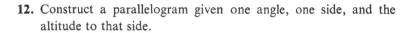

8. Bisect one side of a given triangle. Through this point of bisection, construct a line parallel to another side of the triangle.

9. Draw a circle with radius 3 in. and diameter $\overline{AB}$. Construct a segment of length $\sqrt{5}$ in. [*Hint:* $\sqrt{5} = \sqrt{5(1)}$ and $5 + 1 = 6$.]

10. (a) Copy $\triangle ABC$ and construct its medians.
     (b) Construct a right triangle and its medians.
     (c) Draw an obtuse triangle and construct its medians.
     (d) Do you notice anything?

11. "Trisect" segment $\overline{PQ}$. (*Hint:* See Construction 10.)

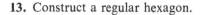

12. Construct a parallelogram given one angle, one side, and the altitude to that side.

13. Construct a regular hexagon.

14. x and y are the radii of internally tangent circles. Construct the circles.

15. PQ is the diameter of a circle used in the construction of tangent $\overleftrightarrow{PT}$. Carry out this construction.

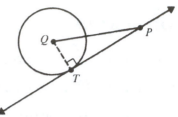

16. Construct an equilateral triangle given the altitude.

17. $\overline{AB}$ represents a plane mirror. A ray of light starting at C is reflected by the mirror to an observer at D.

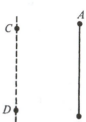

    (a) If $\overline{CD} \parallel \overline{AB}$, construct the point where the light ray strikes the mirror. Recall problem 11 of Problem Set 4.8.

    (b) How would you solve this problem if $\overline{CD}$ was not parallel to $\overline{AB}$?

8.10 CENTERS OF TRIANGLES

In the diagram shown, $\overrightarrow{AP}$ and $\overrightarrow{BP}$ are angle bisectors. Thus P is equidistant from $\overleftrightarrow{AC}$ and $\overleftrightarrow{AB}$ (since it is on the angle bisector of $\angle A$) and equidistant from $\overleftrightarrow{AB}$ and $\overleftrightarrow{BC}$ (since it is on the angle bisector of $\angle B$). Symbolically, we write:

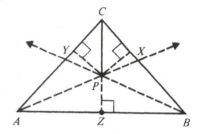

$$PY = PZ$$

and

$$PX = PZ$$

Thus $PX = PY$, which means that P is on the angle bisector of $\angle C$.

DEFINITION 8.17 Two or more lines are *concurrent* if there is a single point that belongs to all of them.

Thus in the discussion preceding the definition, we have derived the following theorem.

THEOREM 98 The angle bisectors of the angles of a triangle are concurrent at a point equidistant from the sides of the triangle.

The derivation of T98 suggests the following construction.

CONSTRUCTION 11   To construct the inscribed circle of a given triangle.

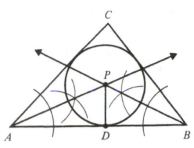

Given: $\triangle ABC$
Construct: The inscribed circle of $\triangle ABC$

Step 1: Bisect $\angle A$ and $\angle B$.
Step 2: Construct $\overline{PD}$, the perpendicular segment, from P, the point of intersection of the above angle bisectors, to $\overline{AB}$.
Step 3: Construct circle P with radius PD.

Theorem 98 assures us that P is equidistant from all three sides of the triangle.

Suppose we consider the perpendicular bisectors of two of the sides of triangle $\triangle ABC$. They intersect at some point P. The Perpendicular Bisector Theorem implies that $PA = PB$ and $PA = PC$. Thus $PB = PC$, which implies that P is on the perpendicular bisector of $\overline{BC}$ as well. The following theorem summarizes these ideas.

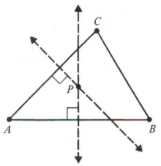

THEOREM 99   The perpendicular bisectors of the sides of a triangle are concurrent at a point equidistant from the vertices of the triangle.

Theorem 99 suggests the technique of constructing the circumscribed circle of a given triangle.

CONSTRUCTION 12   To construct the circumscribed circle of a given triangle.

Given: $\triangle ABC$
Construct: The circumscribed circle of $\triangle ABC$

Step 1: Construct the $\perp$ bisectors of $\overline{AB}$ and $\overline{AC}$.
Step 2: Let P be the point of intersection of the $\perp$ bisectors of $\overline{AB}$ and $\overline{AC}$. Construct circle P with radius PA.

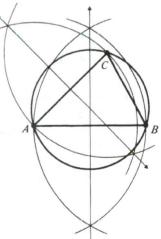

DEFINITION 8.18   The center of the inscribed circle of a triangle is the *incenter* of the triangle. The center of the circumscribed circle of a triangle is the *circumcenter* of the triangle.

Problem 10 of Problem Set 8.9 should have led you to conclude that the first part of the following theorem is true.

THEOREM 100 The medians of a triangle are concurrent at a point which is two thirds of the distance from any vertex to the midpoint of the opposite side.

Proof: We give an outline here. You will be asked to complete parts of the proof as an exercise.

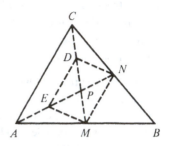

Suppose $\overline{AN}$ and $\overline{CM}$ are medians to $\overline{BC}$ and $\overline{AB}$, respectively, intersecting at P. Let D and E be midpoints of $\overline{CP}$ and $\overline{AP}$, respectively. Then $EMND$ is a parallelogram, and $CD = DP = PM$ and $AE = EP = PN$, from which it follows that $AP = \frac{2}{3}AN$ and $CP = \frac{2}{3}CM$. Let $\overline{AN}$ and $\overline{BL}$, the median from B, intersect at Q. Then by a similar argument, $AQ = \frac{2}{3}AN$ and $CQ = \frac{2}{3}CL$. The Point-Plotting Postulate gives us $Q = P$, that is, concurrency.

DEFINITION 8.19 The intersection of the medians of a triangle is called the *centroid* of the triangle.

In more advanced mathematics we prove that the centroid of a triangle is its "center of gravity" or "balance point." If you cut a triangular region out of a rigid piece of cardboard or sheet metal, you will find the approximate location of the centroid of this figure by finding the point at which it balances on a pencil point.

PROBLEM SET 8.10

1. (a) The point of concurrency of the angle bisectors of a triangle is called the _____ of the triangle.
   (b) The point of concurrency of the perpendicular bisectors of the sides of a triangle is called the _____ of the triangle.
   (c) The point of concurrency of the medians of a triangle is called the _____ of the triangle.

2. Is the incenter of a triangle always in its interior? Explain.

3. Under what conditions is the circumcenter of a triangle
   (a) In the interior of the triangle?
   (b) On the triangle?
   (c) In the exterior of the triangle?

4. In the figure, medians $\overline{AL}$, $\overline{BM}$, $\overline{CN}$ are concurrent at Q.
   (a) If $AP = 6$, find AL
   (b) If $CN = 12$, find PN
   (c) If $PL = 4$, find AL
   (d) If $BM = 16$, find PM

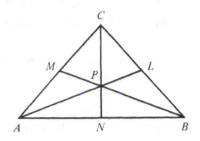

5. How would you go about finding a point that is equidistant from each of three noncollinear points?

6. Construct an equilateral triangle. Construct its inscribed circle and its circumscribed circle.

7. Prove: The perpendicular bisectors of the sides of a right triangle are concurrent at the midpoint of the hypotenuse.

8. Circumscribe a circle about a square.

9. Construct a right triangle given the lengths of the hypotenuse and one leg.

10. Construct an equilateral triangle given the radius of the inscribed circle.

11. In the proof to T100, prove that $EMND$ is a parallelogram.

12. In the figure, P is the centroid of $\triangle ABC$, CH is an altitude. If $PM = 8\frac{2}{3}$, $MH = 5$, find CH.

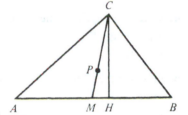

13. Prove: If the angle bisectors of three angles of a parallelogram are concurrent, then that parallelogram is a rhombus.

CHAPTER 8: REVIEW EXERCISES

1. For each of the following statements, indicate whether it is *always* true, *sometimes* but not always true, or *never* true.
   (a) Given circle C. Then $C \in$ circle C.
   (b) If a line intersects the interior of a circle, then it intersects the circle in exactly two points.
   (c) If $\overparen{AB}$ and $\overparen{BC}$ are arcs of the same circle, then $m\,\overparen{AB} + m\,\overparen{BC} = m\,\overparen{ABC}$.
   (d) A diameter of a circle is a secant of that circle.
   (e) If a line intersects a circle in one point, then it intersects the circle in two points.

(f) If $\angle A$ intercepts $\overset{\frown}{BC}$, then $\angle A$ is inscribed in $\overset{\frown}{BAC}$.

(g) If $\angle A$ is inscribed in major arc $\overset{\frown}{BC}$, then $\angle A$ intercepts minor arc $\overset{\frown}{CB}$.

(h) If a point is the midpoint of two chords of a circle, it is the center of the circle.

(i) If two chords of a circle are not congruent, then the shorter chord is closer to the center.

(j) Two tangents to a circle at endpoints of a diameter are parallel.

(k) If in a given circle, $m\,\overset{\frown}{AB} = \tfrac{1}{2}m\,\overset{\frown}{AC}$ (both are minor arcs), then $AB = \tfrac{1}{2}AC$.

(l) If a parallelogram is inscribed in a circle, it is a rectangle.

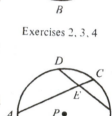

2. Given circle C, in the figure, and $\overline{AB} \perp \overline{ED}$. If $AB = 10$, and $DE = 8$, find FC.

3. Given circle C, in the figure, and $\overline{AB} \perp \overline{ED}$. If $FC = 5$, and $DE = 24$, find AB.

Exercises 2, 3, 4

4. Given circle C, in the figure, $m\,\overset{\frown}{ED} = 60$, $ED = 2$, and $\overline{AB} \perp \overline{ED}$. Find:

   (a) DC (b) AF

5. Given circle P, in the figure, $m\,\overset{\frown}{AB} = 110$, and $m\,\overset{\frown}{DC} = 70$. Find $m\angle AEB$.

6. Given circle P in the figure, $AE = 6$, $EC = 4$, and $DE = 3$. Find EB. (*Hint:* Use similar triangles.)

Exercises 5, 6

7. Given: Circle P and $\overline{AD} \cong \overline{BC}$

   Prove: $\overline{AD} \parallel \overline{BC}$

8. Approximately how far (d in the figure) can a person see from the top of a mountain 3 miles high? Assume the radius of the earth is 4,000 miles.

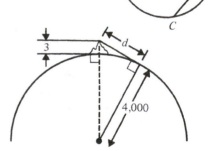

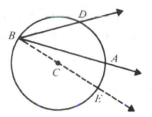

9. Use the figure at the right to complete the proof for Case 3 of T91 (Section 8.5).

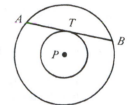

10. Given: Concentric circles with center P, and $\overleftrightarrow{AB}$ is tangent to the inner circle at T and a chord of the outer circle

    Prove: $\overline{AT} \perp \overline{TB}$

11. Given: Circle P and circle Q are tangent at T, and $\overleftrightarrow{RS}$ and $\overleftrightarrow{RV}$ are tangents

    Prove: $\overline{RS} \cong \overline{RV}$

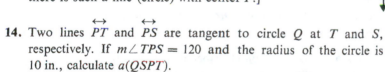

12. Given circle P with tangents $\overleftrightarrow{QT}$ and $\overleftrightarrow{QR}$. If $m\angle TQR = 60$ and the radius of circle P is 10 in., find QT.

13. Prove: No line is a circle. [*Hint:* Assume that there is such a line (circle) with center P.]

14. Two lines $\overleftrightarrow{PT}$ and $\overleftrightarrow{PS}$ are tangent to circle Q at T and S, respectively. If $m\angle TPS = 120$ and the radius of the circle is 10 in., calculate $a(QSPT)$.

15. Find the area of an equilateral triangle inscribed in a circle of circumference equal to 6π.

16. The area of a circle equals the area of a square with side of length 6. Find the radius of the circle.

17. Find the area of a regular hexagon inscribed in a circle of radius 5.

18. In the figure, $\overline{AB}$ and $\overline{CD}$ are parallel chords.

    Prove: $\overparen{AC} \cong \overparen{DB}$

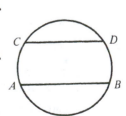

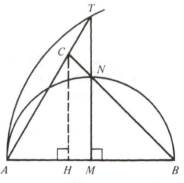

19. In the figure at the right, $\overset{\frown}{ANB}$ is a semicircle with center M. $\overset{\frown}{AT}$ is an arc of the circle with center B and radius $\overline{AB}$. If $\overline{TM} \perp \overline{AB}$ and $AB = 1$, find:
    (a) $m\angle ABC$ (b) $m\angle CAB$
    (c) AT (d) $m\angle ACB$
    (e) AC (*Hint:* Let $AH = x$, and use the fact that $\triangle BHC$ is isosceles.)
    (f) BC (g) $a\triangle ABC$

20. Two circles having radii 10 in. and 20 in. are tangent externally. Find the length of $\overline{AB}$ if $\overleftrightarrow{AB}$ is tangent to both circles and A and B are the points of tangency.

21. Two chords intersect in the interior of a circle. The shorter segment is separated into segments of length $2x$ and $3x$, and the longer segment is separated into segments of length y and $6y$. Find the lengths of the chords if one is 8 in. longer than the other.

22. Prove that the area of a polygon circumscribed about a circle is equal to one-half the product of the perimeter of the polygon and the radius of the circle.

23. Every equiangular polygon inscribed in a circle is regular if it has an odd number of sides. Is this statement true if the polygon has an even number of sides? Explain.

24. Given: Square $ABCD$
    Prove: The area of a circle inscribed in $ABCD$ is $\frac{1}{2}$ the area of a circle circumscribed about $ABCD$.

25. Prove that the area of a regular hexagon inscribed in a circle is the geometric mean of the area of the inscribed equilateral triangle and the area of the circumscribed equilateral triangle.

26. If on a given clock, the length of the hour hand is two-thirds the length of the second hand, find the ratio of the distance traveled by the tip of the second hand to that of the tip of the minute hand in one hour. Is the answer different for a longer period of time? Explain.

27. $\overrightarrow{PA}$ and $\overrightarrow{PB}$ are tangent to a circle at A and B, respectively. If $m \angle APB = 105$ and the radius of the circle is 10 in., find the length of minor arc $\overset{\frown}{AB}$.

28. If a chord 12 in. long is 4 in. from the center of the circle, find the length of another chord which is 3 in. from the center.

29. A circular arch has height 18 ft and spans 48 ft. Find the radius of the circle of which the arch is a part.

30. Suppose one leg of a right triangle is a diameter of a circle. Show that the tangent at the point where the circle intersects the hypotenuse bisects the other leg.

REFERENCES

Brumfiel, Charles F.; Eicholz, Robert E.; and Shanks, Merrill E. *Geometry.* Reading, Mass.: Addison-Wesley Publishing Co., 1960. p. 194.
A readable proof of our assertion that every regular polygon can be inscribed in a circle.

Eves, Howard. *An Introduction to the History of Mathematics.* 2d ed. New York: Holt, Rinehart and Winston, 1953. pp. 90–96.
Man has been attempting to compute π for at least 3600 years. A fascinating, and often amusing, account of some of these attempts is given here.

Moise, Edwin E., and Downs, Floyd L., Jr. *College Geometry.* Reading, Mass.: Addison-Wesley Publishing Co., 1971. p. 423.
The ratio of the circumference of a circle to its diameter is a constant (π). A clear outline of the proof of this assertion is given here.

Polya, G. *Induction and Analogy in Mathematics.* Princeton, N.J.: Princeton University Press, 1954. Chap. 10.
"Of all plane figures of equal perimeter, the circle has the maximum area." Appealing to nature and the reader's experience throughout, Polya provides a proof to this theorem that is an exciting model of how a mathematician does mathematics.

"And what are you taking up here at college?" the Dean asked.
"Space, mostly, sir," John replied.

JOHN BLOW, classmate of the author, 1951

CHAPTER NINE

SPACE GEOMETRY

9.1 INTRODUCTION

Our lives are not lived in a world that is coplanar. The geometry of the plane, or *two-dimensional geometry*, as it is sometimes called, is inadequate to describe much of our noncoplanar existence. All is not lost, however. For as we developed the ideas of the previous chapters, we were careful to leave room for the possibility of noncoplanar geometric figures. Recall the following definition and postulate.

DEFINITION 2.2 *Space* is the set of all points.

POSTULATE 7 Space contains at least four points which are non-collinear and do not lie in the same plane.

These statements set the stage for the study of three-dimensional figures. We call this study *space geometry* or *solid geometry*. It should

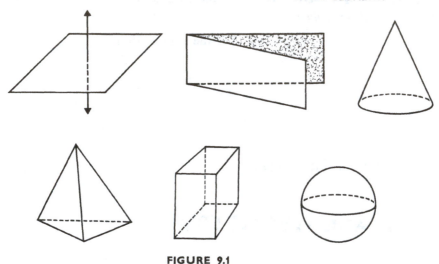

FIGURE 9.1

be viewed as an extension of what has come before because all the definitions, postulates, and theorems of the preceding chapters continue to be applicable.

Figure 9.1 shows some space figures. Now draw some examples of your own. You might begin by drawing figures that represent your own conception of some of the following: parallel planes, a cube, a cylinder, a box with triangular top and bottom.

9.2 LINES, PLANES, AND SEPARATION

In Problem Set 3.2 you were asked to state a Space Separation Postulate analogous to the Plane Separation Postulate. Hopefully you came up with something like the following.

POSTULATE 24   *The Space Separation Postulate:* Given a plane, the set of all points of space that do not belong to that plane is the union of two disjoint sets called *half spaces* such that both the following are true:

1. Each of the sets is convex.
2. If P belongs to one of the sets and Q belongs to the other, the $\overline{PQ}$ intersects the plane.

Each of the walls of a room can be thought of as a plane separating space into disjoint sets. Part (b) of Postulate 24 contains the reason that a doorway has been cut through one of those walls.

From this and earlier postulates we can now prove a number of interesting theorems. Among them,

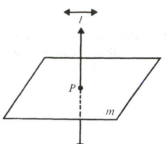

THEOREM 101   If a line intersects a plane not containing it, then the intersection contains a single point.

Proof (indirect): Suppose there were two points in the intersection. Then these two points would be in plane *m*. Thus the line containing them would be in plane *m* (P6). →/←

THEOREM 102   Any point and a line not containing it determine a plane.

Given: Line $\overleftrightarrow{AB}$; point *P*; $P \notin \overleftrightarrow{AB}$
Prove: There is exactly one plane containing $\overleftrightarrow{AB}$ and *P*

Proof: *A*, *B*, and *P* are noncollinear. Therefore, by P5, the Plane Postulate (Section 2.2), they determine a plane. But if *A* and *B* lie in a plane, the entire line lies in that plane [see P6 (Section 2.2)]. Thus the plane determined by *A*, *B*, and *P* is also determined by $\overleftrightarrow{AB}$ and *P*.

The accompanying figure suggests the next postulate.

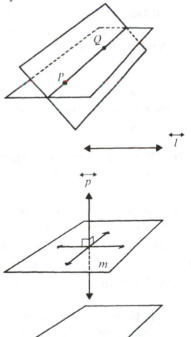

POSTULATE 25   If two planes intersect, then their intersection is a line.

In space geometry, words like "parallel" and "perpendicular" take on expanded meanings that are probably already intuitively clear to you.

DEFINITION 9.1   Two planes, or a line and a plane, are *parallel* if they do not intersect. Polygonal or circular regions are *parallel* if they are subsets of parallel planes. A line is *perpendicular* to a plane if it intersects the plane and is perpendicular to every line of that plane which passes through the point of intersection.

In the figure, $\overleftrightarrow{l}$ is parallel to plane *m*, $\overleftrightarrow{p}$ is perpendicular to plane *m*, and planes *m* and *n* are parallel.

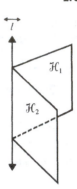

DEFINITION 9.2 The union of a line and two noncoplanar half planes having that line as an edge is called a *dihedral angle*. The line is called the *edge* of the dihedral angle. The union of each half plane with its edge is called a *side* or *face*.

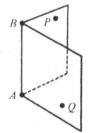

NOTATION RULE 9.1 If $\mathcal{H}_1 \cup \mathcal{H}_2 \cup \overleftrightarrow{AB}$ is a dihedral angle with $P \in \mathcal{H}_1$ and $Q \in \mathcal{H}_2$, then we will denote the angle by

$$\angle P\text{-}AB\text{-}Q$$

DEFINITION 9.3 A *plane angle* of a dihedral angle is the intersection of the dihedral angle with any plane perpendicular to its edge.

In the figure $\angle RST$ is a plane angle of $\angle P\text{-}AB\text{-}Q$. Obviously any given dihedral angle may have infinitely many plane angles. Fortunately, they are all congruent.

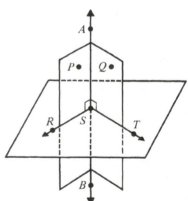

THEOREM 103 Any two plane angles of a dihedral angle are congruent.

In terms of the figure, $\angle D$ and $\angle C$ are plane angles of $\angle P\text{-}AB\text{-}Q$, and we wish to show that $\angle D \cong \angle C$. The proof is left as an exercise. (*Hint:* Pick E, F, G, and H as shown such that $\overline{CE} \cong \overline{DG}$ and $\overline{CF} \cong \overline{DH}$. Then show that $EFHG$ is a parallelogram. Congruent triangles then point the way to the desired result.)

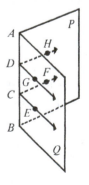

DEFINITION 9.4 The *measure of a dihedral angle* is the measure of any of its plane angles. Acute, obtuse, or right plane angles determine acute, obtuse, or right dihedral angles.

DEFINITION 9.5　Two planes are *perpendicular* if they determine a right dihedral angle.

The preceding definitions lead to the following theorems.

THEOREM 104　If a line is perpendicular to a plane, any plane containing the line is perpendicular to the plane.

THEOREM 105　If two planes are perpendicular, any line in one of them that is perpendicular to their intersection is also perpendicular to the other plane.

THEOREM 106　If two intersecting planes are each perpendicular to a third plane, then their line of intersection is perpendicular to that plane.

The proofs of T104 and T105 are left as exercises. Theorem 106, sometimes called the Revolving-Door Theorem, has a much more tedious proof. The interested student can find a proof in any solid geometry textbook.

PROBLEM SET 9.2

1. Sketch several different three-dimensional geometric figures.

2. Using parallelograms to represent parallel (nonintersecting) planes, sketch two lines which are not parallel and yet which do not intersect.

3. Sketch two intersecting planes and a line parallel to their line of intersection which is not in either plane.

4. Plane *m* and plane *n* have three points in common. What do you conclude?

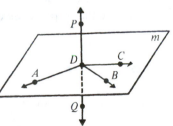

5. Given *A*, *B*, *C*, and *D* in plane *m* and $\overleftrightarrow{PQ} \perp m$ at *D*. Which of the following angles are right angles: $\angle ADP$, $\angle ADB$, $\angle PDC$, $\angle QDB$, $\angle BDC$?

6. If a line is perpendicular to a line in a given plane, is it perpendicular to the plane? Explain.

7. Try to write a definition of the interior of a dihedral angle.

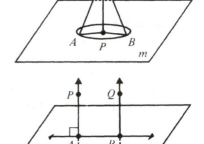

8. Prove T104

9. Given dihedral angle $\angle P\text{-}AB\text{-}Q$. For what real numbers a and b is it always the case that $a < m\angle P\text{-}AB\text{-}Q < b$?

10. Name six dihedral angles determined by the figure.

11. How many planes are determined by the following?
    (a) Three lines that are concurrent but not coplanar.
    (b) Three lines that are parallel but not co-planar.
    (c) Two parallel lines and a point not in the plane of those lines.

12. Given planes l, m, n; $l \cap m \cap n \neq \varnothing$. What are the possibilities?

13. Given: Circle P in plane m, and $\overleftrightarrow{VP} \perp$ plane m
    Prove: $\angle PAV \cong \angle PBV$

14. We know that two parallel lines are coplanar. Use that fact to argue that $\overleftrightarrow{PA}$, $\overleftrightarrow{QB}$, and $\overleftrightarrow{AB}$ in the diagram are coplanar and then prove that if one of two parallel lines is perpendicular to a given plane, then the other line is also perpendicular to that plane.

15. Explain why there is no line that is perpendicular to each of two intersecting planes.

16. Prove T105

17. (a) Prove: Through a point not on a given plane, there exists a line that is perpendicular to that plane. (*Hint:* Use T102.)
    (b) Prove: Through a point not on a given plane, there is at most one line that is perpendicular to that plane.

9.3 POLYHEDRA

The figures shown illustrate what are called polyhedral angles.

DEFINITION 9.6 Let $P_1P_2 \ldots P_n$, $n \geq 3$, be a polygon in plane m. Let V be a point; $V \notin m$. The union of all rays $\overrightarrow{VP}$ such that $P \in P_1P_2 \ldots P_n$ is called a *polyhedral angle*. V is the *vertex* and $\overrightarrow{VP_1}$,

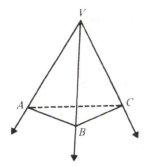

Trihedral angle

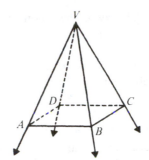

Tetrahedral angle

$\overrightarrow{VP_2}, \ldots, \overrightarrow{VP_n}$ are the *edges*. The angles $\angle P_1VP_2$, $\angle P_2VP_3, \ldots$, $\angle P_{n-1}VP_n$, $\angle P_nVP_1$ are called the *face angles* of the polyhedral angle. The union of each of the face angles with its interior is called a *face* of the polyhedral angle. If $P_1P_2\ldots P_n$ is a convex polygon, then the polyhedral angle is *convex*.

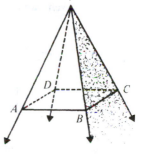

In the accompanying figure, $\overrightarrow{VA}$, $\overrightarrow{VB}$, $\overrightarrow{VC}$, and $\overrightarrow{VD}$ are edges of a tetrahedral angle. The shaded region is a face, and $\angle BVC$ is a face angle.

Each of the shapes in Fig. 9.2 is a polyhedron (*plural:* polyhedra).

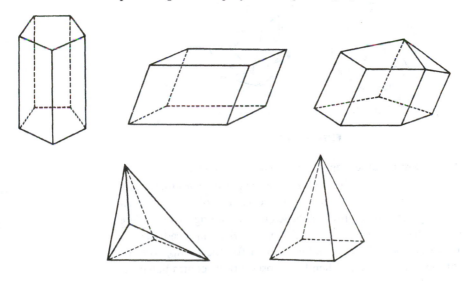

FIGURE 9.2

DEFINITION 9.7   A *polyhedron* is the union of a finite number of polygonal regions such that both the following are true:

1. The interiors of no two regions intersect.
2. Each side of a polygon is a side of exactly one of the other polygons.

Each of the polygonal regions is called a *face* of the polyhedron. The intersection of any two faces is called an *edge*. The intersection of two edges is a *vertex* of the polyhedron. If all the faces of a polyhedron are bounded by regular congruent polygons, then the polyhedron is called a *regular* polyhedron.

Five regular polyhedra are depicted in Fig. 9.3. With the exception of the cube, they are given names that indicate how many faces they have.

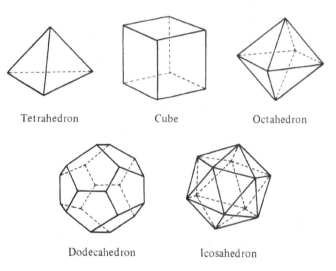

Tetrahedron   Cube   Octahedron

Dodecahedron   Icosahedron

FIGURE 9.3

That there are no other regular polyhedra was known as early as Plato's time. In fact he was so interested in their properties that today they are known as the *Platonic solids*. The existence of only five such figures has impressed many prominent scientists, among them the late sixteenth-century astronomer Johannes Kepler, who employed them to reconstruct what he believed was God's plan for the motions of the planets of our solar system. When later observations contradicted his theory, he willingly gave it up, true scientist that he was.

DEFINITION 9.8   If two of the faces of a polyhedron are bounded by congruent polygons lying in parallel planes, and if each of the other faces is the union of a parallelogram with its interior, then the polyhedron is called a *prism*. The parallel faces are *bases*. The faces joining the bases are called *lateral faces*; their intersections are *lateral edges*. If the lateral edges are perpendicular to a base, the prism is called a right *prism*.

Figure 9.4 shows four prisms with names that are given them. Their names are chosen for the obvious reason that their bases are quadrilaterals, pentagons, triangles, and rectangles, respectively.

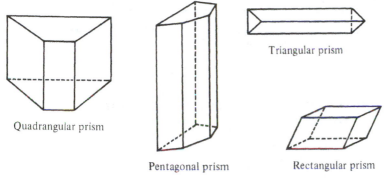

Quadrangular prism

Triangular prism

Pentagonal prism   Rectangular prism

FIGURE 9.4

Among the most familiar of all the polyhedra are two of the Platonic solids: the cube and the tetrahedron. The cube is a rectangular prism. The tetrahedron belongs to a widely known class of polyhedra called *pyramids*. Pyramids, too, are named with reference to their bases. The tetrahedron, for example, is often call a triangular pyramid.

DEFINITION 9.9   Let $P_1 P_2 \ldots P_n$ be a polygon in plane m. Let V be a point in either of the half spaces determined by m. Then the union of polygonal region $P_1 P_2 \ldots P_n$ with the triangular regions bounded by $\triangle V P_1 P_2$, $\triangle V P_2 P_3$, $\ldots$, $\triangle V P_3 P_n$, $\triangle V P_n P_1$, is called a *pyramid* and is denoted pyramid V-$P_1 P_2 \ldots P_n$. The polygonal region bounded by $P_1 P_2 \ldots P_n$ is called the *base* of the pyramid, V is called its *apex*, and the triangular regions are called *lateral faces*. If $P_1 P_2 \ldots P_n$ is a regular polygon and if the perpendicular segment from the apex to m also contains the center of $P_1 P_2 \ldots P_n$, then the pyramid is *regular*. The (length of the) perpendicular segment from the apex to the plane containing the base is called the *altitude* of the pyramid.

Examples   The figures show two types of pyramids.

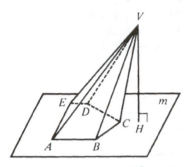

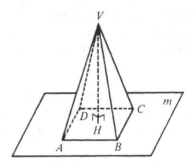

A pentagonal pyramid with its altitude   A regular square pyramid with its altitude

PROBLEM SET 9.3

1. What is the minimum number of faces that a polyhedral angle may have? What is the maximum number?

2. How many dihedral angles are determined by a trihedral angle? A tetrahedral angle?

3. A polyhedral angle is determined if three or more planes intersect at a single point. Name as many physical representations of polyhedral angles as you see in the room you are in.

4. Sketch a pentahedral angle and a hexahedral angle.

5. What is the least number of faces that a polyhedron may have? A prism?

6. Under what conditions would the lateral faces of a prism be bounded by congruent polygons?

7. What is another name for a regular quadrangular prism?

8. Prove: The lateral edges of a regular pyramid are congruent.

9. Sketch a triangular pyramid and a hexagonal pyramid.

10. Name the polygons that bound the faces of each of the five regular polyhedra illustrated in this section.

11. Given the regular pyramid $V\text{-}ABCD$. If $m\angle V\text{-}AD\text{-}B = m\text{-}\angle V\text{-}AB\text{-}C = m\angle V\text{-}BC\text{-}A = m\angle V\text{-}DC\text{-}A = 45$, and if $AB = 2$ in., find the altitude of the pyramid.

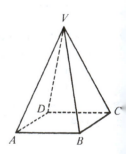

12. Calculate the measure of the dihedral angles determined by intersecting faces of each of the following polyhedra:

    (a) Regular tetrahedron.   (b) Cube.
    (c) Regular octahedron.

13. Use the Pythagorean Theorem (twice) to show that if the edge of a cube has length e, any segment joining two noncoplanar vertices has length $e\sqrt{3}$. Such a segment is called a *diagonal* of the cube. A sketch will prove most useful.

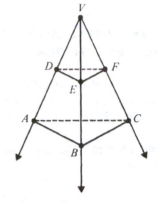

14. Given: Trihedral angle, as in the figure; and D, E, and F are the midpoints of $\overline{VA}$, $\overline{VB}$, and $\overline{VC}$, respectively
    Prove: (a) $\triangle ABC \sim \triangle DEF$
             (b) $a\triangle ABC = 4(a\triangle DEF)$

15. Construct a model of a convex polyhedral angle out of straws or toothpicks and observe what happens when it is flattened out in a plane. Does it "encircle" a point? Make a conjecture about the sum of the measures of the face angles.

16. In the following chart, $V = $ number of vertices, $E = $ number of edges, and $F = $ number of faces.

    (a) Copy and complete the chart.

NAME OF POLYHEDRON	V	E	F	$V - E + F$
tetrahedron				
triangular prism				
cube				
octahedron				
quadrangular pyramid				
dodecahedron			12	
icosahedron		20		

(b) State the conjecture implied by the right-hand column of the above chart. This result is known as Euler's Formula, named after one of its discoverers, Leonhard Euler (1707–1783). It has been used to prove that there are only five regular polyhedra.

9.4 SOME AREAS AND VOLUMES FOR POLYHEDRA

We have previously used the word "area" with respect to coplanar sets of points such as polygonal regions and circular regions. The following definition expands this usage.

DEFINITION 9.10   The *surface area of a polyhedron* is the sum of the areas of its faces.

In practice, the surface area of a polyhedron can be difficult to calculate. We will restrict our attention to some of the less complicated cases.

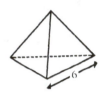

Example 1   Calculate the surface area of a regular tetrahedron with edge of length 6.
Since each of the faces of the tetrahedron is bounded by an equilateral triangle, we recall the formula for the area of an equilateral triangle in terms of the length of one of its sides s.

$$A = \frac{s^2\sqrt{3}}{4}$$

Then the surface area of the above tetrahedron becomes

$$S = 4\left(\frac{6^2\sqrt{3}}{4}\right) = 36\sqrt{3}$$

DEFINITION 9.11   The *lateral surface area of a prism or a pyramid* is the sum of the areas of the lateral faces.

Example 2   The lateral surface area S_ℓ of the tetrahedron of Example 1 is

$$S_\ell = 3\left(\frac{6^2\sqrt{3}}{4}\right) = 27\sqrt{3}$$

Example 3 Find the surface area and the lateral surface area of a right prism with the base as pictured in the figure, and with altitude 20 in. $ABCD$ is a parallelogram.

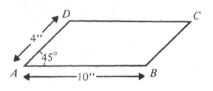

The lateral surface area can be calculated swiftly because each of the lateral faces is a rectangle with altitude 20 in.

$$S_\ell = 2(4 \text{ in.} \times 20 \text{ in.}) + 2(10 \text{ in.} \times 20 \text{ in.})$$

$$= 560 \text{ sq in.}$$

Letting $\overline{AB}$ be the base of $ABCD$ and drawing the altitude from D to $\overline{AB}$, the area of the base becomes

$$A = 10 \text{ in.} \times \frac{4 \text{ in.}}{\sqrt{2}} = 20\sqrt{2} \text{ sq in.}$$

Then the surface area is easily obtained.

$$S = S_\ell + 2A = [560 + 2(20\sqrt{2})] \text{ sq in.}$$

$$= (560 + 40\sqrt{2}) \text{ sq in.}$$

To compute an area of a polygonal region is to assign a number to it. We can compute the volume of a polyhedron in a like manner.

POSTULATE 26 *The Volume Postulate:* To each convex polyhedron there corresponds a unique positive real number.

DEFINITION 9.12 The number assigned to a convex polyhedron by the Volume Postulate is called its *volume*.

If the units of measure for the lengths of the edges are inches, feet, and meters, for example, then the units for volume measure are called cubic (cu) inches, cubic feet, and cubic meters, respectively.

Before we can compute a volume of a prism, we need further theorems and definitions. The proofs of these theorems are very complicated. For this reason, and because the theorems themselves are so obvious, the proofs are omitted.[1]

[1] The interested reader may find the proofs in F. Eugene Seymour and Paul James Smith, *Solid Geometry* (New York: Macmillan Co., 1949).

THEOREM 107   If a line is perpendicular to one of two parallel planes, it is perpendicular to the other.

THEOREM 108   A segment which has its endpoints in each of two parallel planes and is perpendicular to one of them (and hence to both) is the shortest segment joining the two planes.

DEFINITION 9.13   *The distance between two parallel planes* is the length of the shortest segment joining them.

It should be clear from the above definition and T108 that the distance between two parallel planes is the length of the mutually perpendicular segment joining them. Thus the distance between the bases of a prism is the length of such a segment.

DEFINITION 9.14   The *altitude* of a prism is the distance between its bases.

In Chapter 7, we stated a postulate giving the formula for the area of a rectangle which was used in the development of the formulas for the areas of other polygons. Volume formulas can be derived in a like manner.

POSTULATE 27   The volume of a right rectangular prism is equal to the product of the area of its base and its altitude.

Example 4   Suppose the box shown has length 4 ft, width 2 ft, and height 4 ft. Then its volume is given by

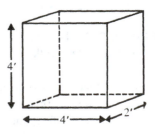

$$V = (4 \text{ ft} \times 2 \text{ ft}) \times 4 \text{ ft}$$
$$= (8 \text{ sq ft}) \times 4 \text{ ft} = 32 \text{ cu ft}$$

If we think of a deck of playing cards as a right rectangular prism, the decks shown suggest that the volume remains unchanged if we allow the deck to lean to one side, since the volume of each card remains constant. The height and area of the base of the leaning prism are equal to those

of the original one. Thus we conclude that the volume of a rectangular prism is the product of its altitude and the area of its base. If we let the area of the base equal B and let its altitude equal h, we can write

$$V = Bh$$

This formula can be generalized even further. Prisms with non-rectangular bases have volumes given by the same formula. Why this works is laid out in Chapter 23 of Edwin E. Moise's book, *Elementary Geometry from an Advanced Standpoint.*[2]

Our discussion may be summarized by a theorem.

THEOREM 109   The volume of a prism is the product of the area of its base and its altitude.

Example 5   Suppose we are given a right triangular prism with isosceles right triangular bases with 1-in. legs. Suppose also that the prism has lateral edges of length 4 in. Then the volume

$$V = Bh = \tfrac{1}{2}(1 \text{ in.} \times 1 \text{ in.}) \times 4 \text{ in.} = 2 \text{ cu in.}$$

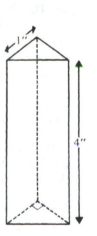

Since this prism is a *right* prism, the length of any one of its congruent lateral edges is its altitude.

The following theorem can be proved using T109. A readable explanation can be found in James M. Moser's *Modern Elementary Geometry.*[3]

THEOREM 110   The volume of a pyramid is one third the product of the area of its base and the altitude.

Example 6   Suppose the base of a regular square pyramid is 14 in. on a side and its altitude is 12 in. Then the volume of the pyramid is given by

$$V = \tfrac{1}{3}(14)^2(12) = 784 \text{ cu in.}$$

[2] (Reading, Mass.: Addison-Wesley Publishing Co., 1963).
[3] (Englewood Cliffs, N.J.: Prentice-Hall, 1971), pp. 234–235.

PROBLEM SET 9.4

1. Find the surface area, lateral surface area, and volume of the right prisms described:

   (a) The base is an equilateral triangle with side of length 6, and the length of a lateral edge is 10.
   (b) The base is a right triangle with legs of length 4 and 5 in., and the altitude is 7 in.
   (c) The base is an isosceles right triangle with hypotenuse 10, and each lateral edge is 12.
   (d) The base is a square with area 144 sq ft and the distance between the bases is $10\frac{1}{2}$ ft.
   (e) The altitude has measure 6 and the base is a regular hexagon with perimeter 60.
   (f) The base is a parallelogram with a 30° angle and consecutive sides of length 6 and 12 ft, and the distance between the bases is 8 ft.

2. Find the surface area, lateral surface area, and volume of a cube if an edge of the cube has length 1.

3. $ABCD$ is the base of the rectangular prism in the figure. The area of $ABCD$ is 18 sq in. Find the volume if

   (a) The distance between the bases is 10 in.
   (b) $m \angle EAB = 60$ and $AE = 5\sqrt{3}$

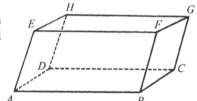

4. How many square feet of material is used in making a topless bin if the bottom measures 6 by 9 ft, and the bin is 7 ft high?

5. Find the surface area, lateral surface area, and volume of the regular pyramid described by problem 1(a).

6. A face diagonal of a cube is $3\sqrt{2}$. Find the surface area and the volume.

7. A rectangular room is 8 by 12 by 16 ft. Find the distance from one corner diagonally to the opposite corner. (*Hint:* Add another auxiliary segment to the figure shown.)

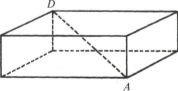

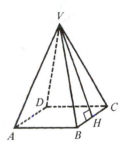

8. The faces of a regular pyramid are bounded by isosceles triangles. The altitude of any one of these triangles is called the *slant height* of the pyramid. (In the figure, *VH* is the slant height.)

  (a) Find the lateral surface area of a regular square pyramid with base 5 in. on a side and slant height 8 in.

  (b) If the slant height of a regular hexagonal pyramid is 5 centimeters (cm), and if the hexagonal base has apothem 3 cm, find the altitude of the pyramid.

  (c) Given a regular square pyramid as in the figure, with altitude *VK*. If slant height $VH = 2$ and $BC = 2$, find the measure of dihedral angle, $\angle V\text{-}BC\text{-}K$.

  (d) Find the volume of the pyramid of part (b).

  (e) Find the volume of the pyramid of part (c).

9. How many units are there in the edge of a cube if the number of square units in its total surface area equals the number of cubic units in its volume?

10. The base of a prism is a rhombus with sides of length 10. One diagonal of the base is 16 units long. If the altitude of the prism is 15 units, find the volume of the prism.

11. A gold ingot is 8 in. long and 3 in. high. Each end of the ingot is in the shape of a trapezoid, with upper and lower bases 3 and 4 in., respectively.

  (a) Find the volume of the ingot in cubic feet.

  (b) How much does the ingot weigh if the weight density of gold is 1210 lb per cu ft?

9.5 SPHERES

The points of a circle are coplanar. If, in space, a circle is rotated about one of its diameters, the figure formed is called a sphere. Few changes need to be made in the definition of a circle to give us the definition of a sphere.

DEFINITION 9.15 The set of all points in space which are a given distance from a given point is called a *sphere*. The given point is the *center* of the sphere and the given distance is called the *radius*. The *interior* of a sphere is the set of points which are a distance less than the radius from the center of the sphere. The *exterior* of a sphere is the set of points which are a distance greater than the radius from the center of the sphere.

THEOREM 111   If a plane intersects the interior of a sphere, its intersection with the sphere is a circle. The center of this circle is the intersection of the plane with a line that passes through the center of the sphere and is perpendicular to the plane.

We will omit the proof of this theorem because of its complexity.

Example 1   Suppose the circle of intersection of plane m and sphere P has a radius 10 and that the radius of the sphere is 15. Theorem 111 implies that the distance from P, the center of the sphere, to plane m is QP. This can be calculated using the Pythagorean Theorem.

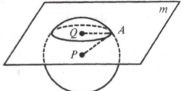

$$(QP)^2 + (QA)^2 = (PA)^2$$
$$(QP)^2 + 10^2 = 15^2$$
$$QP = \sqrt{225 - 100} = \sqrt{125}$$
$$= 5\sqrt{5}$$

DEFINITION 9.16   The circle of intersection of a plane and a sphere is called a *circle of the sphere*. A circle of a sphere is a *great circle* if its plane contains the center of the sphere.

Example 2   In the figure, three of the great circles of sphere P are depicted.

The problem of measuring the surface area and volume of the sphere is solved in a manner similar to that used for the circumference and area of the circle. We will omit the details here.[4]

THEOREM 112   The surface area of a sphere of radius r is $4\pi r^2$.

THEOREM 113   The volume of a sphere of radius r is $\frac{4}{3}\pi r^3$.

Example 3   A given sphere has radius 5 in. To calculate the surface area A, we write:

$$A = 4\pi r^2$$
$$= 4\pi(5)^2$$
$$= 100\pi$$

---

[4] The interested reader can find these details in Seymour and Smith, *Solid Geometry*.

Thus the surface area is 100π sq in.

To calculate the volume V, we write:

$$V = \frac{4}{3}\pi r^3$$

$$= \frac{4}{3}\pi(5)^3$$

$$= \frac{4}{3}\pi(125)$$

$$= \frac{500\pi}{3}$$

Thus the volume is $500\pi/3$ cu in.

Example 4   Lead weighs 710 lb per cu ft. How much does a lead ball 6 in. in diameter weigh?

$$V = \frac{4}{3}\pi r^3$$

$$= \frac{4}{3}\pi(\tfrac{1}{4})^3 \quad (6 \text{ in.} = \tfrac{1}{2} \text{ ft})$$

$$= \frac{\pi}{48} \text{ cu ft}$$

The weight W is given by:

$$W = 710V$$

$$= 710\left(\frac{\pi}{48}\right)$$

$$= 46.15$$

Thus the weight of the ball is approximately 46.15 lb.

PROBLEM SET 9.5

1. How many great circles does a sphere have?

2. On the spherical map of the earth (a globe), there are parallels of latitude and meridians of longitude. Which of these are great circles?

3. A plane is 5 in. from the center of a sphere of radius 13 in. Find the area of the circle of intersection.

4. The area of a circle of a sphere is 189π sq in., and the plane of the circle is 8 in. from the center of the sphere. Find:

   (a) The radius of the sphere.
   (b) The surface area of the sphere.
   (c) The volume of the sphere.

5. Given: P and Q are circles of sphere S, and the planes of P and Q are equidistant from the center of sphere S
   Prove: Circle $P \cong$ circle Q

6. Find the surface area and volume of a sphere if

   (a) Its radius is 6.
   (b) Its diameter is 5.

7. Find the radius of a sphere with surface area 144π.

8. The surface areas of two spheres are in the ratio of 3 to 4. What is the ratio of their radii?

9. What is the diameter of a sphere with volume $179\frac{2}{3}$? (Use $3\frac{1}{7}$ for π.)

10. The volume of a sphere is $3\frac{3}{8}$ times that of another sphere.

   (a) What is the ratio of their radii?   (b) Their areas?

11. A sphere just fits into a cube having a 12-in. edge. Find the volume of the cube.

12. A hollow spherical metal ball has an inside diameter of 6 in. and is 1 in. thick. Find the volume of metal used to make the ball.

13. A solid metal ball, 6 in. in diameter, is to be melted down and cast into a cube. Find, to the nearest tenth of an inch, the length of an edge of the cube.

14. A slab of lead in the shape of a right rectangular prism 3 in. high, 6 in. wide, and 1 ft long is to be used to make spherical buckshot $\frac{1}{4}$ in. in diameter. How many of these buckshot can be made?

15. An old puzzle problem goes as follows. Suppose you have three homes in a row and located across the street from them are three utility companies, water, gas, and electricity, as shown in the figure.

(a) Can you draw a picture connecting each utility with each house without crossing any lines?

(b) What if the diagram were drawn on the surface of a sphere?

16. A hunter walks 10 miles south and then 10 miles east, where he encounters a bear. Frightened, he runs 10 miles north and arrives at his starting point. What color is the bear?

CHAPTER 9: REVIEW EXERCISES

1. For each of the following statements, indicate whether it is *always* true, *sometimes* but not always true, or *never* true.

(a) A dihedral angle is a polyhedral angle.

(b) If $\overline{AB} \perp \overline{CD}$ and $\overline{CD} \in$ plane m, then $\overline{AB} \perp$ plane m.

(c) If two lines are perpendicular to a third line at the same point, then those two lines are coplanar.

(d) Two planes perpendicular to a third plane are parallel.

(e) The bases of a prism are bounded by congruent polygons.

(f) If a prism has only square regions for faces, then it is a cube.

(g) The lateral faces of a prism are triangles.

(h) A polyhedron is a prism.

(i) The lateral faces of a regular pyramid are bounded by congruent isosceles triangles.

(j) If two lateral faces of a right prism intersect, they are in perpendicular planes.

(k) The faces of a regular polyhedron are congruent regular hexagonal regions.

(l) The lateral edges of a prism are congruent.

(m) In a prism, the lateral faces are contained in parallel planes.

(n) If the volumes of two right rectangular prisms are equal, then they have equal altitudes.

(o) If two spheres are congruent, a great circle of one sphere is congruent to a great circle of the other.

(p) Two circles of a sphere intersect in three points.

2. Illustrate how six toothpicks may be placed to form four congruent triangles.

3. Sketch a cube. How many polyhedral angles are determined by the cube? How many dihedral angles?

4. What is the maximum number of mutually disjoint regions into which n distinct planes can separate space if n is

(a) 1   (b) 2   (c) 3

5. Can a line be perpendicular to each of two distinct lines that are not parallel? Explain.

6. Given: $\overline{PQ} \perp$ plane m; $\overline{AQ} \subseteq m$; $\overline{BQ} \subseteq m$; and $\overline{AQ} \cong \overline{BQ}$
Prove: $\overline{PA} \cong \overline{PB}$

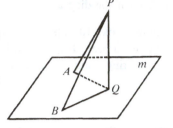

7. (a) Given a point P and a plane m. How many planes containing P are perpendicular to m?

(b) Given a line $\overset{\leftrightarrow}{l}$ and a plane m. How many planes containing $\overset{\leftrightarrow}{l}$ are perpendicular to m?

8. The surface area of a given sphere is numerically equal to the circumference of a great circle of the sphere. Find the diameter of the sphere.

9. How many regular polyhedra are there? Name them.

10. Given the regular square pyramid in the figure with apex V and altitude $\overline{VH}$. $VH = 8$ and $AH = 2\sqrt{2}$, and E is the midpoint of $\overline{AB}$.

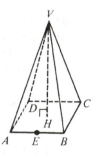

(a) Calculate the area of the base.
(b) What definition of Section 9.2 assures us that $\overline{VH} \perp \overline{HE}$?
(c) Calculate the volume of the pyramid.
(d) Calculate HE and the slant height VE (see problem 8 in Problem Set 9.4).
(e) Calculate the lateral surface area of the pyramid.

11. Prove: The lateral edges of a regular pyramid are congruent.

12. Find the length of the slant height (see problem 8 in Problem Set 9.4) and the altitude of a regular hexagonal pyramid if each lateral edge is 10 in. and each base edge is 6 in. in length.

13. Find the volume of a prism with altitude 12 and a base which is an equilateral triangle with side of length 8.

14. Calculate the volume of a prism with altitude 18 and a base which is a right triangle with legs of length 10 and 16.

15. What is the volume of a pyramid with altitude 9 and a base which is an equilateral triangle 8 units on a side?

16. Find the volume of a square pyramid with altitude 6 if 7 is the measure of each side of the base.

17. Find the surface area of the pyramid of problem 15 if the pyramid is a regular pyramid.

18. Calculate the surface area and volume of a sphere with radius 16.

19. The length of an edge of a regular tetrahedron is 12. Find its

(a) Surface area.   (b) Altitude.   (c) Volume.

20. The radii of two spheres are 3 and 4. Find the ratio of their

(a) Diameters.   (b) Surface areas.   (c) Volumes.

21. The volumes of two spheres are 36π and 288π. Find the ratio of their

(a) Radii.   (b) Diameters.   (c) Surface areas.

22. Given a sphere with radius 10 in., find the area of a circle of this sphere made by a plane 6 in. from its center.

REFERENCES

Moise, Edwin E. *Elementary Geometry from an Advanced Standpoint*. Reading, Mass.: Addison-Wesley Publishing Co., 1963. Chap. 23.
   The author develops volume formulas using a style analogous to the one we used to develop formulas for plane area. Heady stuff at this level, but much of it, particularly "Cavalieri's Principle," is worth looking at.

Moser, James M. *Modern Elementary Geometry*. Englewood Cliffs, N.J.: Prentice-Hall, 1971. pp. 234–235.
   Provided here is a clear explication of how the formula for the volume of a pyramid can be derived from the formula for the volume of a prism.

Seymour, F. Eugene, and Smith, Paul James. *Solid Geometry*. New York: Macmillan Co., 1949.
   This was a popular textbook in the days when solid geometry was taught as a separate course. It is fairly encyclopedic and contains proofs of the theorems we declined to prove in this chapter.

As far as the propositions of mathematics refer to reality they are not certain; and as far as they are certain, they do not refer to reality.

ALBERT EINSTEIN (1879–1955)

CHAPTER TEN

COORDINATE GEOMETRY

10.1 INTRODUCTION

By the early seventeenth century, the vast expansion of science and commerce had raised mathematical problems that, in their geometric formulation, seemed nearly impossible to solve: problems designing lenses for telescopes, problems concerning the motion of projectiles, and the like. Then two Frenchmen, René Descartes and Pierre de Fermat (1601–1665), working independently of each other, realized the potentialities inherent in the algebraic representation of geometric curves. They developed new systematic methods for dealing with curves, and showed how helpful it is to be able to attack geometric problems algebraically and vice versa.

No doubt you recall how, in Chapter 2, the concept of the number line was introduced and geometric methods of solving certain complicated algebraic problems were developed. In this chapter we will extend the notion of a coordinate system to two dimensions—to the plane. Such a study is called *coordinate geometry* or *analytic geometry*.

10.2 COORDINATE SYSTEMS AND DISTANCE

In Section 2.4 we established the existence of a coordinate system, a particular type of one-to-one correspondence between the real numbers and the points on a line. Recall the Ruler Placement Postulate:

POSTULATE 10   Let $\overleftrightarrow{l}$ be a line with $P \in \overleftrightarrow{l}$, $Q \in \overleftrightarrow{l}$, and $P \neq Q$. Then there exists a coordinate system for $\overleftrightarrow{l}$ such that $c(P) = 0$ and $c(Q)$ is positive.

An example of such a coordinate system is depicted in the accompanying figure.

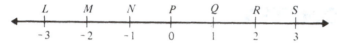

We will now use a method suggested by the work of Descartes to establish a coordinate system for a plane.

First we introduce a line $\overleftrightarrow{x}$ with a given coordinate system. We call this line the *x-axis*, and it is generally drawn horizontally. Next we introduce a second line $\overleftrightarrow{y}$, perpendicular to the first line at the point with coordinate 0. This second line is called the *y-axis*. We establish a coordinate system on the *y*-axis, assigning the coordinate 0 to the point of intersection of the two axes. The positive direction on the *x*-axis is

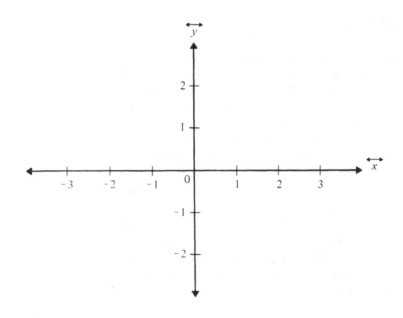

generally taken to be to the right, while on the vertical y-axis it is upward. It is conventional to adopt the same scale on both axes; that is, the unit of distance is the same. Further, we call the intersection of the two axes the *origin*, and usually label it with the letter O.

In Chapter 2, the one-dimensional coordinate system we called the *number line* was explored. Many of the ideas developed there are useful as we build our two-dimensional system. In the one-dimensional system, any point on a line can be completely determined by giving its coordinate. In the two-dimensional system, any point in the plane of the axes can be determined by giving a pair of coordinates associated with it. The scheme for doing this follows.

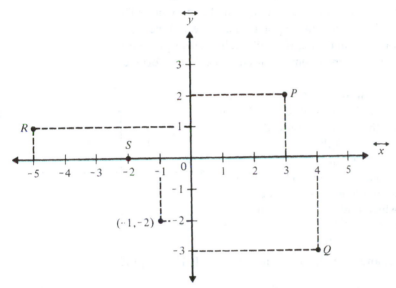

Consider the point P in the plane of $\overleftrightarrow{x}$ and $\overleftrightarrow{y}$ as shown in the figure. From P we draw perpendiculars to the x and y axes. The coordinate on the x-axis is called the *x-coordinate* or *abscissa* of P, and the coordinate of the y-axis is called the *y-coordinate* or *ordinate* of P. In the situation depicted above, 3 is the x-coordinate and 2 is the y-coordinate. Together these numbers are called the *coordinates* of P. It is quite common to compress this information by writing $P(3, 2)$ to mean P is a point with 3 as its x-coordinate and 2 as its y-coordinate. Similarly, we can characterize the points Q, R, and S by writing $Q(4, -3)$, $R(-5, 1)$ and $S(-2, 0)$. It should be clear that the order in which we write the coordinates of a point is important. We will always write the x-coordinate first and the y-coordinate second. Thus we call these number pairs, *ordered pairs*.

Given a point in the plane of $\overleftrightarrow{x}$ and $\overleftrightarrow{y}$, the previous paragraph outlines the process for finding an ordered pair of real numbers associated with it. Conversely, if we are given a pair of coordinates, $(-1, -2)$, for example, we can locate the point corresponding to that pair of numbers by drawing perpendiculars to the x- and y-axes at the points where $x = -1$ and $y = -2$ respectively. The point of intersection of these two perpendiculars is the point we are seeking.

To draw a set of coordinate axes and to draw a set of points which are located in the way that we have been discussing is a task often performed in various branches of applied mathematics. To draw such a set of points is to *plot* that set of points.

The previous paragraphs informally establish the existence of a one-to-one correspondence between the points of a plane and the set of all ordered pairs of real numbers. We call such a correspondence a *Cartesian coordinate system* (from Cartesius, the Latin form of Descartes' name).

Since it is often awkward to refer to "the point with coordinates (a, b)," we will often find it convenient to adopt the more widespread though inexact usage "the point (a, b)."

The coordinate axes separate the plane into four distinct regions known as *quadrants*. The quadrants are named using Roman numerals I–IV, beginning with the region in which both coordinates are positive and then moving around the origin in a counterclockwise direction. Thus the points $(2, 1)$,

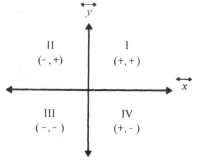

$(-1, -2)$, $(\pi, -\pi)$, and $(-\sqrt{2}, 7)$ are in quadrants I, III, IV, and II, respectively.

Chapter 2 provided a method for computing the distance between two points on a line. We can use this method to derive a formula for finding the distance between two distinct points in a Cartesian coordinate system. The derivation follows.

Suppose $P(x_1, y_1)$ and $Q(x_2, y_2)$ are given as in the accompanying figure. We draw a vertical line through P and a horizontal line through Q that intersect at some point R and which intersect the x- and y-axes at S and T, respectively. We also draw perpendiculars from P and Q to the y- and x-axes, respectively, intersecting them at M and N, as in the figure. The x-coordinates of S and N are x_1 and x_2, and the y-coordinates of M and T are y_1 and y_2, respectively.

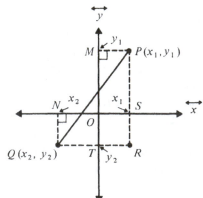

To find PQ, we notice that $\overline{PQ}$ is the hypotenuse of right triangle $\triangle QRP$ and write

$$(PQ)^2 = (QR)^2 + (RP)^2$$

Notice also that $RP = MT$ and $QR = NS$, being opposite sides of rectangles. Therefore,

$$(PQ)^2 = (NS)^2 + (MT)^2$$
$$= |x_2 - x_1|^2 + |y_2 - y_1|^2$$

But the definition of *absolute value* says that

$$|x_2 - x_1|^2 = \begin{cases} (x_2 - x_1)^2 & \text{if } x_2 - x_1 \geq 0 \\ [-(x_2 - x_1)]^2 & \text{if } x_2 - x_1 < 0 \end{cases}$$

which in either case is $(x_2 - x_1)^2$.

Likewise $|y_2 - y_1|^2 = (y_2 - y_1)^2$. So we may write

$$(PQ)^2 = (x_2 - x_1)^2 + (y_2 - y_1)^2$$

from which it follows that

$$PQ = \sqrt{(x_2 - x_1)^2 + (y_2 - y_1)^2}$$

We can restate this result as a theorem.

THEOREM 114　*The Distance Formula:*　Given two distinct points $P(x_1, y_1)$ and $Q(x_2, y_2)$ in a Cartesian coordinate system. Then

$$PQ = \sqrt{(x_2 - x_1)^2 + (y_2 - y_1)^2}$$

Example　To find the distance between $P(3, -1)$ and $Q(-2, 7)$, we write

$$PQ = \sqrt{(-2 - 3)^2 + (7 - [-1])^2}$$
$$= \sqrt{(-5)^2 + (8)^2}$$
$$= \sqrt{25 + 64}$$
$$= \sqrt{89}$$

The careful reader probably noticed that in the derivation of the distance formula, we assumed that P and Q were not on the same horizontal line or the same vertical line. If either is the case, $\triangle QRP$ does not exist. However, if P and Q *are* on the same vertical line, $x_2 = x_1$, and the distance formula becomes

$$PQ = \sqrt{(x_2 - x_1)^2 + (y_2 - y_1)^2}$$

$$= \sqrt{(y_2 - y_1)^2}$$

$$= |y_2 - y_1|$$

Likewise, if P and Q are on the same horizontal line,

$$PQ = |x_2 - x_1|$$

In the following problem set you will be asked to derive the formula for the case when P and Q are on the same horizontal line.

PROBLEM SET 10.2

1. Give, as accurately as you can, the coordinates of each of the points depicted in the figure.

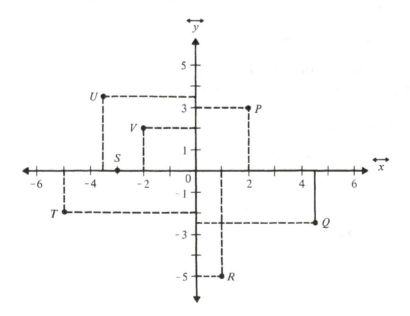

2. Draw and label a set of coordinate axes, and then plot the following points as accurately as you can: $A(-1, 0)$, $B(2, -3)$, $C(\pi, 4)$, $D(-1\frac{1}{2}, -3)$, and $E(5, 1)$.

3. If $P(1, 1)$, $Q(1, 4)$, $R(-1, -3)$, and $S(-2, 5)$ are given, plot quadrilateral $PQSR$.

4. What are the coordinates of the origin?

5. Name the quadrant in which each point of problem 2 lies.

6. (a) What is the abscissa of all of the points lying on the y-axis?
   (b) What is the ordinate of all of the points lying on the x-axis?

7. What are the coordinates of the point that is the intersection of the x-axis and the line drawn perpendicular to the x-axis from $P(-7, 2)$?

8. $A(3, -5)$, $B(3, 2)$ and $C(-6, -5)$ are three of the vertices of a rectangle. Find:

   (a) The coordinates of the fourth vertex.
   (b) The perimeter of the rectangle.
   (c) The area of the rectangle.

9. $P(4, -2)$, $Q(6, -2)$, and $R(5, 3)$ are three of the vertices of a parallelogram. Find:

   (a) The coordinates of the fourth vertex. Is there more than one possibility? Explain.
   (b) The perimeter of the parallelogram. Is there more than one possibility? Explain.
   (c) The area of the parallelogram.

10. Use the distance formula to find the distance between:

   (a) $P(0, 0)$ and $Q(-1, 2)$
   (b) $R(-1, 2)$ and $S(-5, -5)$
   (c) $T(6, -2)$ and $U(9, 2)$
   (d) $V(-3\frac{1}{7}, 4\frac{2}{7})$ and $W(-2\frac{3}{7}, 6)$
   (e) $A(\pi, 2)$ and $B(2, \pi)$
   (f) $C(\pi, 2)$ and $D(\pi, -2)$
   (g) $E(6, 20)$ and $F(15, -20)$
   (h) $G(-3, -8)$ and $H(7, -10)$

11. Find the perimeter of triangle $\triangle ABC$ given $A(4, 7)$, $B(0, 10)$, and $C(-4, -8)$.

12. Show that $\triangle PQR$ is isosceles given $P(2, 2)$, $Q(5, 6)$, and $R(1, 3)$.

13. Given $A(-3, 5)$, $B(5, 6)$, $C(9, -1)$, and $D(1, -2)$. Show that quadrilateral $ABCD$ is a rhombus.

14. Given $P(2, -3)$, $R(5, -2)$, and $S(7, 0)$. Show, without plotting any points, that R is not between P and S.

15. Describe the set of all points for which the abscissa is -2.

16. Describe the set of all points for which the ordinate is 4.

17. Plot the set of all points (x, y) such that x and y are integers if $1 \le x \le 3$ and $-2 < y < 2$.

18. Plot the set of points (x, y) which satisfies the following conditions: $x \le 1$ and $y \ge -2$.

19. Show that $R(-2, 1)$, $S(6, 6)$, and $T(8, -15)$ are vertices of a right triangle.

20. Develop an argument to show that if two points are on the same horizontal line, the distance between them is the absolute value of the difference between their x-coordinates.

10.3   SLOPE AND EQUATIONS FOR LINES

If we draw nonvertical lines or segments or rays in a Cartesian coordinate plane, we notice that they may rise to the right, to the left, or they may be horizontal. We also notice that some lines are steeper than others. In terms of the diagram we see that $\overline{AB}$ is less steep than $\overline{CD}$, and that the ratio of the "rise" to the "run" is less for $\overline{AB}$ than for $\overline{CD}$. We can give a name to this ratio and define it in such a way that it not only indicates how steep a segment is, but also whether it rises to the right or to the left.

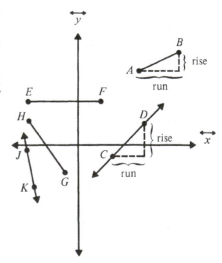

DEFINITION 10.1   Let $P_1(x_1, y_1)$ and $P_2(x_2, y_2)$ be given, $x_1 \ne x_2$. Then m, the *slope* of $\overline{P_1P_2}$, is defined by

$$m = \frac{y_2 - y_1}{x_2 - x_1}$$

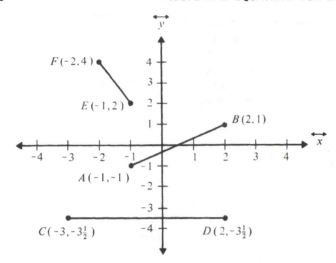

Example 1 In the accompanying figure, we can calculate the slope of each segment.

(a) For $\overline{AB}$,

$$m = \frac{1 - (-1)}{2 - (-1)} = \frac{2}{3}$$

(b) For $\overline{CD}$,

$$m = \frac{-3\frac{1}{2} - (-3\frac{1}{2})}{2 - (-3)} = \frac{0}{5} = 0$$

(c) For $\overline{EF}$,

$$m = \frac{4 - 2}{-2 - (-1)} = \frac{2}{-1} = -2$$

From the above examples and others that you should make up on your own, it should be clear that if a segment rises to the right, it has a positive slope, and if it rises to the left, it has a negative slope. Also, a horizontal segment has 0 for its slope, since $y_2 - y_1$ will always be 0.

The opening sentence of this section limited the discussion to *non-vertical* lines, segments, or rays. Thus the definition of *slope* specifies that $x_1 \neq x_2$. If we allow x_1 to equal x_2, the expression

$$\frac{y_2 - y_1}{x_2 - x_1} \quad \text{becomes} \quad \frac{y_2 - y_1}{0}$$

which is not a *number*! Hence a vertical segment has no slope.

Consider the graph shown here, in which the dashed segments are drawn parallel to the co-ordinate axes. $\angle QPS \cong \angle RQU$ and $\angle QRU \cong \angle PQS$ by T45 (Section 5.3). Therefore, by AA~, we conclude that $\triangle QPS \sim \triangle RPT$ and $\triangle RQU \sim \triangle RPT$. Thus, by transitivity, $\triangle QPS \sim \triangle RQU$. The definition of similarity then allows us to write

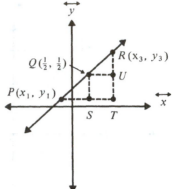

$$\frac{y_2 - y_1}{x_2 - x_1} = \frac{y_3 - y_1}{x_3 - x_1} = \frac{y_3 - y_2}{x_3 - x_2}$$

There is nothing special about the three points that we chose here. In fact, this ratio is the same no matter which pair of points are chosen. We state this useful result as a theorem.

THEOREM 115   All segments on the same nonvertical line have the same slope.

The careful reader probably noticed that the development of T115 would proceed somewhat differently if the line in question did not have a positive slope. You will be asked to consider the two alternative cases in Problem Set 10.3.

Since all segments on the same nonvertical line do have the same slope, it is useful to expand the meaning of the word *slope* as follows.

DEFINITION 10.2   The *slope of a nonvertical line* is the slope of any segment of that line.

Example 2   To calculate the slope of the line through $A(-2, 1)$ and $B(3, -\frac{7}{2})$ we write

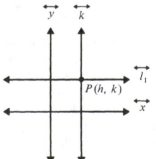

$$m = \frac{y_2 - y_1}{x_2 - x_1} = \frac{-\frac{7}{2} - 1}{3 - (-2)} = \frac{-\frac{9}{2}}{5} = \frac{-9}{10}$$

If a line l_1 is horizontal and contains the point $P(h, k)$, it is clear that every point on that line has k for its y-coordinate. Conversely, if k is the y-co-ordinate of a point, that point is on the horizontal line containing $P(h, k)$. Thus the equation $y = k$ completely characterizes l_1. Using set notation we could write

$$l_1 = \{(x, y) : y = k\}$$

More generally we just say that $y = k$ is the equation of $\overset{\leftrightarrow}{l_1}$. It is even more common (although more imprecise) to refer to $y = k$ as "the line $y = k$."

Using similar reasoning we can conclude that

$$\overset{\leftrightarrow}{l_2} = \{(x, y) : x = h\}$$

or $x = h$ is the equation of $\overset{\leftrightarrow}{l_2}$.

The following theorem is a distillation of these ideas.

THEOREM 116   Every vertical line has an equation of the form $x = h$, and every horizontal line has an equation of the form $y = k$.

Example 3   The equation of $\overset{\leftrightarrow}{l}$, the vertical line three units to the left of the y-axis, is

$$x = -3$$

The equation of $\overset{\leftrightarrow}{m}$, the horizontal line $2\frac{1}{2}$ units above the x-axis, is

$$y = 2\frac{1}{2}$$

Also, $\overset{\leftrightarrow}{l} \cap \overset{\leftrightarrow}{m} = \{(-3, 2\frac{1}{2})\}$.

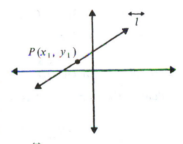

Now that we can algebraically characterize vertical and horizontal lines, we are ready to consider other lines, such as $\overset{\leftrightarrow}{l}$ in the accompanying figure. Suppose we know that $P(x_1, y_1)$ belongs to $\overset{\leftrightarrow}{l}$, and $Q(x, y)$ is any other point on that line. Then the slope of $\overset{\leftrightarrow}{l}$ is given by

$$m = \frac{y - y_1}{x - x_1}$$

which is equivalent to

$$y - y_1 = m(x - x_1)$$

This second equation has an advantage over the first: namely, (x_1, y_1) satisfies it, and since it is also true for any choice of $Q(x, y)$

on $\overleftrightarrow{l}$, it is true for every point on the line. Notice that if $\overleftrightarrow{l}$ were a horizontal line through P, it would still be characterized by

$$y - y_1 = m(x - x_1)$$

since in that case the right-hand side of the equation would become zero and we would have

$$y - y_1 = 0$$

or $$y = y_1$$

These results are formalized in the following theorem.

THEOREM 117 Suppose $\overleftrightarrow{l}$ is a line with slope m and containing $P(x_1, y_1)$. Then for any point $Q(x, y)$ such that $Q \in \overleftrightarrow{l}$, we have the equation

$$y - y_1 = m(x - x_1)$$

Example 4 Suppose a line with slope 3 contains the point $A(-3, 2)$. Then

$$y - 2 = 3[x - (-3)]$$
$$= 3(x + 3)$$

Example 5 Given $\overleftrightarrow{AB}$ with $A(-3, 2)$ and $B(4, 4)$, we can calculate the slope of $\overleftrightarrow{AB}$ as follows:

$$m = \frac{4 - 2}{4 - (-3)} = \frac{2}{7}$$

Then we can write an equation for $\overleftrightarrow{AB}$:

$$y - 2 = \tfrac{2}{7}[x - (-3)] \quad \text{using } A$$

or

$$y - 4 = \tfrac{2}{7}(x - 4) \qquad \text{using } B$$

Theorem 117 tells us that if a point is on $\overleftrightarrow{l}$, its coordinates satisfy the equation. Happily, the converse is also true, as we learn in the following theorem.

THEOREM 118 Suppose $\overleftrightarrow{l}$ is a line with slope m and containing $P(x_1, y_1)$. If the coordinates of $Q(x, y)$ satisfy the equation

$$y - y_1 = m(x - x_1)$$

then $Q \in \overleftrightarrow{l}$.

Proof: Since $\overleftrightarrow{l}$ is not vertical, there must be some point on it with abscissa x; call it (x, y_2). Then by T117, we have

$$y_2 - y_1 = m(x - x_1)$$

But, by hypothesis, $y - y_1 = m(x - x_1)$
Therefore,

$$y_2 - y_1 = y - y_1$$
so that
$$y_2 = y$$

Thus if $(x, y_2) \in \overleftrightarrow{l}$, it follows that $Q(x, y) \in \overleftrightarrow{l}$.

Theorems 117 and 118 can be compressed into the single assertion: The line with slope m and containing $P(x_1, y_1)$ is the set $\{(x, y) : y - y_1 = m(x - x_1)\}$.

Actually, a slight expansion of the vocabulary introduced in Chapter 2 would lead us to say that

$$y - y_1 = m(x - x_1)$$

is a sentence involving two variables, and that

$$\{(x, y) : y - y_1 = m(x - x_1)\}$$

is the *solution set* of this sentence. The *graph of the solution set* is the set of all points in a Cartesian coordinate plane that correspond to the elements of the solution set.

Again we adopt the shorter, though imprecise usage, "graph the line $y - y_1 = m(x - x_1)$," or the even shorter, "graph $y - y_1 = m(x - x_1)$." For obvious reasons, the form

$$y - y_1 = m(x - x_1)$$

is called the *point-slope form* of the equation of a nonvertical line.

After some algebraic juggling, the previous equation can also be written as

$$-mx + y + (mx_1 - y_1) = 0$$

which is of the form

$$Ax + By + C = 0$$

where A, B, and C are real numbers and not both A and B are zero. Thus if $\overset{\leftrightarrow}{l}$ is any line in the plane, then there are numbers A, B, and C, with not both A and B zero, such that every point (x, y) on $\overset{\leftrightarrow}{l}$ has coordinates that satisfy the equation

$$Ax + By + C = 0$$

The converse is also true; that is, all the points with coordinates that satisfy

$$Ax + By + C = 0$$

lie on a single line. The proof for this fact is given as Theorem 15-2 in the reference to Brumfiel, Eicholz, and Shanks at the end of Chapter 9.

DEFINITION 10.3   An equation that can be written in the form $Ax + By + C = 0$, with not both A and B zero, is called a *linear equation*. The form $Ax + By + C = 0$ is called the *standard form* of the linear equation.

Example 6   If we want to write the linear equation

$$y - 3 = 2(x + 7)$$

in standard form, we notice that

$$y - 3 = 2(x + 7)$$

is equivalent to

$$y - 3 = 2x + 14$$

or

$$-2x + y + (-17) = 0$$

which is the desired result.

Example 7   Suppose we want to graph the line $\overleftrightarrow{l}$ characterized by $3x + 2y - 6 = 0$. Since the equation is not in either of the forms $x = h$ or $y = k$, we know the line must intersect both axes. Thus we also know that there are points $P(0, y_1)$ and $Q(x_2, 0)$ belonging to $\overleftrightarrow{l}$.

That $(0, y_1) \in \overleftrightarrow{l}$ implies that

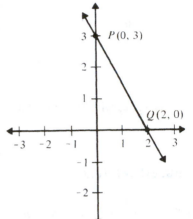

$$3(0) + 2y_1 - 6 = 0$$

Solving for y_1, we obtain $y_1 = 3$. Thus $P(0, 3) \in \overleftrightarrow{l}$.

Likewise, since $(x_2, 0) \in \overleftrightarrow{l}$, we may write

$$3(x_2) + 2(0) - 6 = 0$$

Solving for x_2, we obtain $x_2 = 2$. Thus $Q(2, 0) \in \overleftrightarrow{l}$. Since two points determine a line, we need only plot P and Q and then, using a straight-edge, draw the line containing these two points.

The following example compresses the process outlined in Example 7.

Example 8   To graph the line $3x - 15y = 15$, we form a table by substituting $x = 0$ and $y = 0$ into the equation. This gives us the coordinates of two points on the line, and since two points deter-mine a line, we have enough information to plot the graph as shown. For those of us who make an occasional arithmetic

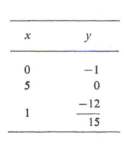

x	y
0	-1
5	0
1	$\dfrac{-12}{15}$

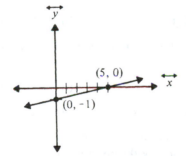

mistake, it is wise to substitute another value of x into the equation and thus obtain the coordinates of a third point. If these three points appear to be collinear, then we know that it is extremely unlikely that we have erred in computing either of the first two points. For example, letting $x = 1$, we obtain

$$3(1) - 15y = 15$$

$$y = \frac{-12}{15}$$

This agrees with the information already obtained.

PROBLEM SET 10.3

In Exercises 1 through 6, find the slope of the line passing through the given points.

1. $(3, -1)$, $(2, 3)$

2. $(5, -5)$, $(-5, 5)$

3. $(-2, 0)$, $(1, 5)$

4. $(-3, -7)$, $(3, -2)$

5. $(0, 0)$, $(4, 3)$

6. (a, a), $(-a, -a)$

7. Find y so that the line determined by the two points will be horizontal.

   (a) $(5, 3)$, $(2, y)$ (b) $(-7, y)$, $(4, -1)$

8. Find x so that the line determined by the two points will be vertical.

   (a) $(x, -7)$, $(3, 5)$ (b) $(-1, 0)$, $(x, -1)$

9. What is the equation of the x-axis? of the y-axis?

10. Write an equation in point-slope form of the line with the given slope and containing the given point:

   (a) $A(-2, 3)$, $m = 4$ (b) $B(1, -4)$, $m = \frac{-3}{4}$

   (c) $C(0, 0)$, $m = 1$ (d) $D(5, 0)$, $m = -3$

11. For each pair of points, write an equation of the line containing them:

    (a) $(1, 3)$ and $(-2, 5)$ (b) $(0, 0)$ and $(5, 6)$
    (c) $(-1, 2)$ and $(1, -2)$ (d) $(3, 4)$ and $(-2, 4)$
    (e) (π, π) and $(-\pi, -\pi)$ (f) $(-2, 7)$ and $(-2, 4)$

12. Express each of the following linear equations in standard form:

    (a) $y - 2 = 3(x + 1)$ (b) $y + 5 = 6x$
    (c) $y = 2x - 7$ (d) $3x = 2y + 5$

13. Find the equation of $\overleftrightarrow{PQ}$ in point-slope form. Rewrite this equation in standard form.

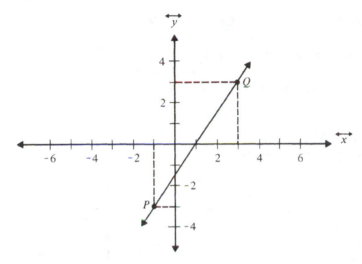

14. For each of the following equations, find the coordinates of two points of its graph and sketch its graph:

    (a) $y - 1 = 3(x - 2)$ (b) $y - \frac{2}{3} = -\frac{1}{3}(x + 6)$
    (c) $2x - 4y + 8 = 0$ (d) $x + 2y = 5$
    (e) $y = x$ (f) $y = -x$

15. What is the graph of the sentence "$x = 2$ and $y = -3$?"

16. (a) What is the equation of the line five units to the left of the y-axis?

    (b) What is the equation of the line four units above the x-axis?

17. The points where the graph of a linear equation intersect the x- and y-axes are called the *x-intercept* and *y-intercept*, respectively.

Find the x- and y-intercepts of the graphs of the following linear equations:

(a) $3x + 2y - 6 = 0$ (b) $x - y = 4$ (c) $\dfrac{x}{2} + \dfrac{y}{3} = 1$

(d) $\dfrac{x}{-2} + \dfrac{y}{4} = 1$ (e) $\dfrac{x}{5} + \dfrac{y}{-3} = 1$

Make a conjecture based on the results of (c), (d), and (e).

18. Graph the equation $y = |x|$. [*Hint:* Consider the definition of absolute value and the graphs for problems 14(e) and 14(f).]

10.4 PARALLELISM AND PERPENDICULARITY

It is obvious that (a) every vertical line is parallel to every other vertical line, and (b) every vertical line is perpendicular to every horizontal line. Since the equations for vertical and horizontal lines are so distinctive ($x = h$, $y = k$), parallelism and perpendicularity are quite recognizable from the equations alone. In the more general case of nonvertical lines, the situation is almost as easy to deal with.

Consider, as in the figure, any nonvertical line, $\overleftrightarrow{AB}$ intersecting the x- and y-axes at $A(a, 0)$ and $B(0, b)$ respectively. These points are called the x-*intercept* and the y-*intercept*. The point-slope form of the equation for this line is

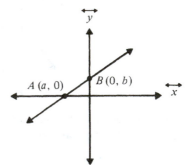

$$y - b = m(x - 0)$$

where m is the slope and $B(0, b)$ is the point used. This equation can be rewritten in the form

$$y = mx + b$$

Since m is the slope and b is the y-coordinate of the y-intercept, this form of a linear equation is called the *slope-intercept* form.

Example 1 The equation of the line with slope 3 and y-intercept $B(0, -2)$ is

$$y = 3x + 2$$

Example 2 We can rewrite

$$3x - 2y + 7 = 0$$

in *slope-intercept* form by solving for y as follows:

$$3x - 2y + 7 = 0$$
$$2y = 3x + 7$$
$$y = \tfrac{1}{2}(3x + 7)$$
$$= \tfrac{3}{2}x + \tfrac{7}{2}$$

Slope-intercept form is useful in establishing the following important result.

THEOREM 119 Two distinct nonvertical lines are parallel iff they have the same slope.

Proof: Suppose

$$y = m_1x + b_1 \text{ is the equation of } \overleftrightarrow{l_1}$$
$$y = m_2x + b_2 \text{ is the equation of } \overleftrightarrow{l_2}$$

Then we have two things to prove:

(i) If $m_1 = m_2$, then $\overleftrightarrow{l_1} \parallel \overleftrightarrow{l_2}$

(ii) If $\overleftrightarrow{l_1} \parallel \overleftrightarrow{l_2}$, then $m_1 = m_2$

The proof of (i) is indirect, as follows. Suppose $m_1 = m_2$ and $\overleftrightarrow{l_1} \nparallel \overleftrightarrow{l_2}$. Then $\overleftrightarrow{l_1}$ and $\overleftrightarrow{l_2}$ must intersect at some point, the coordinates of which satisfy the equation of both lines. Therefore, for that point (x, y), we can write

$$y = m_1x + b_1 \quad \text{and} \quad y = m_1x + b_2$$

or

$$b_1 = y - m_1x = b_2$$

This cannot be, however, for in such a case the equations of the two lines would be identical and we would be considering only one line, not two. $\rightarrow\!/\!\leftarrow$ Therefore, if $m_1 = m_2$, then $\overleftrightarrow{l_1} \parallel \overleftrightarrow{l_2}$.

The proof of (ii) is as follows. Suppose $\overset{\leftrightarrow}{l_1} \parallel \overset{\leftrightarrow}{l_2}$. If we perform the construction of the right triangles indicated in the figure, it is obvious that

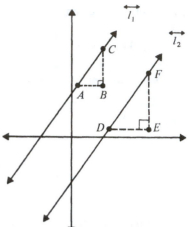

$$\triangle ABC \sim \triangle DEF$$

and

$$m_1 = \frac{BC}{AB} = \frac{FE}{DE} = m_2$$

Therefore, if $\overset{\leftrightarrow}{l_1} \parallel \overset{\leftrightarrow}{l_2}$, then $m_1 = m_2$.

Example 3 Suppose the equation of $\overset{\leftrightarrow}{l_1}$ is $3x + 2y = 7$; and suppose $\overset{\leftrightarrow}{l_1} \parallel \overset{\leftrightarrow}{l_2}$ and $P(2, -3) \in \overset{\leftrightarrow}{l_2}$. We can find the equation of $\overset{\leftrightarrow}{l_2}$ by noting that the equation of $\overset{\leftrightarrow}{l_1}$ may be written

$$y = -\tfrac{3}{2}x + \tfrac{7}{2}$$

so that the slope of $\overset{\leftrightarrow}{l_2}$ is $-\tfrac{3}{2}$. Therefore, in *point-slope* form, the equation of $\overset{\leftrightarrow}{l_2}$ is

$$y + 3 = \frac{-3}{2}(x - 2)$$

Example 4 The lines characterized by the linear equations

$$2x + 3y - 5 = 0$$

and

$$4x = 7 - 6y$$

are parallel since they may be rewritten (in slope-intercept form) as

$$y = \frac{-2}{3}x + \frac{5}{2}$$

and

$$y = \frac{-2}{3}x + \frac{7}{6}$$

respectively.

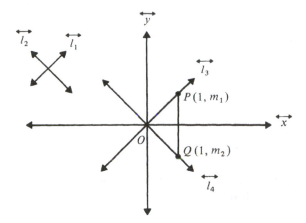

Next we consider slopes of perpendicular lines. Suppose $\overleftrightarrow{l_1}$ and $\overleftrightarrow{l_2}$ are lines in the Cartesian coordinate plane and $\overleftrightarrow{l_1} \perp \overleftrightarrow{l_2}$. Let the slopes of $\overleftrightarrow{l_1}$ and $\overleftrightarrow{l_2}$ be m_1 and m_2, respectively. We can construct lines $\overleftrightarrow{l_3}$ and $\overleftrightarrow{l_4}$ that are parallel to (or perhaps identical with) $\overleftrightarrow{l_1}$ and $\overleftrightarrow{l_2}$, respectively, and which pass through the origin. Obviously $\overleftrightarrow{l_3} \perp \overleftrightarrow{l_4}$. Using slope-intercept form, we arrive at the following equations for $\overleftrightarrow{l_3}$ and $\overleftrightarrow{l_4}$:

$$y = m_1 x$$

and

$$y = m_2 x$$

Since their coordinates satisfy the above equations, $P(1, m_1)$ and $Q(1, m_2)$ lie on $\overleftrightarrow{l_3}$ and $\overleftrightarrow{l_4}$, respectively. It should be clear that $\triangle OPQ$ is a right triangle. Then the Pythagorean Theorem implies that

$$(OP)^2 + (OQ)^2 = (PQ)^2$$

Using the Distance Formula (T114), we can write

$$(m_1 - 0)^2 + (1 - 0)^2 + (m_2 - 0)^2 + (1 - 0)^2 =$$
$$(1 - 1)^2 + (m_1 - m_2)^2$$

or

$$m_1^2 + 1 + m_2^2 + 1 = m_1^2 - 2m_1 m_2 + m_2^2$$

Therefore,

$$2 = -2m_1m_2$$

$$1 = -m_1m_2$$

$$m_1 = -\frac{1}{m_2}$$

Thus we have the following theorem.

THEOREM 120   If two lines with slopes m_1 and m_2 are perpendicular, then $m_1 = -1/m_2$; that is, each of these slopes is the negative reciprocal of the other.

Example 5   Suppose $\overset{\leftrightarrow}{l}$ is perpendicular to the line $y = 3x + 5$ at $P(-1, 2)$. Then the equation of $\overset{\leftrightarrow}{l}$ is

$$y - 2 = \frac{-1}{3}[x - (-1)]$$

$$y - 2 = \frac{-1}{3}(x + 1)$$

By retracing the steps of the previous proof we learn that its converse is also true.

THEOREM 121   If the slopes of two lines are negative reciprocals of one another, then the lines are perpendicular.

Example 6   If we are given $A(-3, 0)$, $B(3, -6)$, and $C(5, 8)$, we can use the converse of the Pythagorean Theorem to show that $\triangle ABC$ is a right triangle. The result is more easily obtained by using T121. Consider the following calculations:

$$\text{Slope of } \overline{AB} = \frac{-6 - 0}{3 - (-3)} = -1$$

$$\text{Slope of } \overline{AC} = \frac{8 - 0}{5 - (-3)} = 1$$

These numbers are negative reciprocals of each other. Therefore, $\overline{AB} \perp \overline{AC}$, which implies that $\triangle ABC$ is a right triangle.

PROBLEM SET 10.4

1. Rewrite each of the following linear equations in slope-intercept form. Identify the slope and the coordinates of the y-intercept in each case.

   (a) $x - y = 1$ (b) $y - 1 = 3(x + 2)$
   (c) $3x + 4y - 12 = 0$ (d) $y - x = 0$
   (e) $x = 3 - y$

2. (a) Write the equation of the line with x- and y-intercepts $A(1, 0)$ and $B(0, 3)$, respectively, in slope-intercept form.
   (b) Write the equation of the line with no y-intercept and containing the point $P(3, 8)$.
   (c) Write the equation of the line with no x-intercept and containing the point $Q(-2, -1)$.

3. Find an equation of the line through $A(-1, 4)$ that is parallel to the line characterized by $5x - 2y = 7$.

4. Find an equation of the line through $B(7, 2)$ that is perpendicular to the line characterized by $3x - 7y = 20$.

5. Given the points $A(-1, 5)$, $B(2, -3)$, and $C(5, 7)$, find:

   (a) An equation of a line through A and parallel to $\overline{BC}$.
   (b) An equation of a line through B and perpendicular to $\overline{AC}$.
   (c) An equation of the altitude to side $\overline{AB}$ of $\triangle ABC$.

6. Find an equation of the line that passes through $A(2, -3)$ and

   (a) Passes through the origin.
   (b) Has slope -3.
   (c) Is parallel to the line characterized by $2x - 7 = 4y$.

7. Which pairs of segments determined by the following four points are parallel? $A(3, 6)$, $B(5, 9)$, $C(8, 2)$, and $D(6, -1)$.

8. Show that $ABCD$ is a rectangle, given $A(3, -2)$, $B(6, 2)$, $C(2, 5)$, and $D(-1, 1)$.

9. Without using the Distance Formula, show that $ABCD$ is a parallelogram, given $A(-2, -5)$, $B(1, -2)$, $C(2, 1)$, and $D(-1, -2)$.

10. If the slope of a line is $-\frac{1}{3}$, what is the equation of a line perpendicular to it if the two lines intersect at $R(5, 7)$?

11. Find the equation of the line $\overleftrightarrow{l}$ through $A(-2, 7)$ and satisfying the following condition:

(a) $\overleftrightarrow{l}$ is parallel to the line $3x - 2y - 1 = 0$

(b) $\overleftrightarrow{l}$ is perpendicular to the line $x - 60y = 5$

12. In a given plane, $\overleftrightarrow{l_1}$ is perpendicular to the line with equation $2x - 5y - 6 = 0$. Suppose $\overleftrightarrow{l_1} \perp \overleftrightarrow{l_2}$ and $(3, 4) \in l_2$. What is the equation of $\overleftrightarrow{l_2}$?

13. Given $A(3, 2)$, $B(-2, 5)$, and $C(k, -1)$. Find a value of k so that $\overline{AB} \perp \overline{CA}$.

14. Find a value of k so that the lines characterized by the following equations are (a) parallel, and (b) perpendicular: $3x - ky + 3 = 0;\ 5x + 2y = 4$.

15. Under what conditions on A, B, and C will the line $Ax + By + C = 0$

(a) Pass through the origin?

(b) Have no x-intercept?

(c) Have no y-intercept?

10.5   CIRCLES: EQUATIONS AND GRAPHS

In Chapter 8 we learned that the set of all points in a plane which are a given distance from a given point in that plane is called a circle.

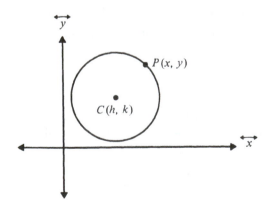

Suppose we are given a circle with center $C(h, k)$ and radius r. Then for any point $P(x, y)$ belonging to the circle we have

$$PC = r$$

Using the Distance Formula (T114), this becomes

$$\sqrt{(x - h)^2 + (y - k)^2} = r$$

Squaring both sides yields the equivalent statement

$$(x - h)^2 + (y - k)^2 = r^2$$

Therefore, we conclude:

THEOREM 122   The circle with center $C(h, k)$ and radius r is the the graph of the equation

$$(x - h)^2 + (y - k)^2 = r^2$$

Example 1   Graph the circle with equation

$$(x - 3)^2 + (y + 4)^2 = 9$$

Applying Theorem 122, we conclude that the circle we are looking for has center $C(3, -4)$ and radius 3. The graph of this circle is as shown.

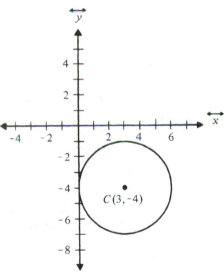

Consider the equation of the circle with radius 4 and with center at $C(-2, 1)$:

$$(x + 2)^2 + (y - 1)^2 = 16$$

This can be rewritten as

$$x^2 + 4x + 4 + y^2 - 2y + 1 = 16$$
$$x^2 + 4x + y^2 - 2y - 11 = 0$$

Since these steps are reversible, it should be clear that we can examine *any* equation of the form

$$x^2 + y^2 + Dx + Ey + F = 0$$

and come up with a graphical interpretation.

Example 2   Suppose we are given the equation

$$x^2 + y^2 - 4x + 6y - 12 = 0$$

We can group terms as follows

$$(x^2 - 4x \qquad) + (y^2 + 6y \qquad) = 12$$

"Completing the square" to form perfect square trinomials, we obtain

$$(x^2 - 4x + 4) + (y^2 + 6y + 9) = 12 + 4 + 9$$

or
$$(x - 2)^2 + (y + 3)^2 = 25$$

which is the equation of a circle with center $C(2, -3)$ and radius 5.

Performing the process outlined in Example 2 may lead us to conclude that the graph of an equation of this form is not a circle at all. Suppose we "complete the square" and obtain something like

$$(x - h)^2 + (y - k)^2 = 0$$

Then the graph is just the point $C(h, k)$. If the equation obtained after "completing the square" has a *negative* constant on the right-hand side, there are no ordered pairs satisfying the equation. Thus the graph is the empty set.

Example 3   The graph of $x^2 + (y - 1)^2 = 0$ is the point that is zero units from $C(0, 1)$, namely, $C(0, 1)$.

Example 4   Consider the equation

$$x^2 - x + y^2 + 3y + 3 = 0$$

We "complete the square" and obtain

$$(x^2 - x + \tfrac{1}{4}) + (y^2 + 3y + \tfrac{9}{4}) + 3 - \tfrac{1}{4} - \tfrac{9}{4} = 0$$
$$(x - \tfrac{1}{2})^2 + (y + \tfrac{3}{2})^2 = -\tfrac{1}{2}$$

Since no "squares" have a negative sum, the solution set and graph of this equation are the empty set.

We can summarize the above discussion and examples with a theorem.

THEOREM 123   The graph of an equation of the form

$$x^2 + y^2 + Dx + Ey + F = 0$$

is a circle or a point or the empty set.

Of the two forms of the equation for a circle,

$$(x - h)^2 + (y - k)^2 + r^2$$

and

$$x^2 + y^2 + Dx + Ey + F = 0$$

which do you feel is more helpful?

If we know the center of a circle and a point on that circle, we can determine its equation.

Example 5   Suppose a circle has center $(2, -3)$ and passes through $(3, 1)$. Then

$$(x - 2)^2 + (y + 3)^2 = r^2$$

But since it passes through $(3, 1)$, we may write

$$(3 - 2)^2 + (1 + 3)^2 = r^2$$

or

$$1 + 16 = r^2$$

Therefore, $r^2 = 17$, and we have

$$(x - 2)^2 + (y + 3)^2 = 17$$

In Chapter 8 we learned that a tangent to a circle is perpendicular to a radius drawn to a point of tangency. We can put this knowledge to work in the following example.

Example 6   To find the equation of the line $\overleftrightarrow{l}$ tangent to the circle

$$(x + 1)^2 + (y - 2)^2 = 25$$

at the point $P(3, 5)$, we use a fact from Chapter 8.

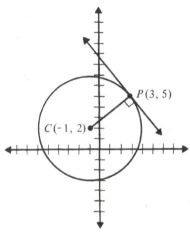

As the accompanying diagram suggests, we need to find the equation of the line containing $P(3, 5)$ and perpendicular to $\overline{CP}$. We can write this equation in point-slope form if we know the slope of the tangent.

The slope of $\overline{CP}$ is given by

$$\frac{5 - 2}{3 - (-1)} = \frac{3}{4}$$

Therefore, the slope of the tangent is $-\frac{4}{3}$, and its equation becomes

$$y - 5 = -\tfrac{4}{3}(x - 3)$$

PROBLEM SET 10.5

1. Sketch the graphs of the following equations.

 (a) $(x - 2)^2 + (y + 3)^2 = 4$
 (b) $(x + 4)^2 + y^2 = 16$
 (c) $(x + 1)^2 + (y - 1)^2 = \frac{9}{4}$
 (d) $x^2 + y^2 = 25$

2. Find the equation of the circle with center C and radius r, given

 (a) $C(3, 4)$ and $r = 2$
 (b) $C(-2, -3)$ and $r = 3$
 (c) $C(0, -1)$ and $r = \frac{1}{2}$
 (d) $C(a, b)$ and $r = k$

3. Obtain the equation of the circle that has center $C(4, 0)$ and is tangent to the y-axis.

4. Find the equation of the circle that is tangent to the line $y = 3$ and has center $C(7, -2)$.

5. Find the equation of the circle with center at the origin and passing through $P(3, 4)$.

6. If the graph of the equation is a circle, find its center and radius. If it is not a circle, tell whether the graph is a single point or the empty set.

   (a) $x^2 + 2x + y^2 - 4y = 4$
   (b) $x^2 + y^2 + 4x + 2y = -1$
   (c) $x^2 - 6x + y^2 - 4y = 3$
   (d) $x^2 + y^2 + 5x + \frac{9}{4} = 6y$
   (e) $x^2 + y^2 + 6x + 12y + 45 = 0$

7. Give an algebraic relationship between D, E, and F that ensures that
   $$x^2 + y^2 + Dx + Ey + F = 0$$

   is the equation of a circle. (*Hint:* Rewrite this equation in another form by completing the square.)

8. Show that the line $2x - 3y = 5$ is tangent to the circle $(x + 1)^2 + (y - 2)^2 = 13$, at $P(1, -1)$.

9. Find the equation of the tangent to the circle $x^2 + y^2 = 25$, at $(-3, 4)$.

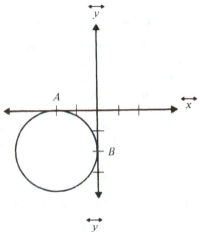

10. Find the equation of the tangent to the circle $(x - 3)^2 + (y - 2)^2 = 25$, at $P(6, -6)$.

11. The circle in the figure is tangent to the coordinate axes at $A(-2, 0)$ and $B(0, -2)$. Find the radius and the coordinates of the center of the circle and then write the equation of the circle.

12. Circles P and Q are externally tangent circles. Circle P is congruent to circle Q. If $P(-2, 2)$ and $Q(4, -\frac{1}{2})$ are given, find the equations of both circles.

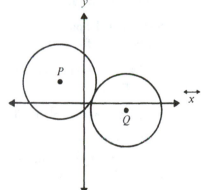

13. Use the Distance Formula to find an equation for the graph of the set of all points that are twice as far from the origin as from $P(3, 0)$. Square both sides of this equation and reduce to the simplest equivalent equation that you can. Sketch the graph.

CHAPTER 10: REVIEW EXERCISES

1. For each of the following statements, indicate whether it is *always* true, *sometimes* but not always true, or *never* true.

   (a) If P is a point in the fourth quadrant, then the abscissa of P is positive.

   (b) If $\overleftrightarrow{l}$ and $\overleftrightarrow{m}$ have the same slope and $\overleftrightarrow{l} \neq \overleftrightarrow{m}$, then $\overleftrightarrow{l}$ and $\overleftrightarrow{m}$ are parallel.

   (c) If $\overleftrightarrow{l}$ and $\overleftrightarrow{m}$ are parallel, then $\overleftrightarrow{l}$ and $\overleftrightarrow{m}$ have the same slope.

   (d) If $\overleftrightarrow{l}$ intersects the negative portion of the y-axis and the positive portion of the x-axis, then the slope of $\overleftrightarrow{l}$ is a positive number.

   (e) If $\overleftrightarrow{l} = \{(x, y) : x = h\}$, then the slope of $\overleftrightarrow{l}$ is zero.

   (f) If $\overleftrightarrow{l}$ and $\overleftrightarrow{m}$ are perpendicular lines, then the product of their slopes is -1.

   (g) If the product of the slopes of two lines is -1, then the lines are perpendicular.

   (h) The graph of $\{(x, y) : Ax + By + C = 0\}$ is a line.

   (i) The graph of an equation of the form $x^2 + y^2 + Dx + Ey + F = 0$, where D, E, and F are real numbers, is a circle.

   (j) The graph of an equation of the form $(x - h)^2 + (y - k)^2 = r^2$, where h, k, and r are real numbers, is a circle.

2. Find the slopes of the lines joining the following pairs of points:

   (a) $(2, 7)$ and $(5, -1)$ (b) $(2, 1)$ and $(7, 4)$

3. Show that the points $A(2, 1)$, $B(7, 4)$, $C(2, 6)$, and $D(-3, 3)$ are the vertices of a parallelogram.

4. Find an equation of the line with slope $-\frac{3}{4}$ that passes through the point $P(-2, 3)$.

5. Find an equation of the line through the points $A(7, -1)$ and $B(\frac{3}{5}, -\frac{1}{2})$.

6. Find an equation of the line through $A(2, -3)$ that is parallel to the line $3x + 5y = 7$.

7. Prove that the points $A(3, -4)$, $B(5, 1)$ and $C(9, 11)$ are collinear.

8. Find a value of k for which the line $2x - ky = 14$ is

(a) Parallel to the line $y = 2x$.

(b) Perpendicular to the line $x + y = 6$.

9. What is the graph of the following equation?

$$x^2 + y^2 + 3x - 5y + 18 = 0$$

10. (a) Use the Distance Formula to derive an equation for the set of all points in the Cartesian coordinate plane which are equidistant from $A(2, 1)$ and $B(4, 4)$.

(b) How is this graph related to $\overline{AB}$?

11. Given $\triangle ABC$ with $A(-5, 0)$, $B(3, 4)$, and $C(5, 0)$. What can you conclude about $\triangle ABC$?

12. Sketch the graph of $(x - 2)^2 + (y - 3)^2 = 25$.

13. How would you describe the set of lines characterized by equations of the form $y = x + k$, where k is a constant?

14. (a) Find the length of the altitude from A to $\overline{BC}$ in $\triangle ABC$, given $A(5, 2)$, $B(-1, 1)$, $C(2, 0)$.

(b) Find the area of the triangle.

15. Find the equation of the circle that is tangent to the y-axis and with center $C(3, -4)$.

16. Find the equation of the tangent to the circle

$$(x + 2)^2 + (y - 3)^2 = 16$$

at $P(-2, 7)$.

17. Write the equation of the circle with center at $C(4, -3)$ and passing through the origin.

18. How would you describe the set of lines characterized by equations of the form $y = mx$?

19. What can you say about a point if the product of its coordinates is

(a) Negative?   (b) Zero?   (c) Positive?

20. Use the diagram of the rectangle and methods of coordinate geometry to prove that the diagonals of a rectangle are congruent.

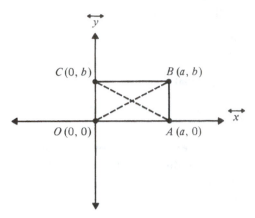

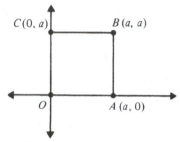

21. Use the diagram of the rectangle and the methods of coordinate geometry to prove that the diagonals of a square are perpendicular.

REFERENCES

Descartes, René. *La Géométrie*. New York: Dover Publications, 1954.
This volume contains the original French as well as an English translation of the work that revolutionized method in geometry. It was originally published in 1637 as an appendix to another work, *Discourse on Method*, that was intended to revolutionize method in philosophy.

Kline, Morris. *Mathematics, A Cultural Approach*. Reading, Mass.: Addison-Wesley Publishing Co., 1962. pp. 271–276.
Kline gives a clear picture of the reasons for the timely appearance of analytic geometry in the early seventeenth century.

APPENDIX A

PROPERTIES OF THE REAL NUMBER SYSTEM

The set of real numbers and the operation of addition and multiplication comprise what mathematicians call a *field*. The laws that define this field are given on page 322. We will let letters $a, b, c, \ldots$ represent elements of the set of real numbers, and will assume that there are at least two such elements.

In addition to the field properties, we also assume the following properties of the equals relation. These properties are true for all elements of the set of real numbers.

Reflexive law: $a = a$

Symmetric law: If $a = b$, then $b = a$

Transitive law: If $a = b$ and $b = c$, then $a = c$

Addition law: If $a = b$ and $c = d$, then $a + c = b + d$

Multiplication law: If $a = b$ and $c = d$, then $ac = bd$

Division law: If $a = b$ and $c = d \neq 0$, then $\dfrac{a}{c} = \dfrac{b}{d}$

329

	Addition	*Multiplication*
Closure law:	$a + b$ is a real number for all a and b	ab is a real number for all a and b
Associative law:	For all $a, b, c,$ $(a + b) + c = a + (b + c)$	For all $a, b, c,$ $(ab)c = a(bc)$
Existence of identity elements:	There is an element 0 such that $a + 0 = a,$ for all a	There is an element 1 such that $a \cdot 1 = a,$ for all a
Existence of inverses:	For each element a there is an element $(-a)$ such that $a + (-a) = 0$	For each element a, $a \neq 0$, there is an element a^{-1} such that $a \cdot a^{-1} = 1$

And linking these two operations, we have the *Distributive law:*
$$a(b + c) = ab + ac$$

The following useful generalization of the transitive law is used in many places in the text.

> *Substitution law:* If a and b are two names for the same real number, then either may be substituted for the other in any mathematical sentence without altering the truth value of the sentence.

Other familiar algebraic properties that can be derived from those above are used often in this book as well as other applications of algebra. Some of these are as follows:

Cancellation law for addition: If $a + b = a + c$, then $b = c$

Cancellation law I for multiplication: If $ab = ac$ and $a \neq 0$, then $b = c$

Cancellation law II for multiplication: $\dfrac{ab}{ac} = \dfrac{b}{c}$

Fraction addition law: $\dfrac{a}{b} + \dfrac{c}{d} = \dfrac{ad + bc}{bd}$

Laws of exponents: 1. $a^m \cdot a^n = a^{m+n}$, where m and n are integers
2. $(a^m)^n = a^{mn}$, where m and n are integers
3. $a^0 = 1$
4. $(a + b)^2 = a^2 + 2ab + b^2$

Quadratic formula: If $ax^2 + bx + c = 0$, $a \neq 0$, then

$$x = \frac{-b \pm \sqrt{b^2 - 4ac}}{2a}$$

Laws for radical expressions 1. If $a > 0$ and $b > 0$, then

$$\sqrt{(a)(b)} = \sqrt{a} \cdot \sqrt{b}$$

2. If $a > 0$ and $b > 0$, then

$$\sqrt{\frac{a}{b}} = \frac{\sqrt{a}}{\sqrt{b}}$$

TABLE OF SQUARES
AND SQUARE ROOTS

Table of Squares and Square Roots of the Numbers from 1 to 300

NUMBER	SQUARE	SQUARE ROOT	NUMBER	SQUARE	SQUARE ROOT
1	1	1.000	11	121	3.317
2	4	1.414	12	144	3.464
3	9	1.732	13	169	3.606
4	16	2.000	14	196	3.742
5	25	2.236	15	225	3.873
6	36	2.449	16	256	4.000
7	49	2.646	17	289	4.123
8	64	2.828	18	324	4.243
9	81	3.000	19	361	4.359
10	100	3.162	20	400	4.472

Table of Squares and Square Roots (continued)

NUMBER	SQUARE	SQUARE ROOT	NUMBER	SQUARE	SQUARE ROOT
21	441	4.583	62	3,844	7.874
22	484	4.690	63	3,969	7.937
23	529	4.796	64	4,096	8.000
24	576	4.899	65	4,225	8.062
25	625	5.000	66	4,356	8.124
26	676	5.099	67	4,489	8.185
27	729	5.196	68	4,624	8.246
28	784	5.292	69	4,761	8.307
29	841	5.385	70	4,900	8.367
30	900	5.477	71	5,041	8.426
31	961	5.568	72	5,184	8.485
32	1,024	5.657	73	5,329	8.544
33	1,089	5.745	74	5,476	8.602
34	1,156	5.831	75	5,625	8.660
35	1,225	5.916	76	5,776	8.718
36	1,296	6.000	77	5,929	8.775
37	1,369	6.083	78	6,084	8.832
38	1,444	6.164	79	6,241	8.888
39	1,521	6.245	80	6,400	8.944
40	1,600	6.325	81	6,561	9.000
41	1,681	6.403	82	6,724	9.055
42	1,764	6.481	83	6,889	9.110
43	1,849	6.557	84	7,056	9.165
44	1,936	6.633	85	7,225	9.220
45	2,025	6.708	86	7,396	9.274
46	2,116	6.782	87	7,569	9.327
47	2,209	6.856	88	7,744	9.381
48	2,304	6.928	89	7,921	9.434
49	2,401	7.000	90	8,100	9.487
50	2,500	7.071	91	8,281	9.539
51	2,601	7.141	92	8,464	9.592
52	2,704	7.211	93	8,649	9.644
53	2,809	7.280	94	8,836	9.695
54	2,916	7.348	95	9,025	9.747
55	3,025	7.416	96	9,216	9.798
56	3,136	7.483	97	9,409	9.849
57	3,249	7.550	98	9,604	9.899
58	3,364	7.616	99	9,801	9.950
59	3,481	7.681	100	10,000	10.000
60	3,600	7.746	101	10,201	10.050
61	3,721	7.810	102	10,404	10.100

Table of Squares and Square Roots (*continued*)

NUMBER	SQUARE	SQUARE ROOT	NUMBER	SQUARE	SQUARE ROOT
103	10,609	10.149	144	20,736	12.000
104	10,816	10.198	145	21,025	12.042
105	11,025	10.247	146	21,316	12.083
106	11,236	10.296	147	21,609	12.124
107	11,449	10.344	148	21,904	12.166
108	11,664	10.392	149	22,201	12.207
109	11,881	10.440	150	22,500	12.247
110	12,100	10.488	151	22,801	12.288
111	12,321	10.536	152	23,104	12.329
112	12,544	10.583	153	23,409	12.369
113	12,769	10.630	154	23,716	12.410
114	12,996	10.776	155	24,025	12.450
115	13,225	10.724	156	24,336	12.490
116	13,456	10.770	157	24,649	12.530
117	13,689	10.817	158	24,964	12.570
118	13,924	10.863	159	25,281	12.610
119	14,161	10.909	160	25,600	12.649
120	14,400	10.954	161	25,921	12.689
121	14,641	11.000	162	26,244	12.728
122	14,884	11.045	163	26,569	12.767
123	15,129	11.091	164	26,896	12.806
124	15,376	11.136	165	27,225	12.845
125	15,625	11.180	166	27,556	12.884
126	15,876	11.225	167	27,889	12.923
127	16,129	11,269	168	28,224	12.961
128	16,384	11.314	169	28,561	13.000
129	16,641	11.358	170	28,900	13.038
130	16,900	11.402	171	29,241	13.077
131	17,161	11.446	172	29,584	13.115
132	17,424	11.489	173	29,929	13.153
133	17,689	11.533	174	30,276	13.191
134	17,956	11.576	175	30,625	13.229
135	18,225	11.619	176	30,976	13.266
136	18,496	11.662	177	31,329	13.304
137	18,769	11.705	178	31,684	13.342
138	19,044	11.747	179	32,041	13.379
139	19,321	11.790	180	32,400	13.416
140	19,600	11.832	181	32,761	13.454
141	19,881	11.874	182	33,124	13.491
142	20,164	11.916	183	33,489	13.528
143	20,449	11.958	184	33,856	13.565

Table of Squares and Square Roots (continued)

NUMBER	SQUARE	SQUARE ROOT	NUMBER	SQUARE	SQUARE ROOT
185	34,225	13.601	226	51,076	15.033
186	34,596	13.638	227	51,529	15.067
187	34,969	13.675	228	51,984	15.100
188	35,344	13.711	229	52,441	15.133
189	35,721	13.748	230	52,900	15.166
190	36,100	13.784	231	53,361	15.199
191	36,481	13.820	232	53,824	15.232
192	36,864	13.856	233	54,289	15.264
193	37,249	13.892	234	54,756	15.297
194	37,636	13.928	235	55,225	15.330
195	38,025	13.964	236	55,696	15.362
196	38,416	14.000	237	56,169	15.395
197	38,809	14.036	238	56,644	15.427
198	39,204	14.071	239	57,121	15.460
199	39,601	14.107	240	57,600	15.492
200	40,000	14.142	241	58,081	15.524
201	40,401	14.177	242	58,564	15.556
202	40,804	14.213	243	59,049	15.588
203	41,209	14.248	244	59,536	15.620
204	41,616	14.283	245	60,025	15.652
205	42,025	14.318	246	60,516	15.684
206	42,436	14.353	247	61,009	15.716
207	42,849	14.387	248	61,504	15.748
208	43,264	14.422	249	62,001	15.780
209	43,681	14.457	250	62,500	15.811
210	44,100	14.491	251	63,001	15.843
211	44,521	14.526	252	63,504	15.875
212	44,944	14.560	253	64,009	15.906
213	45,369	14.595	254	64,516	15.937
214	45,796	14.629	255	65,025	15.969
215	46,225	14.663	256	65,536	16.000
216	46,656	14.697	257	66,049	16.031
217	47,089	14.731	258	66,564	16.062
218	47,524	14.765	259	67,081	16.093
219	47,961	14.799	260	67,600	16.125
220	48,400	14.832	261	68,121	16.155
221	48,841	14.866	262	68,644	16.186
222	49,284	14.900	263	69,169	16.217
223	49,729	14.933	264	69,696	16.248
224	50,176	14.967	265	70,225	16.279
225	50,625	15.000	266	70,756	16.310

Table of Squares and Square Roots (*continued*)

NUMBER	SQUARE	SQUARE ROOT	NUMBER	SQUARE	SQUARE ROOT
267	71,289	16.340	284	80,656	16.852
268	71,824	16.371	285	81,225	16.882
269	72,361	16.401	286	81,796	16.912
270	72,900	16.432	287	82,369	16.941
271	73,441	16.462	288	82,944	16.971
272	73,984	16.492	289	83,521	17.000
273	74,529	16.523	290	84,100	17.029
274	75,076	16.553	291	84,681	17.059
275	75,625	16.583	292	85,264	17.088
276	76,176	16.613	293	85,849	17.117
277	76,729	16.643	294	86,436	17.146
278	77,284	16.673	295	87,025	17.176
279	77,841	16.703	296	87,616	17.205
280	78,400	16.733	297	88,209	17.234
281	78,961	16.763	298	88,804	17.263
282	79,524	16.793	299	89,401	17.292
283	80,089	16.823	300	90,000	17.321

APPENDIX C

LIST OF POSTULATES, THEOREMS, AND COROLLARIES

POSTULATE 1   There exist at least two lines.

POSTULATE 2   Every line is a set of points and contains at least two distinct points.

POSTULATE 3   *The Line Postulate*: If P and Q are distinct points, then there is exactly one line that contains both of them.

POSTULATE 4   A plane is a set of points and contains at least three non-collinear points.

POSTULATE 5   *The Plane Postulate*: If P, Q, and R are three noncollinear points, then there is exactly one plane that contains all of them.

POSTULATE 6   If two points of a line lie in a plane, then the line lies in that plane.

POSTULATE 7   Space contains at least four points which are noncollinear and do not lie in the same plane.

POSTULATE 8   The set of points on a line and the set of real numbers are matching sets.

POSTULATE 9   *The Ruler Postulate*: Every line has a coordinate system.

POSTULATE 10   *The Ruler Placement Postulate*: Let $\overleftrightarrow{l}$ be a line with $P \in \overleftrightarrow{l}$, $Q \in \overleftrightarrow{l}$, and $P \neq Q$. Then there exists a coordinate system for $\overleftrightarrow{l}$ such that $c(P) = 0$ and $c(Q)$ is positive.

POSTULATE 11   *The Point-Plotting Postulate* (PPP): Suppose $\overrightarrow{AB}$ and real number r, $r > 0$ are given. Then there is exactly one point C, $C \in \overrightarrow{AB}$ such that $AC = r$.

POSTULATE 12   *The Plane Separation Postulate* (PSP): Given a line and a plane containing it, the set of all points of that plane that do not lie on the line is the union of two disjoint sets such that
   1. Each of the sets is convex.
   2. If P belongs to one of the sets and Q belongs to the other, then $\overline{PQ}$ intersects the line.

POSTULATE 13   *The Angle Measurement Postulate*: To every angle there corresponds a unique real number k, $0 < k < 180$.

POSTULATE 14   *The Angle Construction Postulate*: Let $\overrightarrow{AB}$ be a ray on the edge of half plane $\mathcal{H}$, and let k be a real number such that $0 < k < 180$; then there is exactly one ray $\overrightarrow{AP}$ with $P \in \mathcal{H}$ and such that $m \angle BAP = k$.

POSTULATE 15   *The Angle Addition Postulate* (AAP): Suppose $\angle BAC$ is given. If $P \in \mathcal{I} \angle BAC$, then $m \angle BAP + m \angle PAC = m \angle BAC$.

POSTULATE 16   *The Supplement Postulate*: If two angles are a linear pair, then they are supplementary.

POSTULATE 17   *The Side-Angle-Side Postulate* (SAS): Given two triangles, or one triangle and itself, and a one-to-one correspondence between their vertices such that two sides and the included angle in one triangle are congruent to the corresponding parts of the other triangle, then the triangles are congruent.

POSTULATE 18   *The Parallel Postulate*: Through a given point not on a given line, there is at most one line parallel to the given line.

POSTULATE 19   *The Area Postulate*: To each polygonal region there corresponds a unique positive real number.

POSTULATE 20   *The Area Addition Postulate*: If two polygonal regions do not intersect other than in a finite number of segments and vertices, then the area of their union is the sum of their areas.

POSTULATE 21   If two triangles are congruent, then the triangular regions determined by them have the same area.

POSTULATE 22   The area of a rectangular region is the product of the lengths of a pair of consecutive sides.

POSTULATE 23   *The Two-Circle Postulate*:   Given circles P and Q with radii a and b, respectively. Let c be the distance between their centers. If each of a, b, and c is less than the sum of the other two, then the circles intersect in exactly two points, and the two points of intersection lie on opposite sides of the line containing their centers.

POSTULATE 24   *The Space Separation Postulate*: Given a plane, the set of all points of space that do not belong to that plane is the union of two disjoint sets called *half spaces* such that both the following are true:
  1. Each of the sets is convex.
  2. If P belongs to one of the sets and Q belongs to the other, the $\overline{PQ}$ intersects the plane.

POSTULATE 25   If two planes intersect, then their intersection is a line.

POSTULATE 26   *The Volume Postulate*: To each convex polyhedron there corresponds a unique positive real number.

POSTULATE 27   The volume of a right rectangular prism is equal to the product of the area of its base and its altitude.

THEOREM 1   If two lines intersect, their intersection contains exactly one point.

THEOREM 2   If A and C are two points, then there is a point B such that A-B-C.

THEOREM 3   If A and B are two points, then there is a point C such that B is between A and C.

THEOREM 4   *The Midpoint Theorem*: Every segment has one and only one midpoint.

THEOREM 5   Suppose M is the midpoint of $\overline{AB}$ and $c(A) = a$, $c(M) = m$, and $c(B) = b$. Then

$$m = \frac{a + b}{2}$$

THEOREM 6   Segment congruence is reflexive. Symbolically, if $\overline{AB}$ is a segment, then $\overline{AB} \cong \overline{AB}$.

THEOREM 7   Segment congruence is symmetric; that is, if $\overline{AB} \cong \overline{CD}$, then $\overline{CD} \cong \overline{AB}$.

THEOREM 8   Segment congruence is transitive; that is, if $\overline{AB} \cong \overline{CD}$ and $\overline{CD} \cong \overline{EF}$, then $\overline{AB} \cong \overline{EF}$.

THEOREM 9   *Segment Addition Theorem* (SAT): If A-B-C, R-S-T, $\overline{AB} \cong \overline{RS}$, and $\overline{BC} \cong \overline{ST}$, then $\overline{AC} \cong \overline{RT}$.

THEOREM 10   *Segment Subtraction Theorem* (SST): If A-B-C, R-S-T, $\overline{AB} \cong \overline{RS}$, and $\overline{AC} \cong \overline{RT}$, then $\overline{BC} \cong \overline{ST}$.

THEOREM 11   *Segment Bisection Theorem*: If $\overline{AC} \cong \overline{RT}$, B is the midpoint of $\overline{AC}$, and S is the midpoint of $\overline{RT}$, then $\overline{AB} \cong \overline{RS}$.

THEOREM 12   The intersection of two convex sets is a convex set.

THEOREM 13   Let $\angle A$ be given, then $\angle A \cong \angle A$ (reflexive property).

THEOREM 14   If $\angle A \cong \angle B$, then $\angle B \cong \angle A$ (symmetric property).

THEOREM 15   If $\angle A \cong \angle B$ and $\angle B \cong \angle C$, then $\angle A \cong \angle C$ (transitive property).

THEOREM 16A   *The Angle Addition Theorem*: Let $\angle ABC$ and $\angle DEF$ be given. Suppose $P \in \angle ABC$ and $Q \in \angle DEF$. If $\angle ABP \cong \angle DEQ$ and $\angle PBC \cong \angle QEF$, then $\angle ABC \cong \angle DEF$.

THEOREM 16B   *The Angle Subtraction Theorem*: Let $\angle ABC \cong \angle DEF$. Suppose $P \in \angle ABC$, $Q \in \angle DEF$, and $\angle PBC \cong \angle QEF$; then $\angle PBA \cong \angle QED$.

THEOREM 17   *The Angle Bisection Theorem*: Suppose that $\angle ABC \cong \angle DEF$, $\overrightarrow{BP}$ bisects $\angle ABC$, and $\overrightarrow{EQ}$ bisects $\angle DEF$. Then $\angle PBC \cong \angle QEF$.

THEOREM 18   Right angles are congruent.

THEOREM 19   Complements of congruent angles are congruent.

THEOREM 20   If two angles are supplementary and congruent, then each is a right angle.

THEOREM 21   Supplements of congruent angles are congruent.

THEOREM 22   If two lines are perpendicular, then they form four right angles.

THEOREM 23   *The Vertical Angle Theorem* (VAT): If two angles are vertical angles, then they are congruent.

THEOREM 24   (a) Triangle congruence is reflexive.
              (b) Triangle congruence is symmetric.
              (c) Triangle congruence is transitive.

THEOREM 25   *The Isosceles Triangle Theorem* (ITT): If two sides of a triangle are congruent, then the angles opposite them are congruent.

THEOREM 26   The bisector of the vertex angle of an isosceles triangle bisects and is perpendicular to the base of the triangle.

THEOREM 27   *The Angle-Side-Angle Theorem* (ASA): Given two triangles, or a triangle and itself, and a one-to-one correspondence between their vertices. If two angles and the included side of the first triangle are congruent to the corresponding parts of the second triangle, then the triangles are congruent.

THEOREM 28   If two angles of a triangle are congruent, then the sides opposite those angles are congruent.

THEOREM 29   *The Side-Side-Side Theorem* (SSS): Given two triangles, or a triangle and itself, and a one-to-one correspondence between their vertices. If the three sides of one triangle are congruent to the corresponding sides of the other triangle, then the triangles are congruent.

THEOREM 30   Let $\overleftrightarrow{l}$ be given, $\overleftrightarrow{l}$ in plane m. Let $P \in \overleftrightarrow{l}$. Then there is one and only one line in plane m that is perpendicular to $\overleftrightarrow{l}$ at P.

THEOREM 31   The median from the vertex of an isosceles triangle is an altitude and an angle bisector.

THEOREM 32   *The Exterior Angle Theorem* (EAT): The measure of an exterior angle of a triangle is greater than the measures of either of the remote interior angles.

THEOREM 33   The lengths of two sides of a triangle are unequal if and only if the measures of the angles opposite them are unequal in the same order.

THEOREM 34   *The Triangle Inequality*: The sum of the lengths of two sides of a triangle is greater than the length of the third side.

THEOREM 35   Given a line and a point not on the line, there is one and only one line perpendicular to the given line and which contains the given point.

THEOREM 36   *The Hypotenuse-Leg Theorem*: Given two right triangles, or a right triangle and itself, and a correspondence between their vertices. If one leg and the hypotenuse of one of the triangles are congruent to the corresponding parts of the second right triangle, then the triangles are congruent.

THEOREM 37   *The Perpendicular Bisector Theorem* (PBT): In a plane, the perpendicular bisector of a segment is the set of all points that are equidistant from the endpoints of the segment.

THEOREM 38   In a plane, if two lines are perpendicular to the same line, then they are parallel.

THEOREM 39   Let $\overleftrightarrow{l}$ be a line and let P be a point not on $\overleftrightarrow{l}$. Then there is a line, containing P, that is parallel to $\overleftrightarrow{l}$.

THEOREM 40   If two lines are cut by a transversal so that one pair of alternate interior angles are congruent, then the other pair of alternate interior angles are congruent.

THEOREM 41   *The Alternate Interior-Parallel Theorem* (AIP): If two lines are cut by a transversal so that a pair of alternate interior angles are congruent, then the lines are parallel.

THEOREM 42   *The Corresponding-Parallel Theorem*: If two lines are cut by a transversal so that a pair of corresponding angles are congruent, then the lines are parallel.

THEOREM 43   If two lines are cut by a transversal so that a pair of interior angles on the same side of the transversal are supplementary, then the lines are parallel.

THEOREM 44   If two parallel lines are cut by a transversal, then alternate interior angles are congruent.

THEOREM 45   If two parallel lines are cut by a transversal, each pair of corresponding angles are congruent.

THEOREM 46   If two parallel lines are cut by a transversal, interior angles on the same side of the transversal are supplementary.

THEOREM 47   In a plane, if a line intersects one of two parallel lines in a single point, then it intersects the other.

THEOREM 48   In a plane, if a line is perpendicular to one of two parallel lines, it is perpendicular to the other.

THEOREM 49   In a plane, two lines parallel to the same line are parallel to each other.

THEOREM 50   *The Angle Sum Theorem* (AST): The sum of the measures of the angles of a triangle is 180.

THEOREM 51   *The Side-Angle-Angle Theorem* (SAA): Given a correspondence between two triangles. If two angles and a side of one triangle are congruent to the corresponding parts of the other, then the triangles are congruent.

THEOREM 52   Parallel lines are everywhere equidistant.

THEOREM 53   If three parallel lines intercept congruent segments on one transversal, then they intercept congruent segments on any transversal parallel to the given one.

THEOREM 54   If three parallel lines intercept congruent segments on one transversal, then they intercept congruent segments on any transversal.

THEOREM 55   If two triangles are congruent, they are similar.

THEOREM 56:   *The Basic Proportionality Theorem* (BPT): A line parallel to one side of a triangle and intersecting the interiors of the other two sides separates these sides proportionally.

THEOREM 57   If a line intersects two sides of a triangle, separating them into segments which are proportional to these sides, then it is parallel to the third side.

THEOREM 58   *The Angle-Angle-Angle Similarity Theorem* (AAA~): Given a one-to-one correspondence between the vertices of two triangles, or a triangle and itself. If the three angles of one of the triangles are congruent to the corresponding angles of the other, then the triangles are similar.

THEOREM 59   *The Side-Angle-Side Similarity Theorem* (SAS~): Given a one-to-one correspondence between the vertices of two triangles, or a triangle and itself. If two sides of one triangle are proportional to the corresponding sides of the other and the included angles are congruent, then the triangles are similar.

THEOREM 60   *The Side-Side-Side Similarity Theorem* (SSS~): Given a one-to-one correspondence between the vertices of two triangles, or a triangle and itself. If the corresponding sides are proportional, then the triangles are similar.

THEOREM 61   The altitude to the hypotenuse of a right triangle forms two triangles, each of which is similar to the other and to the original right triangle.

THEOREM 62   Given a right triangle and an altitude to the hypotenuse:
   (a) The altitude to the hypotenuse is the geometric mean between the segments into which it separates the hypotenuse.
   (b) Each leg is the geometric mean between the hypotenuse and the segment determined by the altitude to the hypotenuse which is adjacent to that leg.

THEOREM 63   *The Pythagorean Theorem* (PT): The square of the length of the hypotenuse of a right triangle is equal to the sum of the squares of the lengths of the legs.

THEOREM 64   *Converse of the Pythagorean Theorem* (CPT): If a triangle has sides of lengths a, b, and c, and $a^2 + b^2 = c^2$, then the triangle is a right triangle and the side of length c is the hypotenuse.

THEOREM 65   One leg of a right triangle is half as long as the hypotenuse if and only if the angle opposite that leg has measure 30.

THEOREM 66   The hypotenuse of an isosceles right triangle is $\sqrt{2}$ times as long as a leg.

THEOREM 67   The consecutive angles of a parallelogram are supplementary.

THEOREM 68   The nonconsecutive angles of a parallelogram are congruent.

THEOREM 69   Each diagonal separates a parallelogram into two congruent triangles.

THEOREM 70   The nonconsecutive sides of a parallelogram are congruent.

THEOREM 71   The diagonals of a parallelogram bisect each other.

THEOREM 72   If both pairs of nonconsecutive angles of a quadrilateral are congruent, the quadrilateral is a parallelogram.

THEOREM 73   If both pairs of nonconsecutive sides of a quadrilateral are congruent, the quadrilateral is a parallelogram.

THEOREM 74   If two sides of a quadrilateral are parallel and congruent, the quadrilateral is a parallelogram.

THEOREM 75   If the diagonals of a quadrilateral bisect each other, the quadrilateral is a parallelogram.

THEOREM 76   The area of a triangle is half the product of any altitude and the corresponding base.

THEOREM 77   The area of a trapezoid is half the product of the altitude and the sum of the bases.

THEOREM 78   The area of a parallelogram is the product of the length of a side and altitude to that side.

THEOREM 79   If two triangles are similar, then the ratio of their areas is the square of the ratios of any two corresponding sides.

THEOREM 80   If two polygons are similar, then the ratio of their perimeters is the ratio of any pair of corresponding sides.

THEOREM 81   If two polygons are similar, then the ratio of their areas is the square of the ratios of any pair of corresponding sides.

THEOREM 82 Each tangent to a circle is perpendicular to the radius drawn to the point of tangency.

THEOREM 83 If a line is perpendicular to a radius at its outer end, it is tangent to the circle.

THEOREM 84 The perpendicular segment from the center of a circle to a chord of that circle bisects the chord.

THEOREM 85 If a segment bisecting a chord which is not a diameter passes through the center of the circle, then that segment is perpendicular to the chord.

THEOREM 86 In the plane of a given circle, the perpendicular bisector of a chord passes through the center of the circle.

THEOREM 87 Every circle is congruent to itself.

THEOREM 88 Chords of congruent circles are congruent iff they are equidistant from their centers.

THEOREM 89 If a line and a circle are coplanar and if the line intersects the interior of the circle, then the line contains two points of the circle.

THEOREM 90 *The Arc Addition Theorem*: If X is an interior point of $\overarc{AB}$, then $m\,\overarc{AXB} = m\,\overarc{AX} + m\,\overarc{XB}$.

THEOREM 91 The measure of an inscribed angle is one-half the measure of its intercepted arc.

THEOREM 92 If a secant ray and tangent ray have a common endpoint, then the measure of the angle formed is one half the measure of the intercepted arc.

THEOREM 93 In the same circle or congruent circles, if two chords that are not diameters are congruent, then the corresponding minor arcs are congruent.

THEOREM 94 In the same circle or congruent circles, if two arcs are congruent, then the corresponding chords are congruent.

THEOREM 95 The area of a regular polygon is the product of one half its apothem and its perimeter.

THEOREM 96 The ratio of the circumference to the diameter is the same for all circles.

THEOREM 97 The area of a circle with radius r is πr^2.

THEOREM 98 The angle bisectors of the angles of a triangle are concurrent at a point equidistant from the sides of the triangle.

THEOREM 99   The perpendicular bisectors of the sides of a triangle are concurrent at a point equidistant from the vertices of the triangle.

THEOREM 100   The medians of a triangle are concurrent at a point which is two-thirds of the distance from any vertex to the midpoint of the opposite side.

THEOREM 101   If a line intersects a plane not containing it, then the intersection contains a single point.

THEOREM 102   Any point and a line not containing it determine a plane.

THEOREM 103   Any two plane angles of a dihedral angle are congruent.

THEOREM 104   If a line is perpendicular to a plane, any plane containing the line is perpendicular to the plane.

THEOREM 105   If two planes are perpendicular, any line in one of them that is perpendicular to their intersection is also perpendicular to the other plane.

THEOREM 106   If two intersecting planes are each perpendicular to a third plane, then their line of intersection is perpendicular to that plane.

THEOREM 107   If a line is perpendicular to one of two parallel planes, it is perpendicular to the other.

THEOREM 108   A segment which has its endpoints in each of two parallel planes and is perpendicular to one of them (and hence to both) is the shortest segment joining the two planes.

THEOREM 109   The volume of a prism is the product of the area of its base and its altitude.

THEOREM 110   The volume of a pyramid is one-third the product of the area of its base and the altitude.

THEOREM 111   If a plane intersects the interior of a sphere, its intersection with the sphere is a circle. The center of this circle is the intersection of the plane with a line that passes through the center of the sphere and is perpendicular to the plane.

THEOREM 112   The surface area of a sphere of radius r is $4\pi r^2$.

THEOREM 113   The volume of a sphere of radius r is $\frac{4}{3}\pi r^3$.

THEOREM 114   *The Distance Formula*: Given two distinct points $P(x_1, y_1)$ and $Q(x_2, y_2)$ in a Cartesian coordinate system. Then

$$PQ = \sqrt{(x_2 - x_1)^2 \, (y_2 - y_1)^2}$$

THEOREM 115   All segments on the same nonvertical line have the same slope.

THEOREM 116   Every vertical line has an equation of the form $x = h$, and every horizontal line has an equation of the form $y = k$.

THEOREM 117   Suppose $\overleftrightarrow{l}$ is a line with slope m and containing $P(x_1, y_1)$. Then for any point $Q(x, y)$ such that $Q \in \overleftrightarrow{l}$, we have the equation

$$y - y_1 = m(x - x_1)$$

THEOREM 118   Suppose $\overleftrightarrow{l}$ is a line with slope m and containing $P(x_1, y_1)$. If the coordinates of $Q(x, y)$ satisfy the equation

$$y - y_1 = m(x - x_1)$$

then $Q \in \overleftrightarrow{l}$.

THEOREM 119   Two distinct nonvertical lines are parallel iff they have the same slope.

THEOREM 120   If two lines with slopes m_1 and m_2 are perpendicular, then $m_1(-1/m_2)$; that is, each of these slopes is the negative reciprocal of the other.

THEOREM 121   If the slopes of two lines are negative reciprocals of one another, then the lines are perpendicular.

THEOREM 122   The circle with center $C(h, k)$ and radius r is the graph of the equation

$$(x - h)^2 + (y - k)^2 = r^2$$

THEOREM 123   The graph of an equation of the form

$$x^2 + y^2 + Dx + Ey + F = 0$$

is a circle or a point or the empty set.

COROLLARY P11   *The Segment Construction Theorem* (SCT): Let $\overline{AB}$ be any segment and $\overrightarrow{PQ}$ any ray. Then there is exactly one point S, $S \in \overline{PQ}$, such that $\overline{AB} \cong \overline{PS}$.

COROLLARY P14   *The Angle Construction Theorem* (ACT): Let $\angle A$ be given. Let $\overrightarrow{PB}$ be a ray on the edge of half plane $\mathcal{H}$. Then there exists exactly one ray $\overrightarrow{PQ}$, $Q \in \mathcal{H}$ such that $\angle BPQ \cong \angle A$.

COROLLARY P17   If two legs of one right triangle are congruent to the corresponding legs of another right triangle, then the triangles are congruent.

COROLLARY T31   If a line in the plane of an isosceles triangle bisects and is perpendicular to the base, that line contains the opposite vertex.

COROLLARY T35(a)   No triangle has two right angles.

COROLLARY T35(b)   The altitude to the base of an isosceles triangle is also a median.

COROLLARY T36   *The Angle Bisector Theorem*: Let $P \in \mathscr{I} \angle BAC$. P is equidistant from the sides of $\angle BAC$ if and only if P is on the angle bisector of $\angle BAC$.

COROLLARY T37   *The Perpendicular Bisector Corollary*: In a plane, if two points are each equidistant from the endpoints of a segment, then the line determined by them is the perpendicular bisector of that segment.

COROLLARY T50(a)   *The Third Angle Corollary*: Given a correspondence between two triangles. If two pairs of corresponding angles are congruent, then the third pair of corresponding angles are also congruent.

COROLLARY T50(b)   The acute angles of a right triangle are complementary.

COROLLARY T50(c)   *The Exterior Angle Corollary* (EAC): The measure of an exterior angle of a triangle is equal to the sum of the measures of the two remote interior angles.

COROLLARY T50(d)   In an isosceles right triangle, the measure of each base angle is 45.

COROLLARY T50(e)   The measure of each angle of an equilateral triangle is 60.

COROLLARY T54(a)   *The Egg-Slicer Corollary*: If three or more parallel lines intercept congruent segments on one transversal, then they intercept congruent segments on any transversal.

COROLLARY T54(b)   If a line bisects one side of a triangle and is parallel to a second side, then it bisects the third side.

COROLLARY T56   *The Basic Proportionality Corollary* (BPC): A line parallel to one side of a triangle and intersecting the interiors of the other two sides separates them into segments which are proportional to these sides.

COROLLARY T58(a)   *The Angle-Angle Corollary* (AA~): If two angles of one triangle are congruent to two angles of a second triangle, then the triangles are similar.

COROLLARY T58(b)   A line parallel to one side of a triangle and intersecting the interiors of the other two sides determines a triangle similar to the given one.

COROLLARY T65(a)   Given an equilateral triangle with side of length s. The length of each altitude is

$$\frac{s\sqrt{3}}{2}$$

COROLLARY T65(b)   Given a 30-60-90 right triangle. The length of the side opposite the 60° angle is $\sqrt{3}/2$ times the length of the hypotenuse and $\sqrt{3}$ times the length of the side opposite the 30° angle.

COROLLARY T67   A rectangle has four right angles.

COROLLARY T70(a)   All the sides of a rhombus are congruent.

COROLLARY T70(b)   The diagonals of a rhombus perpendicularly bisect each other.

COROLLARY P22   Let s be the length of a side of a square. Then the area of that square (region) is s^2.

COROLLARY T76   The area of an equilateral triangle with side of length s is given by

$$\frac{s^2\sqrt{3}}{4}$$

COROLLARY T91(a)   An angle inscribed in a semicircle is a right angle.

COROLLARY T91(b)   Angles inscribed in the same arc or in congruent arcs are congruent.

ANSWERS AND HINTS FOR SELECTED PROBLEMS

Chapter 1

Problem Set 1.2   1. T, 1   3. F, 2   5. F, 4   7. Mathematics is not interesting. 9. No Marxists live in California.   11. Nuts cannot be found in grocery stores; or nuts cannot be found on nut trees.

13.(a) _____

$\sim p$	$(\sim p) \wedge q$
F	F
F	F
T	T
T	F

(b) _____

$\sim q$	$p \wedge (\sim q)$
F	F
T	T
F	F
T	F

Problem Set 1.3   1.(a) If 2 is an even integer, then James Joyce wrote *Death of a Sales-man*. F   (c) F   (e) T   (g) F.

2.(a)

p	q	$\sim q$	$p \to (\sim q)$
T	T	F	F
T	F	T	T
F	T	F	T
F	F	T	T

(c)

$(\sim q) \vee (\sim p)$
F
T
T
T

(e)

$(\sim p) \leftrightarrow q$
F
T
T
F

(g)

$p \to (q \to p)$
T
T
T
T

3. a,b,c, and d; e and f, g and h.

Problem Set 1.4
1.(a)

		$a1$	$a2$	c
p	q	$p \wedge q$	$\sim q$	p
T	T	T	F	T
T	F	F	T	T
F	T	F	F	F
F	F	F	T	F

valid.   (c) Invalid, see third row.   (e) invalid.

4.(a) m.t.   (c) h.s.   (e) h.s.   (f) no conclusion.

Problem Set 1.5   1.(a) Converse: If it is cold, it rains. Inverse: If it doesn't rain, it isn't cold. Contrapositive: If it isn't cold, then it doesn't rain.   3.(a) No   (b) No   (c) Yes.   5. They are equivalent.   7.(a) (ii).

Problem Set 1.6   1.(a) $\{0,1,2,3, \ldots\}$, $\{x : x$ is a whole number$\}$, $\{$The whole numbers$\}$.   2.(a) $\{0,2,4,6, \ldots\}$ (c) $\{0,1\}$.   3. a,d,e.   5.(c) W and $\{3n : n \in W\}$.   6.(b) 2   7. Set up a one-to-one correspondence.

Problem Set 1.7   1.(a) $\{-1,0,1,2,3\}$ (c) J (e) D (g) $\varnothing$ (i) U.   3. $\varnothing$, $(A \cap B) \cap C, A \cap B, A \cup B, (A \cup B) \cup C, A \cup (B \cup C), U$, is one possibility.   5. a,b, c,d,h,i,j,k.   7. 75.

Chapter 1 Review Exercises

1.(a) T   (c) T   (e) T   (g) T   (i) T   (k) F   (m) T   (o) F.   2.(a) Socrates was mortal. m.p.   (c) This sentence is not an implication. m.t.   (e) No conclusion.

(g) No conclusion.   (i) h.s.   3.(a) Try, for example, a case in which $C \subseteq B \subseteq A$. 5.(a) Row 1 of your truth table will show this argument is invalid.   (c) Valid.   7. Let $B = \{1,2,3, \ldots\}$, $A = \{1,3,5, \ldots\}$.   9. $A \cup B = \{x: (x \in A) \vee (x \in B)\}$.   11.(a) $A \cup C$ (c) $C \cap (\sim D)$ (e) $C \cap \sim(D \cup B)$.

Chapter 2

Problem Set 2.2   1.(a) One   (e) 6.   3. 6.   5. P3.   7.

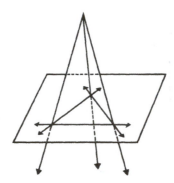

9. P4.

Problem Set 2.3   1.(a) 3   (b) 6.   3. $|x - y| = |y - x|$ for all x and y.   5.(a) (i) 5, (ii) 4,   (iii) $7\frac{1}{2}$.   (c) $|x - y| \geq |x| - |y|$.   7.(a) $\{x: -2 < x < 2$ and $x \in J\}$. 9. $\{x: x \leq 2\}$.

Problem Set 2.4   1.(a) -2 (c) 2.   3.(a) 2   (c) 2   (e) 4   (g) 4   (i) 6.   5.(a) 2 (b) 15   (c) $2\frac{3}{4}$ (d) 12.   7.(a) Absolute value is nonnegative;   (b) They are names for the same point.   8. (c).

Problem Set 2.5   1.

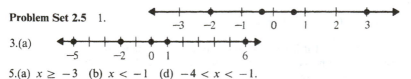

3.(a)

5.(a) $x \geq -3$ (b) $x < -1$ (d) $-4 < x < -1$.

6.(a) $\{6\}$ (c) $\{x: x < 7\}$.   7.(a)

(c)   9. $(x > 3)$ and $(x < 5)$.

11.(b) $-2 \leq x \leq 2$.

Problem Set 2.6   1. 5.   3. $4 + 6 = 10$ or $BC + AC = BC$. Thus A-C-B.   5. Begin with: If $x < y$, then $x + x < x + y$, by the addition law. Then mimic the proof of other half of the lemma in the text.   7. *Hint:* there is a point C such that A-C-B, and a point D such that A-D-C, and a point E such that A-E-D, and so on. 9.(a) $\{x: 1 < x < 5\}$

(c) $\{x: x \leq 1$ or $x \geq 5\}$.

11. $x < -1$ or $x > 1$, $|x| > 1$.

Problem Set 2.7   1.(a)

$\overrightarrow{AB} \cap \overrightarrow{BA} = \overline{AB}$

(c)

$\overrightarrow{BA} \cap \overrightarrow{BC} = \{B\}$

4.(a) $\overleftrightarrow{AB}$ (c) $\varnothing$ (e) $\overrightarrow{AB}$.   5.(a) $b = \dfrac{-1 + 3}{2} = 1$ (c) There are two possibilities.

8. $1\frac{1}{2}$.   11. $\dfrac{\frac{1}{3} + (-11)}{2} = \dfrac{-10\frac{2}{3}}{2} = -5\frac{1}{3}$.

Problem Set 2.8   1.(a) T6, segment congruence is reflexive   (c) T7, segment congruence is symmetric   (e) SST   (g) Definitions of between and distance.   3. 14.
5.(1) T6   (2) SST   (3) Hypothesis.

Chapter 2 Review Exercises

1.(a) S   (c) S   (e) A   (g) S   (i) N   (k) S.   2.(a)

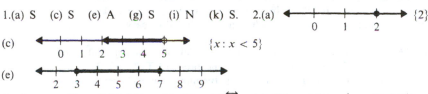

$\{2\}$

(c)

$\{x : x < 5\}$

(e)

$\{x : 3 \le x \le 7\}$.   3.(a) $\{B\}$ (c) $\varnothing$ (e) $\overleftrightarrow{T}$ (g) BC.   4.(a) $-\frac{1}{3}$.   5.(a) T6,
Reflexive Theorem   (c) Definition of congruence   (e) Line Postulate.

Chapter 3

Problem Set 3.2   1.(a) Yes   (c) Yes   (e) No   (g) No   (i) No.   3. P6.   5. Yes,
maybe, yes, yes.   7.(a) Given a point and a line containing it, ...   (b) Given a plane

in space, ...   9.(a)

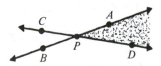

(c)

11. Not necessarily.

Problem Set 3.3   1. a,d,e.   3. 5.   5.(a) No.   A number that is not between 0 and
180 does not qualify.   (b) Only true if $x + y < 180$.   (c)(1) $-144 < x < -36$.
9.(a) 3.   11. $\angle BAP$ and $\angle PAC$.

Problem Set 3.4   1.(a) 45   (c) 80   (e) 115   (g) 105.   3. The Angle Subtraction Theorem can be used.   5.(a) Symmetric Theorem   (c) The Angle Addition Theorem   (e) Definition of bisect.   7.(a) 145.   9. Let $m \angle A = a$. Then, by the Angle Construction Postulate, there is exactly one ray $\overrightarrow{PQ}$, $Q \in H$ such that $m \angle BPQ = a$. Therefore, by the definition of congruence, $\angle BPQ \cong \angle A$.   11. Use the definition of bisect and the Angle Subtraction Theorem.

Problem Set 3.5   1.(a) 55   (c) $90 - x$ (e) $78 - x$.   3. A ray opposite to $\overrightarrow{AC}$.   5. 36.   6. $m \angle A = 72$, $m \angle B = 18$.

STATEMENTS	REASONS
1. $\angle BAC$ and $\angle CAD$ are adjacent and complementary	1. Hypothesis
2. $m \angle BAC + m \angle CAD = m \angle BAD$	2. AAP
3. $m \angle BAC + m \angle CAD = 90$	3. Definition of complementary $\angle s$
4. $m \angle BAD = 90$	4. Substitution, S2 and S3
5. $\overrightarrow{AD} \perp \overrightarrow{AB}$	5. Definition of $\perp$

7.

8.(b) Not necessarily. Suppose $B \in \mathcal{I} \angle CAD$?   9. From the definition of "less than" it follows that $m \angle A < 90$.   10. Begin by naming two right $\angle s$, $\angle A$ and $\angle B$ for example, and then specify the measure of each. The rest should be easy.

Problem Set 3.6   1. There are two pairs.   3.(a) 45   (c) $180 - x$ (e) $90 + x$.   4.(a) $10 < x < 190$.   5. 70 and 110.   7. Mimic the proof of T19.   9. Use the definition of a linear pair, P15 and T20.   11. Follow the hint given in the text.

Chapter 3 Review Exercises

1.(a) S   (c) S   (e) A   (g) N   (i) S   (k) N   3.   5. Use the reflexive theorem for angle congruence and the Angle Subtraction Theorem.   6. 6.   8.(a) 45, 135   (c) 15, 105.   9.(a) $-7 < x < 38$.   11. Use the Vertical Angle Theorem and T15 (transitive property).

Chapter 4

Problem Set 4.2   1.(a) Vertices   (b) $\mathcal{I} \triangle ABC$, $\mathcal{E} \triangle ABC$ (c) Between   (e) $\angle A$ (g) $\angle A$, $\angle C$.   3. No. Explain!   5.(1) Reflexive Theorem for segment congruence. (2) Reflexive Theorem for angle congruence.   (3) Definition of congruent $\triangle$.   7.(a) $\angle A \leftrightarrow \angle D$, $\angle B \leftrightarrow \angle E$, $\angle C \leftrightarrow \angle F$, $\overline{AB} \leftrightarrow \overline{DE}$, $\overline{BC} \leftrightarrow \overline{EF}$, $\overline{AC} \leftrightarrow \overline{DF}$.

Problem Set 4.3   1.(a) $\triangle APB \cong \triangle DPC$, SAS.   (c) None   (e) None.   3. Use ITT, the Supplement Postulate, and T21.   5. ITT implies that $\angle EBC \cong \angle ECB$, which in turn leads to $\angle ABE \cong \angle DCE$ (you must justify this). Then $\triangle ABE \cong \triangle DCE$ and the rest of the proof is easy.   6. Use VAT and SAS.   8. Use Segment Bisection Theorem, Reflexive Theorem for $\angle$ congruence, and SAS.   9. Use T18 and SAS.   10. Use the Reflexive Theorem for $\angle$ congruence and SAS for starters.

Problem Set 4.4   1.(a) $\triangle ABC \cong \triangle EDF$, SAS   (c) $\triangle ADC \cong \triangle BDC$, ASA
(e) $\triangle ABE \cong \triangle DCE$, ASA or SAS or SSS and $\triangle ACE \cong \triangle DBE$, SAS or ASA or
SSS   (g) None.   3. First use ITT and SAS to show $\triangle ACE \cong \triangle DBE$. Then $BE =$
CE.   5. For easy proof, use SSS and CPCTC.   7. Use definition of midpoint, ITT,
SAS, and CPCTC.   9.(a) $\triangle ADC \cong \triangle ABC$, SSS (The Line Postulate assures us that
$\overline{AC}$ exists.)

Problem Set 4.5

1.(a)   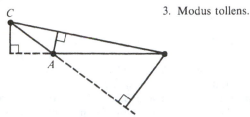   3. Modus tollens.

5. Suppose $\angle A \cong \angle B$. Then by the converse of ITT, $\overline{AC} \cong \overline{BC}$. $\rightarrow/\leftarrow$.   7. The
bisector of the vertex angle of an isosceles triangle is a median and an altitude.   9. Sup-
pose it does not. Then consider the line that does and arrive at a contradiction of T30.
11. Suppose $A \neq \emptyset$. Then there are two cases to consider. First, if $A \subseteq B$, then
$A \cap B = A$ which means $A \cap B \neq \emptyset$, a contradiction of the hypothesis. Second, if
$A \cap B \neq \emptyset, \ldots$

Problem Set 4.6   1.(a) $\angle DCA$ (c) $m\angle FBA > m\angle CAB$.   3.(a) 90   (c) $= 60$
(e) $a < 90$, $b < 90$, $d > 90$.   4. a,c,d.   5. $\angle C$, $\angle A$, $\angle B$.   7. Suppose A-H-B.
Then right $\angle CHB$ is an exterior $\angle$ of $\triangle AHC$. Therefore, by EAT, $m\angle CHB > m\angle A$
$\rightarrow/\leftarrow$.   9. $x = 1, y = 11$.   13. Use the fact that an exterior angle at A is a right $\angle$,
and therefore $m\angle B < 90$ and $m\angle C < 90$. Then apply the Addition Law for inequalities.

Problem Set 4.7   1. A line and a point not on it *determine* a plane.   3. Suppose there
were two perpendiculars as in the figure. Then $m\angle PXA = m\angle PYX \rightarrow/\leftarrow$.

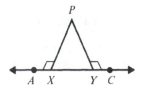

5. $\triangle ABD \cong \triangle CBD$ by HL.   7. See T26.   9. $\triangle AED \cong \triangle BEC$ by HL.   11. Another application of HL.

Problem Set 4.8   1. A point is on the perpendicular bisector of a segment iff it is
equidistant from the endpoints of the segment.   3.(a) 1   (b) an unlimited number.
5. $\overleftrightarrow{CD}$ is $\perp$ bisector of $\overline{AB}$. Therefore by PBT, $CA = CB$.   7.(a) There are four
such lines   (b) A,C,D,E,H,I,K,M,N,O,T,U,V,W,X.   9. Use PBT and transitivity.
11.(a) Since $\overleftrightarrow{m}$ is an axis of symmetry, it is the $\perp$ bisector of $\overline{BB'}$. Therefore, if we
draw a line through B and $\perp \overleftrightarrow{m}$, at X for example, we need only measure off a distance
equal to XB on the ray opposite $\overrightarrow{XB}$ to locate B' such that $\overrightarrow{AB'} \cap \overleftrightarrow{m} = \{C\}$.

Chapter 4 Review Exercises

1.(a) S   (c) A   (e) S   (g) A   (i) A   (k) A   (m) A   (o) S.   3. a.   5.(a) If $\overleftrightarrow{AC}$ is a line and P is a point not on $\overleftrightarrow{AC}$, then there is a line through P that is perpendicular to $\overleftrightarrow{AC}$ (b) Follow proof in Section 4.7.   7. Use the fact that each of the sides of the triangle is the hypotenuse (longest side) of a smaller triangle having an altitude as a leg.   9. Derive a contradiction of Corollary T35a.

Chapter 5

Problem Set 5.2   1.(a) 3 and 6, 4 and 5, 3 and 10, 4 and 9, 7 and 10, 8 and 9   (c) 3 and 9, 4 and 10.   2.(a) $\angle 3$ and $\angle 2$, $\angle 4$ and $\angle 1$ (c) $\angle 2$ and $\angle B$, $\angle 4$ and $\angle B$, $\angle 3$ and $\angle D$, $\angle 1$ and $\angle D$, $\angle A$ and $\angle B$, $\angle B$ and $\angle C$, $\angle C$ and $\angle D$.   4.(a) Interior $\angle$ s on same side of transversal   (c) Vertical $\angle$ s   (e) Corresponding $\angle$ s   (g) Alternate interior $\angle$ s.   5. Use the Supplement Postulate and T21.   7. Impossible here! 9. Use HL, CPCTC, and AIP.   11. Begin by using the Supplement Postulate. 13.(b) Use VAT and transitivity.

Problem Set 5.3   1. 50, 130, 50, 50, 130, 130, 50.   3.(a) 50, 70, 50   (b) 180.   5. Use T45, ITT, transitivity, and the converse of ITT.   7. Use T47 and then use T45. 9. *Hint:* use T38 and PAI to show $\triangle AEB \cong \triangle DEC$.   11.(a) T49   (c) Yes. If $\overleftrightarrow{l} \parallel \overleftrightarrow{m}$ then $\overleftrightarrow{m} \parallel \overleftrightarrow{l}$.   13. Use AIP and T42.   15.(a) 200/3   (b) 20.

Problem Set 5.4   1.(a) 90   (c) 1   (e) 120.   3.(a) 30   (b) 77.   5. 52.5.   7. 33, 67, 80.   9. 55.   11. Definition of linear pair, Supplement Postulate, definition of supplementary, Substitution, Subtraction law.   13. Use T39, PAI, and T45.   15. 72. 17. Show that $\angle B$ is the complement of $\angle A$ and of $\angle BCD$.

Problem Set 5.5   1. 20.   3. Introduce a line through B and parallel to $\overline{AC}$.   5. Introduce a line through C and parallel to $\overline{AB}$ and show that F is the midpoint. $\rightarrow/\leftarrow$ 7. Use T54, twice.   9. Introduce lines through A, B, and C and parallel to $\overline{DE}$.

Chapter 5 Review Exercises

1.(a) S   (c) A   (e) N   (g) A   (i) N   (k) A   (m) S.   2.(a) 50.   3. 70, 30, 50. 5. 70.   7. Use ITT, transitivity, and AIP.   9. Use definition of right $\triangle$ and SAA.

Chapter 6

Problem Set 6.2   1.(a) 6   (b) 4, 9   (c) 90, 10   (d) 288, 21, 3.5.   3.(a) 2/3   (b) b/a (c) 2   (d) 5.   5. a and f; b, d, and g; c and e.   6.(a) 6   (c) 10/3   (e) Factoring a quadratic equation tells us that $x = 2$.   7.(a) 8   (c) 12   (e) $2\sqrt{2}$ (g) 1/6. 8.(a) 25.   9.(a) If $a/b = c/d$, then $(a/b) \cdot (bd) = (c/d) \cdot (bd)$. Therefore $ad = bc$.   (c) If $a/b = c/d$ then $ad = bc$. Therefore $ad \cdot 1/ac = bc \cdot 1/ac$, or $ad/ac = bc/ac$. Thus $d/c = b/a$, or $b/a = d/c$.   11. b and d might be zero, in which case Lemma 6.1(b) would be nonsense.   13.(a) Use Lemma 6.1(d)   (c) Lemma 6.1(e) is involved here.

Problem Set 6.3   1. $DE = 6$, $EF = 6\frac{2}{3}$, $k = \frac{2}{3}$.   3. 3, 4.5.   5.(a) $AB = \frac{4}{3}DE$. Therefore $DE = \frac{3}{4}AB = 4\frac{1}{2}$. $EF = 7\frac{1}{2}$. $DF = 9$.   (b) $BC = 30/DE$ (c) 30   (d) 7 and 7.
7.(a) 16/3   (c) 48/7   (e) 2.   9. 8, 6, 36/7, 48/7.   11. Use $AC/CE = (BC + CE)/$
$(AC - AD)$ so that $AC = 20$.   13. Use definition of similar $\triangle$ and the Third Angle
Corollary.

Problem Set 6.4   1.(b) $\triangle ADC \sim \triangle BDC$ by AA$\sim$ (d) $\triangle ABC \sim \triangle DEB$ by SAS$\sim$
(e) $\triangle ABC \sim \triangle ADC$ by SSS$\sim$ (f) $\triangle GHJ \sim \triangle KML$ by SSS$\sim$ (g) $\triangle AEB \sim$
$\triangle CED$ by SAS$\sim$ (h) $\triangle ABC \sim \triangle ACD \sim \triangle CBD$ by AA$\sim$.   5.(a)
$\triangle ABC \sim \triangle DEB$. Hint: Extend $\overline{DE}$ to intersect $\overline{AC}$ and then show $\angle D \cong \angle A$ and
$\angle BED \cong \angle CBA$.   7.(a) $\triangle ABE \sim \triangle DBC$ (b) AA$\sim$ (c) 80/9.   9. Hint: Use
SAS$\sim$, definition of $\sim \triangle$, and AIP.   11. Hint: First show triangles formed by a
pair of corresponding altitudes are similar.   12. Yes, it is true since it is part of the
definition of $\sim \triangle$.   15. Use converse of ITT, Segment Bisection Theorem and SAS$\sim$.
17. (5) SAS$\sim$ (6) Definition of $\sim \triangle$.

Problem Set 6.5   1.(a) 15   (c) 8   (e) 4   2. 15, 20, 25.   3. a, b, c, e, g.   5.(a) 10
(b) $6\sqrt{3}$ (c) $a\sqrt{2}$ (d) 1.   7. $60\sqrt{2}$ miles.   9. Hint: $h^2 + x^2 = 49$, $h^2 + (3 +$
$x)^2 = 81$.   11. Approximately 72 ft.   13. 10 in., 24 in., 26 in.,   15. 9 and 16.
17.(a) In the third row, for example, $a = 30$, $b = 16$, $c = 34$.

Problem Set 6.6   1.(a) $x = 2$, $y = 2\sqrt{3}$ (c) $x = 3$, $z = 6$ (e) $x = \frac{4}{3}\sqrt{3}$,
$z = \frac{8}{3}\sqrt{3}$.   2.(a) $z = y = 2$ (c) $y = z = \frac{3}{2}\sqrt{2}$.   3. $3\sqrt{3}$.   5.(a) $12\sqrt{2}$
(b) $10(\sqrt{2} - 1) \approx 4.14$.   7.(a) $2\sqrt{3}$ (b) $\sqrt{3}$.   9.(a) 100 yd   (b) $100\sqrt{3}$ yd.
11.(a) $3\sqrt{2}$ (b) 3.   13. $3\sqrt{2}$ in.

Chapter 6 Review Exercises

1.(a) S   (c) S   (e) A   (g) N   (i) A   (k) N.   2.(a) $\frac{4}{3}$.   3.(a) 24   (b) 9   (c) 3.

4. 4.5.   5.   □   ⬦   7. $5\sqrt{3}$.   8.(a) 6.   9. 60.   11. 12, 8 and 9, 6.

13. Show $\triangle ABC \sim \triangle ACD$ and then use the definition of similar $\triangle$ and definition of
$\perp$.   15. $h = 10$. Legs have lengths $5\sqrt{5}$ and $10\sqrt{5}$.   17. $\sqrt{3}/2$.

Chapter 7

Problem Set 7.2   1.(a) hexagon, $ABCDEF$ (b) $\angle A$ and $\angle F$, $\angle A$ and $\angle B$, $\angle B$ and
$\angle C$ (c) $\overline{AB}$ and $\overline{DC}$, $\overline{BC}$ and $\overline{ED}$, $\overline{BC}$ and $\overline{EF}$ (d) $\overline{AC}$, $\overline{AD}$, $\overline{AE}$ (e) $AB + BC +$
$CD + DE + EF + FA$.   3. No. A diagonal of a quadrilateral, for example, is not
a subset of the quadrilateral.   5. Equilateral, equiangular.   7. 360, 540, 720.
9.(a) 360   (b) 360   (c) 360   (d) This should be easy.   11.(a) $10 + 5\sqrt{3}$ (b) 12
(c) $17 + 5\sqrt{2} + 5\sqrt{3}$.   13.(a) Given a correspondence between two polygons (or a
polygon and itself). . . .

Problem Set 7.3   1. 130, 50, 130.   3. Introduce $\overline{AC}$ and use BPT (twice).   5. Use
definition of ▱, PAI, and ASA.   7.(a) $m\angle A = 63$, $m\angle B = 117$, $m\angle C = 63$,
$m\angle D = 117$ (b) $m\angle A = 50$, $m\angle B = 130$, etc.   9. If the diagonals intersect at P,

show $\triangle DPC \cong \triangle BPA$.   11. Show $\triangle DCN \cong \triangle EBN$ to get $DC = EB$.   13. Use
SSS and AIP.   15. For a rectangle, b, c, d and e; for a rhombus, a, b and g; for a
square, all.   17. 20.   19.(a) Use SAS and CPCTC   (b) *Hint*: $m\angle CGH + m\angle HGF$
$+ m\angle FGB = 180$.

Problem Set **7.4**
1.(a) For example,   2. 37, 84.   3. 12, 9.   4. 16, 13, 16.

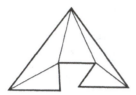

5.(a) 48 ft²   (c) 18 ft²   (e) $\frac{2}{15}$ mi²   6. $280.   7.(a) 16   (c) 4.5   (e) 12.
9.(a) doubled   (b) quadrupled, multiplied by 9   (c) quadrupled, multiplied by 9.
11. 336.   13.(a) 6   (b) 12   (c) $6 + 8\sqrt{3}$ (d) $\frac{25}{4}\sqrt{3}$.   15. 5.   17. $(a + b)^2 = a^2$
$+ 2ab + b^2$.   19. Use Case 1 and the Area Addition Postulate.

Problem Set 7.5   1. 40.   3. 105.   5. 200.   7. 50.   9.(a) 50   (b) 100.   11. 3/1.
13. 1.5, 3.75, 4.5, 6.75, 7.5.   15. $\frac{3}{4}$.

Chapter 7 Review Exercises

1.(a) S   (c) S   (e) A   (g) N   (i) S   (k) A   (m) S   (o) N.   2.(a) 32   (c) 16
(e) 48   (g) 30   (i) 220.   3. 24, 18$\sqrt{3}$.   5. 60.   7. Let $\overline{AB}$ be the base of both triangles. Then their altitudes are equal, since two parallel lines are everywhere equidis-
tant. Therefore . . .   9. $\dfrac{x}{1} = \dfrac{1}{x - 1}$, $x = \dfrac{1 + \sqrt{5}}{2}$.   11. *Hint:* You might show that
a pair of triangles formed are congruent and then use AIP and the theorem preceding
this one.   13. Yes. No. $\overline{AC}$ intersects $\overline{DB}$ at an interior point.   15. Follow the hint
given in the text.   17. Yes. No.

Chapter 8

Problem Set 8.2   1.(a) Diameter   (b) Center of circle P (c) Chord   (d) Secant
(e) Tangent   (f) Point of tangency   (g) Radius   (h) Radius.

3.   5. 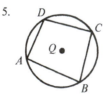   7.(a) 40   (b) 6$\sqrt{5}$ (c) 16/3.   9. The

Trichotomy law.   11. Actually two theorems are involved here. One of them is the existence part of T30. What is the other?   13. *Hint:* If two coplanar lines are ⊥ to the same line, then what?   15. A secant.

Problem Set 8.3   1. 13.   3. $2\sqrt{33}$ in.   5. *Hint:* Use HL.   7. Externally.   9. *Hint:* Use T88.   11. $r = 2\sqrt{5}$.   13. Let the points be A, B, C and let the center of the circle be P. Then $PA = PB = PC$. Sketch this situation and then derive a contradiction of EAT.   15.(c) *Hint:* Use Segment Bisection Theorem and HL.   17. *Hint:* Radii of congruent circles are congruent.

Problem Set 8.4   1.(a) $\angle APC$, $\angle CPB$ (c) $\widehat{ACB}$ and $\widehat{AXB}$ (e) 50   (g) 130.   3.(a) 60   (b) 120   (c) 240   (d) 180.   5.(a) $\widehat{FG}$ and $\widehat{EB}$ (c) $\widehat{BHG}$ (e) $\widehat{BH}$.   7. 130 and 230.   9. Use Definition 8.8.   10. 90.   13. *Hint:* Right $\angle$s should be involved.   15. You might use Definition 8.8 and SAS.

Problem Set 8.5   1.(a) 140   (b) 60, 70, 50   (c) 140, 80, 220   (d) 70, 70.   3. 80, 74, 100, 106.   5. Use AAP.   7.(a) 72   (b) 36.   9. Follow the hint given in the text.   11. *Hint:* Use Case 1 and subtract from 360.   13. Use T93 and the Arc Addition Theorem.   15. 240 and 120.   17. *Hint:* $\triangle TQS$ and $\triangle TPR$ are isosceles and similar, and since $TQ = \frac{1}{2}TP$, it follows that $TS = \frac{1}{2}TR$.

Problem Set 8.6   1.(a) 60   (b) 120   (c) 30   (d) $(5\sqrt{3})/2$ (e) $(75/4)\sqrt{3}$.   3. 3.   5. $\sqrt{13}$.   7. $4\sqrt{2}$.   9. $54\sqrt{3}$.   11. $\frac{1}{2}$.   13. $\sqrt{2}/1$.

Problem Set 8.7   1.(a) 4π, 4π (b) 6π, 9π (c) 8π, 16π (d) 2, $1/\pi$.   2.(a) 5   (c) 4.   3.(a) 81.64 in.   (b) Approximately 776.5.   4. $\frac{1}{4}$.   5.(a) $\frac{2}{3}$ (b) $\frac{4}{5}$.   7. $6\pi\sqrt{2}$, 18π.   9.(a) $4\pi\sqrt{3}$ (b) 8π.   10. 2.   11. 724π. *Hint:* The end of the rope sweeps out $\frac{3}{4}$ of a circle of radius 30, and $\frac{1}{4}$ of a circle of radius 14.   13.(a) 2.5π in.   (b) 4π in.   (c) $12/\pi$ in.   (d) $2\pi/3$ in.

Problem Set 8.8   1. If only a compass and straightedge are used, and in accordance with the Euclidean rules, a drawing is called a construction.   3. Construct arcs with radius AB with centers at A and B. Where these arcs intersect is the third vertex of such a triangle.   5.(d) Where do the $\angle$ bisectors intersect?   7. See the hint for problem 6 and bisect that angle.   9. An intersection of circular arcs with radius 2 in. and centers at the endpoints of the base is the third vertex.   11. At the ends of the lower base construct 30° angles (see problem 7) and then use the fact that an altitude from an upper vertex will intersect the lower base at a point $\frac{1}{4}$ in. from a lower vertex.   13. Construct the perpendicular to $\overset{\leftrightarrow}{PX}$ at X.

Problem Set 8.9   1. Follow the hint or bisect a right $\angle$.   3.(d) Where do they intersect?   5. Draw any line through P and construct the ⊥ bisector of the resulting chord. The vertices of the square should be apparent!   7. Recall that a rhombus is equilateral and then use Construction 9.   9. Use Construction 10 and recall T62.   13. Begin by constructing a sequence of six adjacent 60° $\angle$s with the same vertex.   15. Given circle Q and point P, construct a circle with $\overline{QP}$ as a diameter.   17.(a) The easiest method: construct ⊥ bisector of $\overline{CD}$.   (b) Construct a line through C that is ⊥ to $\overline{AB}$ at X. Then follow the method of Problem Set 4.8.

Problem Set 8.10   1.(a) incenter   (b) circumcenter   (c) centroid.   3.(a) acute $\triangle$
(b) right $\triangle$ (c) obtuse $\triangle$.   5. Construct the $\perp$ bisectors of two of the segments
determined by the three points.   7. Use the corollary following the Egg-slicer Corollary.   9. Construct $\perp$ lines and on one line lay off the leg. One of the endpoints of
this leg is the vertex of an acute angle. Use this vertex as the center of an arc with
radius equal to the length of the hypotenuse.   11. Introduce $\overline{BP}$ and use T56.

Chapter 8 Review Exercises

1.(a) N   (c) S   (e) S   (g) A   (i) N   (k) N.   3. 26.   5. 90.   7. $\overline{AD} \cong \overline{BC}$
implies that $m\,\widehat{AD} = m\,\widehat{BC}$. Then use the Arc Addition Theorem to show that $m\,\widehat{DB} = m\,\widehat{AC}$. Therefore $\angle DAB \cong \angle CBA$. Then use AIP.   9. Use Case 1 and the Arc
Addition Theorem.   11. Transitivity is the key idea.   13. Follow the hint for
Problem 13 of Problem Set 8.3.   15. $(27\sqrt{3})/4$.   17. $(75\sqrt{3})/2$.   19.(a) 45   (b) 60
(Hint: $AT = AB$)   (c) 1   (d) 75   (e) $\sqrt{3} - 1$ (j) $\frac{1}{2}(3\sqrt{2} - \sqrt{6})$ (g) Hint:
$CH = AC \cdot \sqrt{3}/2$.

Chapter 9

Problem Set 9.2
2.

4. There are two possibilities.   5. $\angle ADP$, $\angle PDC$, $\angle QDB$.

7. Hint: It is the intersection of two half-spaces.   9. $a = 0$, $b = 180$.   11.(a) 3
(b) 3   (c) 3.   13. Use Definition 9.1 and SAS.   15. There are two cases to consider:
(1) The line is $\perp$ to both of them at their line of intersection; (2) The line is $\perp$ to them
at distinct points.   17.(a) Follow the hint.   (b) Suppose there were two.

Problem Set 9.3   1. 3, unlimited.   3. Corners of room, desks, books, and so on.
5. 4, 5.   7. Cube.   8. Use Definition 9.6 and CPCTC.   10. Equilateral $\triangle$, squares,
regular pentagons.   11. 1 in.   13. A diagonal of the base has length $e\sqrt{2}$.   Then
use: $d^2 = e^2 + (e\sqrt{2})^2$.   14.(a) Use the second corollary of Section 6.4 and SSS $\sim$.

Problem Set 9.4   1.(a) $180 + 18\sqrt{3}$, 180, $90\sqrt{3}$ (c) $170 + 120\sqrt{2}$, $120 + 120\sqrt{2}$,
300   (e) $360 + 50\sqrt{3}$, 360, $150\sqrt{3}$.   3.(a) 180 cu. in.   (b) 135.   5. $9\sqrt{91} + 9\sqrt{3}$,
$9\sqrt{91}$, $30\sqrt{3}$.   7. $4\sqrt{29}$ ft.   8.(a) 80 sq in.   9. 6.   11.(a) 7/96 ft.³   (b) Approximately 88.24 lb.

Problem Set 9.5   1. Unlimited number.   3. 144π.   5. Hint: Use HL and CPCTC
to show they have $\cong$ radii.   7. 6.   9. 3.5.   11. 288π cu in.   13. 10.6 in.   14. 26,391
using $\pi = 3\frac{1}{7}$.

Chapter 9 Review Exercises

1.(a) N   (c) A   (e) A   (g) N   (i) A   (k) N   (m) S   (o) A.   3. 8, 12.
5. Yes.   Sketch a cube.   7.(a) Unlimited number   (b) Depends on whether or not
$\overleftrightarrow{T} \perp m$.   9. See Section 9.3.   11. Write an informal proof based on the fact that the
center of the base is equidistant from its vertices.   13. $192\sqrt{3}$.   15. $48\sqrt{3}$.
17. $16\sqrt{3} + 4\sqrt{793}$.   19.(a) $144\sqrt{3}$ (b) $4\sqrt{6}$ (c) $144\sqrt{2}$.   21.(a) $\frac{1}{2}$ (b) $\frac{1}{2}$
(c) $\frac{1}{4}$

Chapter 10

Problem Set 10.2   1. $P(2,3)$, $Q(4\frac{1}{2}, -2\frac{1}{2})$, $R(1, -5)$, $S(-3, 0)$ $T(-5, -2)$, $U(-3\frac{1}{2}, 3\frac{1}{2})$,
$V(-2, 2)$ 3.   5. A, none; B, IV; C, I; D, III; E, I.

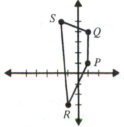

7. $(-7, 0)$.   9.(a) Either $(3, 3)$ or $(6, 3)$ (b) $4 + 2\sqrt{26}$ (c) 10.   11. $22 + 2\sqrt{85}$.
13. $AB = BC = CD = DA = \sqrt{65}$.   15. A line parallel to and 2 units to the left
of the y-axis.   17.

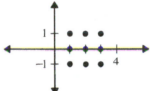

19. Show that $(RS)^2 + (RT)^2 = (ST)^2$.

Problem Set 10.3   1. -4.   3. $\frac{5}{3}$.   5. $\frac{3}{4}$.   7.(a) $y = 3$ (b) $y = -1$.   9. $y = 0$,
$x = 0$.   11.(a) $y - 3 = -\frac{2}{3}(x - 1)$ (c) $y - 2 = -2(x + 1)$ (e) $y = x$.
13. $y - 3 = \frac{3}{2}(x - 3)$, $3x - 2y - 3 = 0$.   15. A single point.   17.(a) $(2, 0)$ and
$(0, 3)$ (b) $(4, 0)$ and $(0, -4)$ (d) $(-2, 0)$ and $(0, 4)$.

Problem Set 10.4   1.(a) $y = x - 1$, $m = 1$, intercept $(0, -1)$ (b) $y = 3x + 7$,
$m = 3$, intercept $(0, 7)$ (c) $y = -\frac{3}{4}x + 3$, $m = -\frac{3}{4}$, intercept $(0, 3)$ (d) $y = x$,
$m = 1$, intercept $(0, 0)$ (e) $y = -x + 3$, $m = -1$, intercept $(0, 3)$.   3. $y - 4 =$
$\frac{5}{2}(x + 1)$.   5.(a) $y - 5 = \frac{10}{3}(x + 1)$ (b) $y + 3 = -3(x - 2)$ (c) $y - 7 =$
$-\frac{8}{3}(x - 5)$ 7. $\overline{AB} \| \overline{CD}$, $\overline{BC} \| \overline{AD}$.   9. Show $\overline{AB} \| \overline{CD}$ and $\overline{AD} \| \overline{BC}$.   11. (a) $y - 7$
$= \frac{3}{2}(x + 2)$ (b) $y - 7 = -60(x + 2)$.   13. $k = 4.8$.   15.(a) $C = 0$ (b) $A = 0$
(c) $B = 0$.

Problem Set 10.5

1.(a)

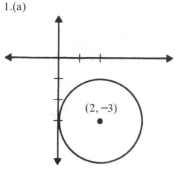

(c)

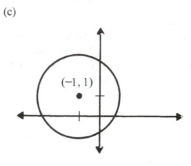

3. $(x - 4)^2 + y^2 = 16$.   5. $x^2 + y^2 = 25$.   7. $D^2 + E^2 > 4F$.   9. $y - 4 = \frac{3}{4}(x + 3)$.   11. $(x + 2)^2 + (y + 2)^2 = 4$.   13. You should obtain a circle with center at $(4, 0)$ and $r = 2$.

Chapter 10 Review Exercises

1.(a) A   (c) S   (e) N   (g) A   (i) S.   3. Show $\overline{AB} \| \overline{CD}$ and $\overline{AD} \| \overline{BC}$.   5. $y + 1 = (-5/64)(x - 7)$.   7. Show that $AB + BC = AC$.   9. $\varnothing$.   11. $\triangle ABC$ is a right $\triangle$, with right angle $\angle B$.   13. The set of lines intersecting the x-axis at a 45° angle. 15. $(x - 3)^2 + (y + 4)^2 = 9$.   17. $(x - 4)^2 + (y + 3)^2 = 25$.   19.(a) In quadrant II or IV   (b) On an axis   (c) In quadrant I or III.   21. Slope of $\overline{OB} = 1$, slope of $\overline{AC} = -1$; therefore $\overline{OB} \perp \overline{AC}$.

INDEX

ABOUT THE AUTHOR

A native of Detroit, Michigan, Edward T. Walsh (Tom to his friends and colleagues) received his A.B. in Physics from the University of California at Berkeley in 1959. After several years spent teaching secondary school science and mathematics he accepted a National Science Foundation grant to study at the University of Montana where he earned an M.A. in mathematics in 1965. He then taught mathematics for 36 years at City College of San Francisco. In 1996, the College Bookstore published *A Liberal Arts Sampler,* a text he wrote to accompany a course for liberal arts students that he conceived and taught for many years. His mathematical interests outside the classroom have included study of the retrospective astronomical calculations related to the megalithic monuments of Britain and Brittany and applications of mathematics to the environmental sciences. He resides with his wife, Maura FitzGerald, in Oakland, California, where he is an avid runner, cyclist, and political gadfly.

A CATALOG OF SELECTED

DOVER BOOKS
IN SCIENCE AND MATHEMATICS

Mathematics–Geometry and Topology

PROBLEMS AND SOLUTIONS IN EUCLIDEAN GEOMETRY, M. N. Aref and William Wernick. Based on classical principles, this book is intended for a second course in Euclidean geometry and can be used as a refresher. More than 200 problems include hints and solutions. 1968 edition. 272pp. 5 3/8 x 8 1/2.   0-486-47720-7

TOPOLOGY OF 3-MANIFOLDS AND RELATED TOPICS, Edited by M. K. Fort, Jr. With a New Introduction by Daniel Silver. Summaries and full reports from a 1961 conference discuss decompositions and subsets of 3-space; n-manifolds; knot theory; the Poincaré conjecture; and periodic maps and isotopies. Familiarity with algebraic topology required. 1962 edition. 272pp. 6 1/8 x 9 1/4.   0-486-47753-3

POINT SET TOPOLOGY, Steven A. Gaal. Suitable for a complete course in topology, this text also functions as a self-contained treatment for independent study. Additional enrichment materials make it equally valuable as a reference. 1964 edition. 336pp. 5 3/8 x 8 1/2.   0-486-47222-1

INVITATION TO GEOMETRY, Z. A. Melzak. Intended for students of many different backgrounds with only a modest knowledge of mathematics, this text features self-contained chapters that can be adapted to several types of geometry courses. 1983 edition. 240pp. 5 3/8 x 8 1/2.   0-486-46626-4

TOPOLOGY AND GEOMETRY FOR PHYSICISTS, Charles Nash and Siddhartha Sen. Written by physicists for physics students, this text assumes no detailed background in topology or geometry. Topics include differential forms, homotopy, homology, cohomology, fiber bundles, connection and covariant derivatives, and Morse theory. 1983 edition. 320pp. 5 3/8 x 8 1/2.   0-486-47852-1

BEYOND GEOMETRY: Classic Papers from Riemann to Einstein, Edited with an Introduction and Notes by Peter Pesic. This is the only English-language collection of these 8 accessible essays. They trace seminal ideas about the foundations of geometry that led to Einstein's general theory of relativity. 224pp. 6 1/8 x 9 1/4.   0-486-45350-2

GEOMETRY FROM EUCLID TO KNOTS, Saul Stahl. This text provides a historical perspective on plane geometry and covers non-neutral Euclidean geometry, circles and regular polygons, projective geometry, symmetries, inversions, informal topology, and more. Includes 1,000 practice problems. Solutions available. 2003 edition. 480pp. 6 1/8 x 9 1/4.   0-486-47459-3

TOPOLOGICAL VECTOR SPACES, DISTRIBUTIONS AND KERNELS, François Trèves. Extending beyond the boundaries of Hilbert and Banach space theory, this text focuses on key aspects of functional analysis, particularly in regard to solving partial differential equations. 1967 edition. 592pp. 5 3/8 x 8 1/2.

0-486-45352-9

INTRODUCTION TO PROJECTIVE GEOMETRY, C. R. Wylie, Jr. This introductory volume offers strong reinforcement for its teachings, with detailed examples and numerous theorems, proofs, and exercises, plus complete answers to all odd-numbered end-of-chapter problems. 1970 edition. 576pp. 6 1/8 x 9 1/4.   0-486-46895-X

FOUNDATIONS OF GEOMETRY, C. R. Wylie, Jr. Geared toward students preparing to teach high school mathematics, this text explores the principles of Euclidean and non-Euclidean geometry and covers both generalities and specifics of the axiomatic method. 1964 edition. 352pp. 6 x 9.   0-486-47214-0

Browse over 9,000 books at www.doverpublications.com

Mathematics–History

THE WORKS OF ARCHIMEDES, Archimedes. Translated by Sir Thomas Heath. Complete works of ancient geometer feature such topics as the famous problems of the ratio of the areas of a cylinder and an inscribed sphere; the properties of conoids, spheroids, and spirals; more. 326pp. 5 3/8 x 8 1/2. 0-486-42084-1

THE HISTORICAL ROOTS OF ELEMENTARY MATHEMATICS, Lucas N. H. Bunt, Phillip S. Jones, and Jack D. Bedient. Exciting, hands-on approach to understanding fundamental underpinnings of modern arithmetic, algebra, geometry and number systems examines their origins in early Egyptian, Babylonian, and Greek sources. 336pp. 5 3/8 x 8 1/2. 0-486-25563-8

THE THIRTEEN BOOKS OF EUCLID'S ELEMENTS, Euclid. Contains complete English text of all 13 books of the Elements plus critical apparatus analyzing each definition, postulate, and proposition in great detail. Covers textual and linguistic matters; mathematical analyses of Euclid's ideas; classical, medieval, Renaissance and modern commentators; refutations, supports, extrapolations, reinterpretations and historical notes. 995 figures. Total of 1,425pp. All books 5 3/8 x 8 1/2.

Vol. I: 443pp. 0-486-60088-2
Vol. II: 464pp. 0-486-60089-0
Vol. III: 546pp. 0-486-60090-4

A HISTORY OF GREEK MATHEMATICS, Sir Thomas Heath. This authoritative two-volume set that covers the essentials of mathematics and features every landmark innovation and every important figure, including Euclid, Apollonius, and others. 5 3/8 x 8 1/2.

Vol. I: 461pp. 0-486-24073-8
Vol. II: 597pp. 0-486-24074-6

A MANUAL OF GREEK MATHEMATICS, Sir Thomas L. Heath. This concise but thorough history encompasses the enduring contributions of the ancient Greek mathematicians whose works form the basis of most modern mathematics. Discusses Pythagorean arithmetic, Plato, Euclid, more. 1931 edition. 576pp. 5 3/8 x 8 1/2.

0-486-43231-9

CHINESE MATHEMATICS IN THE THIRTEENTH CENTURY, Ulrich Libbrecht. An exploration of the 13th-century mathematician Ch'in, this fascinating book combines what is known of the mathematician's life with a history of his only extant work, the Shu-shu chiu-chang. 1973 edition. 592pp. 5 3/8 x 8 1/2.

0-486-44619-0

PHILOSOPHY OF MATHEMATICS AND DEDUCTIVE STRUCTURE IN EUCLID'S ELEMENTS, Ian Mueller. This text provides an understanding of the classical Greek conception of mathematics as expressed in Euclid's Elements. It focuses on philosophical, foundational, and logical questions and features helpful appendixes. 400pp. 6 1/2 x 9 1/4. 0-486-45300-6

BEYOND GEOMETRY: Classic Papers from Riemann to Einstein, Edited with an Introduction and Notes by Peter Pesic. This is the only English-language collection of these 8 accessible essays. They trace seminal ideas about the foundations of geometry that led to Einstein's general theory of relativity. 224pp. 6 1/8 x 9 1/4. 0-486-45350-2

HISTORY OF MATHEMATICS, David E. Smith. Two-volume history – from Egyptian papyri and medieval maps to modern graphs and diagrams. Non-technical chronological survey with thousands of biographical notes, critical evaluations, and contemporary opinions on over 1,100 mathematicians. 5 3/8 x 8 1/2.

Vol. I: 618pp. 0-486-20429-4
Vol. II: 736pp. 0-486-20430-8

Engineering

FUNDAMENTALS OF ASTRODYNAMICS, Roger R. Bate, Donald D. Mueller, and Jerry E. White. Teaching text developed by U.S. Air Force Academy develops the basic two-body and n-body equations of motion; orbit determination; classical orbital elements, coordinate transformations; differential correction; more. 1971 edition. 455pp. 5 3/8 x 8 1/2. 0-486-60061-0

INTRODUCTION TO CONTINUUM MECHANICS FOR ENGINEERS: Revised Edition, Ray M. Bowen. This self-contained text introduces classical continuum models within a modern framework. Its numerous exercises illustrate the governing principles, linearizations, and other approximations that constitute classical continuum models. 2007 edition. 320pp. 6 1/8 x 9 1/4. 0-486-47460-7

ENGINEERING MECHANICS FOR STRUCTURES, Louis L. Bucciarelli. This text explores the mechanics of solids and statics as well as the strength of materials and elasticity theory. Its many design exercises encourage creative initiative and systems thinking. 2009 edition. 320pp. 6 1/8 x 9 1/4. 0-486-46855-0

FEEDBACK CONTROL THEORY, John C. Doyle, Bruce A. Francis and Allen R. Tannenbaum. This excellent introduction to feedback control system design offers a theoretical approach that captures the essential issues and can be applied to a wide range of practical problems. 1992 edition. 224pp. 6 1/2 x 9 1/4. 0-486-46933-6

THE FORCES OF MATTER, Michael Faraday. These lectures by a famous inventor offer an easy-to-understand introduction to the interactions of the universe's physical forces. Six essays explore gravitation, cohesion, chemical affinity, heat, magnetism, and electricity. 1993 edition. 96pp. 5 3/8 x 8 1/2. 0-486-47482-8

DYNAMICS, Lawrence E. Goodman and William H. Warner. Beginning engineering text introduces calculus of vectors, particle motion, dynamics of particle systems and plane rigid bodies, technical applications in plane motions, and more. Exercises and answers in every chapter. 619pp. 5 3/8 x 8 1/2. 0-486-42006-X

ADAPTIVE FILTERING PREDICTION AND CONTROL, Graham C. Goodwin and Kwai Sang Sin. This unified survey focuses on linear discrete-time systems and explores natural extensions to nonlinear systems. It emphasizes discrete-time systems, summarizing theoretical and practical aspects of a large class of adaptive algorithms. 1984 edition. 560pp. 6 1/2 x 9 1/4. 0-486-46932-8

INDUCTANCE CALCULATIONS, Frederick W. Grover. This authoritative reference enables the design of virtually every type of inductor. It features a single simple formula for each type of inductor, together with tables containing essential numerical factors. 1946 edition. 304pp. 5 3/8 x 8 1/2. 0-486-47440-2

THERMODYNAMICS: Foundations and Applications, Elias P. Gyftopoulos and Gian Paolo Beretta. Designed by two MIT professors, this authoritative text discusses basic concepts and applications in detail, emphasizing generality, definitions, and logical consistency. More than 300 solved problems cover realistic energy systems and processes. 800pp. 6 1/8 x 9 1/4. 0-486-43932-1

THE FINITE ELEMENT METHOD: Linear Static and Dynamic Finite Element Analysis, Thomas J. R. Hughes. Text for students without in-depth mathematical training, this text includes a comprehensive presentation and analysis of algorithms of time-dependent phenomena plus beam, plate, and shell theories. Solution guide available upon request. 672pp. 6 1/2 x 9 1/4. 0-486-41181-8

Browse over 9,000 books at www.doverpublications.com

HELICOPTER THEORY, Wayne Johnson. Monumental engineering text covers vertical flight, forward flight, performance, mathematics of rotating systems, rotary wing dynamics and aerodynamics, aeroelasticity, stability and control, stall, noise, and more. 189 illustrations. 1980 edition. 1089pp. 5 5/8 x 8 1/4.   0-486-68230-7

MATHEMATICAL HANDBOOK FOR SCIENTISTS AND ENGINEERS: Definitions, Theorems, and Formulas for Reference and Review, Granino A. Korn and Theresa M. Korn. Convenient access to information from every area of mathematics: Fourier transforms, Z transforms, linear and nonlinear programming, calculus of variations, random-process theory, special functions, combinatorial analysis, game theory, much more. 1152pp. 5 3/8 x 8 1/2.   0-486-41147-8

A HEAT TRANSFER TEXTBOOK: Fourth Edition, John H. Lienhard V and John H. Lienhard IV. This introduction to heat and mass transfer for engineering students features worked examples and end-of-chapter exercises. Worked examples and end-of-chapter exercises appear throughout the book, along with well-drawn, illuminating figures. 768pp. 7 x 9 1/4.   0-486-47931-5

BASIC ELECTRICITY, U.S. Bureau of Naval Personnel. Originally a training course; best nontechnical coverage. Topics include batteries, circuits, conductors, AC and DC, inductance and capacitance, generators, motors, transformers, amplifiers, etc. Many questions with answers. 349 illustrations. 1969 edition. 448pp. 6 1/2 x 9 1/4.

0-486-20973-3

BASIC ELECTRONICS, U.S. Bureau of Naval Personnel. Clear, well-illustrated introduction to electronic equipment covers numerous essential topics: electron tubes, semiconductors, electronic power supplies, tuned circuits, amplifiers, receivers, ranging and navigation systems, computers, antennas, more. 560 illustrations. 567pp. 6 1/2 x 9 1/4.   0-486-21076-6

BASIC WING AND AIRFOIL THEORY, Alan Pope. This self-contained treatment by a pioneer in the study of wind effects covers flow functions, airfoil construction and pressure distribution, finite and monoplane wings, and many other subjects. 1951 edition. 320pp. 5 3/8 x 8 1/2.   0-486-47188-8

SYNTHETIC FUELS, Ronald F. Probstein and R. Edwin Hicks. This unified presentation examines the methods and processes for converting coal, oil, shale, tar sands, and various forms of biomass into liquid, gaseous, and clean solid fuels. 1982 edition. 512pp. 6 1/8 x 9 1/4.   0-486-44977-7

THEORY OF ELASTIC STABILITY, Stephen P. Timoshenko and James M. Gere. Written by world-renowned authorities on mechanics, this classic ranges from theoretical explanations of 2- and 3-D stress and strain to practical applications such as torsion, bending, and thermal stress. 1961 edition. 560pp. 5 3/8 x 8 1/2.   0-486-47207-8

PRINCIPLES OF DIGITAL COMMUNICATION AND CODING, Andrew J. Viterbi and Jim K. Omura. This classic by two digital communications experts is geared toward students of communications theory and to designers of channels, links, terminals, modems, or networks used to transmit and receive digital messages. 1979 edition. 576pp. 6 1/8 x 9 1/4.   0-486-46901-8

LINEAR SYSTEM THEORY: The State Space Approach, Lotfi A. Zadeh and Charles A. Desoer. Written by two pioneers in the field, this exploration of the state space approach focuses on problems of stability and control, plus connections between this approach and classical techniques. 1963 edition. 656pp. 6 1/8 x 9 1/4.

0-486-46663-9

Browse over 9,000 books at www.doverpublications.com

Mathematics–Bestsellers

HANDBOOK OF MATHEMATICAL FUNCTIONS: with Formulas, Graphs, and Mathematical Tables, Edited by Milton Abramowitz and Irene A. Stegun. A classic resource for working with special functions, standard trig, and exponential logarithmic definitions and extensions, it features 29 sets of tables, some to as high as 20 places. 1046pp. 8 x 10 1/2. 0-486-61272-4

ABSTRACT AND CONCRETE CATEGORIES: The Joy of Cats, Jiri Adamek, Horst Herrlich, and George E. Strecker. This up-to-date introductory treatment employs category theory to explore the theory of structures. Its unique approach stresses concrete categories and presents a systematic view of factorization structures. Numerous examples. 1990 edition, updated 2004. 528pp. 6 1/8 x 9 1/4. 0-486-46934-4

MATHEMATICS: Its Content, Methods and Meaning, A. D. Aleksandrov, A. N. Kolmogorov, and M. A. Lavrent'ev. Major survey offers comprehensive, coherent discussions of analytic geometry, algebra, differential equations, calculus of variations, functions of a complex variable, prime numbers, linear and non-Euclidean geometry, topology, functional analysis, more. 1963 edition. 1120pp. 5 3/8 x 8 1/2. 0-486-40916-3

INTRODUCTION TO VECTORS AND TENSORS: Second Edition--Two Volumes Bound as One, Ray M. Bowen and C.-C. Wang. Convenient single-volume compilation of two texts offers both introduction and in-depth survey. Geared toward engineering and science students rather than mathematicians, it focuses on physics and engineering applications. 1976 edition. 560pp. 6 1/2 x 9 1/4. 0-486-46914-X

AN INTRODUCTION TO ORTHOGONAL POLYNOMIALS, Theodore S. Chihara. Concise introduction covers general elementary theory, including the representation theorem and distribution functions, continued fractions and chain sequences, the recurrence formula, special functions, and some specific systems. 1978 edition. 272pp. 5 3/8 x 8 1/2. 0-486-47929-3

ADVANCED MATHEMATICS FOR ENGINEERS AND SCIENTISTS, Paul DuChateau. This primary text and supplemental reference focuses on linear algebra, calculus, and ordinary differential equations. Additional topics include partial differential equations and approximation methods. Includes solved problems. 1992 edition. 400pp. 7 1/2 x 9 1/4. 0-486-47930-7

PARTIAL DIFFERENTIAL EQUATIONS FOR SCIENTISTS AND ENGINEERS, Stanley J. Farlow. Practical text shows how to formulate and solve partial differential equations. Coverage of diffusion-type problems, hyperbolic-type problems, elliptic-type problems, numerical and approximate methods. Solution guide available upon request. 1982 edition. 414pp. 6 1/8 x 9 1/4. 0-486-67620-X

VARIATIONAL PRINCIPLES AND FREE-BOUNDARY PROBLEMS, Avner Friedman. Advanced graduate-level text examines variational methods in partial differential equations and illustrates their applications to free-boundary problems. Features detailed statements of standard theory of elliptic and parabolic operators. 1982 edition. 720pp. 6 1/8 x 9 1/4. 0-486-47853-X

LINEAR ANALYSIS AND REPRESENTATION THEORY, Steven A. Gaal. Unified treatment covers topics from the theory of operators and operator algebras on Hilbert spaces; integration and representation theory for topological groups; and the theory of Lie algebras, Lie groups, and transform groups. 1973 edition. 704pp. 6 1/8 x 9 1/4. 0-486-47851-3

Browse over 9,000 books at www.doverpublications.com

A SURVEY OF INDUSTRIAL MATHEMATICS, Charles R. MacCluer. Students learn how to solve problems they'll encounter in their professional lives with this concise single-volume treatment. It employs MATLAB and other strategies to explore typical industrial problems. 2000 edition. 384pp. 5 3/8 x 8 1/2.   0-486-47702-9

NUMBER SYSTEMS AND THE FOUNDATIONS OF ANALYSIS, Elliott Mendelson. Geared toward undergraduate and beginning graduate students, this study explores natural numbers, integers, rational numbers, real numbers, and complex numbers. Numerous exercises and appendixes supplement the text. 1973 edition. 368pp. 5 3/8 x 8 1/2.   0-486-45792-3

A FIRST LOOK AT NUMERICAL FUNCTIONAL ANALYSIS, W. W. Sawyer. Text by renowned educator shows how problems in numerical analysis lead to concepts of functional analysis. Topics include Banach and Hilbert spaces, contraction mappings, convergence, differentiation and integration, and Euclidean space. 1978 edition. 208pp. 5 3/8 x 8 1/2.   0-486-47882-3

FRACTALS, CHAOS, POWER LAWS: Minutes from an Infinite Paradise, Manfred Schroeder. A fascinating exploration of the connections between chaos theory, physics, biology, and mathematics, this book abounds in award-winning computer graphics, optical illusions, and games that clarify memorable insights into self-similarity. 1992 edition. 448pp. 6 1/8 x 9 1/4.   0-486-47204-3

SET THEORY AND THE CONTINUUM PROBLEM, Raymond M. Smullyan and Melvin Fitting. A lucid, elegant, and complete survey of set theory, this three-part treatment explores axiomatic set theory, the consistency of the continuum hypothesis, and forcing and independence results. 1996 edition. 336pp. 6 x 9.   0-486-47484-4

DYNAMICAL SYSTEMS, Shlomo Sternberg. A pioneer in the field of dynamical systems discusses one-dimensional dynamics, differential equations, random walks, iterated function systems, symbolic dynamics, and Markov chains. Supplementary materials include PowerPoint slides and MATLAB exercises. 2010 edition. 272pp. 6 1/8 x 9 1/4.   0-486-47705-3

ORDINARY DIFFERENTIAL EQUATIONS, Morris Tenenbaum and Harry Pollard. Skillfully organized introductory text examines origin of differential equations, then defines basic terms and outlines general solution of a differential equation. Explores integrating factors; dilution and accretion problems; Laplace Transforms; Newton's Interpolation Formulas, more. 818pp. 5 3/8 x 8 1/2.   0-486-64940-7

MATROID THEORY, D. J. A. Welsh. Text by a noted expert describes standard examples and investigation results, using elementary proofs to develop basic matroid properties before advancing to a more sophisticated treatment. Includes numerous exercises. 1976 edition. 448pp. 5 3/8 x 8 1/2.   0-486-47439-9

THE CONCEPT OF A RIEMANN SURFACE, Hermann Weyl. This classic on the general history of functions combines function theory and geometry, forming the basis of the modern approach to analysis, geometry, and topology. 1955 edition. 208pp. 5 3/8 x 8 1/2.   0-486-47004-0

THE LAPLACE TRANSFORM, David Vernon Widder. This volume focuses on the Laplace and Stieltjes transforms, offering a highly theoretical treatment. Topics include fundamental formulas, the moment problem, monotonic functions, and Tauberian theorems. 1941 edition. 416pp. 5 3/8 x 8 1/2.   0-486-47755-X

Browse over 9,000 books at www.doverpublications.com

Mathematics–Logic and Problem Solving

PERPLEXING PUZZLES AND TANTALIZING TEASERS, Martin Gardner. Ninety-three riddles, mazes, illusions, tricky questions, word and picture puzzles, and other challenges offer hours of entertainment for youngsters. Filled with rib-tickling drawings. Solutions. 224pp. 5 3/8 x 8 1/2. 0-486-25637-5

MY BEST MATHEMATICAL AND LOGIC PUZZLES, Martin Gardner. The noted expert selects 70 of his favorite "short" puzzles. Includes The Returning Explorer, The Mutilated Chessboard, Scrambled Box Tops, and dozens more. Complete solutions included. 96pp. 5 3/8 x 8 1/2. 0-486-28152-3

THE LADY OR THE TIGER?: and Other Logic Puzzles, Raymond M. Smullyan. Created by a renowned puzzle master, these whimsically themed challenges involve paradoxes about probability, time, and change; metapuzzles; and self-referentiality. Nineteen chapters advance in difficulty from relatively simple to highly complex. 1982 edition. 240pp. 5 3/8 x 8 1/2. 0-486-47027-X

SATAN, CANTOR AND INFINITY: Mind-Boggling Puzzles, Raymond M. Smullyan. A renowned mathematician tells stories of knights and knaves in an entertaining look at the logical precepts behind infinity, probability, time, and change. Requires a strong background in mathematics. Complete solutions. 288pp. 5 3/8 x 8 1/2.
0-486-47036-9

THE RED BOOK OF MATHEMATICAL PROBLEMS, Kenneth S. Williams and Kenneth Hardy. Handy compilation of 100 practice problems, hints and solutions indispensable for students preparing for the William Lowell Putnam and other mathematical competitions. Preface to the First Edition. Sources. 1988 edition. 192pp. 5 3/8 x 8 1/2. 0-486-69415-1

KING ARTHUR IN SEARCH OF HIS DOG AND OTHER CURIOUS PUZZLES, Raymond M. Smullyan. This fanciful, original collection for readers of all ages features arithmetic puzzles, logic problems related to crime detection, and logic and arithmetic puzzles involving King Arthur and his Dogs of the Round Table. 160pp. 5 3/8 x 8 1/2.
0-486-47435-6

UNDECIDABLE THEORIES: Studies in Logic and the Foundation of Mathematics, Alfred Tarski in collaboration with Andrzej Mostowski and Raphael M. Robinson. This well-known book by the famed logician consists of three treatises: "A General Method in Proofs of Undecidability," "Undecidability and Essential Undecidability in Mathematics," and "Undecidability of the Elementary Theory of Groups." 1953 edition. 112pp. 5 3/8 x 8 1/2. 0-486-47703-7

LOGIC FOR MATHEMATICIANS, J. Barkley Rosser. Examination of essential topics and theorems assumes no background in logic. "Undoubtedly a major addition to the literature of mathematical logic." – *Bulletin of the American Mathematical Society.* 1978 edition. 592pp. 6 1/8 x 9 1/4. 0-486-46898-4

INTRODUCTION TO PROOF IN ABSTRACT MATHEMATICS, Andrew Wohlgemuth. This undergraduate text teaches students what constitutes an acceptable proof, and it develops their ability to do proofs of routine problems as well as those requiring creative insights. 1990 edition. 384pp. 6 1/2 x 9 1/4. 0-486-47854-8

FIRST COURSE IN MATHEMATICAL LOGIC, Patrick Suppes and Shirley Hill. Rigorous introduction is simple enough in presentation and context for wide range of students. Symbolizing sentences; logical inference; truth and validity; truth tables; terms, predicates, universal quantifiers; universal specification and laws of identity; more. 288pp. 5 3/8 x 8 1/2. 0-486-42259-3

Mathematics–Algebra and Calculus

VECTOR CALCULUS, Peter Baxandall and Hans Liebeck. This introductory text offers a rigorous, comprehensive treatment. Classical theorems of vector calculus are amply illustrated with figures, worked examples, physical applications, and exercises with hints and answers. 1986 edition. 560pp. 5 3/8 x 8 1/2.   0-486-46620-5

ADVANCED CALCULUS: An Introduction to Classical Analysis, Louis Brand. A course in analysis that focuses on the functions of a real variable, this text introduces the basic concepts in their simplest setting and illustrates its teachings with numerous examples, theorems, and proofs. 1955 edition. 592pp. 5 3/8 x 8 1/2.   0-486-44548-8

ADVANCED CALCULUS, Avner Friedman. Intended for students who have already completed a one-year course in elementary calculus, this two-part treatment advances from functions of one variable to those of several variables. Solutions. 1971 edition. 432pp. 5 3/8 x 8 1/2.   0-486-45795-8

METHODS OF MATHEMATICS APPLIED TO CALCULUS, PROBABILITY, AND STATISTICS, Richard W. Hamming. This 4-part treatment begins with algebra and analytic geometry and proceeds to an exploration of the calculus of algebraic functions and transcendental functions and applications. 1985 edition. Includes 310 figures and 18 tables. 880pp. 6 1/2 x 9 1/4.   0-486-43945-3

BASIC ALGEBRA I: Second Edition, Nathan Jacobson. A classic text and standard reference for a generation, this volume covers all undergraduate algebra topics, including groups, rings, modules, Galois theory, polynomials, linear algebra, and associative algebra. 1985 edition. 528pp. 6 1/8 x 9 1/4.   0-486-47189-6

BASIC ALGEBRA II: Second Edition, Nathan Jacobson. This classic text and standard reference comprises all subjects of a first-year graduate-level course, including in-depth coverage of groups and polynomials and extensive use of categories and functors. 1989 edition. 704pp. 6 1/8 x 9 1/4.   0-486-47187-X

CALCULUS: An Intuitive and Physical Approach (Second Edition), Morris Kline. Application-oriented introduction relates the subject as closely as possible to science with explorations of the derivative; differentiation and integration of the powers of x; theorems on differentiation, antidifferentiation; the chain rule; trigonometric functions; more. Examples. 1967 edition. 960pp. 6 1/2 x 9 1/4.   0-486-40453-6

ABSTRACT ALGEBRA AND SOLUTION BY RADICALS, John E. Maxfield and Margaret W. Maxfield. Accessible advanced undergraduate-level text starts with groups, rings, fields, and polynomials and advances to Galois theory, radicals and roots of unity, and solution by radicals. Numerous examples, illustrations, exercises, appendixes. 1971 edition. 224pp. 6 1/8 x 9 1/4.   0-486-47723-1

AN INTRODUCTION TO THE THEORY OF LINEAR SPACES, Georgi E. Shilov. Translated by Richard A. Silverman. Introductory treatment offers a clear exposition of algebra, geometry, and analysis as parts of an integrated whole rather than separate subjects. Numerous examples illustrate many different fields, and problems include hints or answers. 1961 edition. 320pp. 5 3/8 x 8 1/2.   0-486-63070-6

LINEAR ALGEBRA, Georgi E. Shilov. Covers determinants, linear spaces, systems of linear equations, linear functions of a vector argument, coordinate transformations, the canonical form of the matrix of a linear operator, bilinear and quadratic forms, and more. 387pp. 5 3/8 x 8 1/2.   0-486-63518-X

Browse over 9,000 books at www.doverpublications.com

Mathematics–Probability and Statistics

BASIC PROBABILITY THEORY, Robert B. Ash. This text emphasizes the probabilistic way of thinking, rather than measure-theoretic concepts. Geared toward advanced undergraduates and graduate students, it features solutions to some of the problems. 1970 edition. 352pp. 5 3/8 x 8 1/2. 0-486-46628-0

PRINCIPLES OF STATISTICS, M. G. Bulmer. Concise description of classical statistics, from basic dice probabilities to modern regression analysis. Equal stress on theory and applications. Moderate difficulty; only basic calculus required. Includes problems with answers. 252pp. 5 5/8 x 8 1/4. 0-486-63760-3

OUTLINE OF BASIC STATISTICS: Dictionary and Formulas, John E. Freund and Frank J. Williams. Handy guide includes a 70-page outline of essential statistical formulas covering grouped and ungrouped data, finite populations, probability, and more, plus over 1,000 clear, concise definitions of statistical terms. 1966 edition. 208pp. 5 3/8 x 8 1/2. 0-486-47769-X

GOOD THINKING: The Foundations of Probability and Its Applications, Irving J. Good. This in-depth treatment of probability theory by a famous British statistician explores Keynesian principles and surveys such topics as Bayesian rationality, corroboration, hypothesis testing, and mathematical tools for induction and simplicity. 1983 edition. 352pp. 5 3/8 x 8 1/2. 0-486-47438-0

INTRODUCTION TO PROBABILITY THEORY WITH CONTEMPORARY APPLICATIONS, Lester L. Helms. Extensive discussions and clear examples, written in plain language, expose students to the rules and methods of probability. Exercises foster problem-solving skills, and all problems feature step-by-step solutions. 1997 edition. 368pp. 6 1/2 x 9 1/4. 0-486-47418-6

CHANCE, LUCK, AND STATISTICS, Horace C. Levinson. In simple, non-technical language, this volume explores the fundamentals governing chance and applies them to sports, government, and business. "Clear and lively ... remarkably accurate." – *Scientific Monthly*. 384pp. 5 3/8 x 8 1/2. 0-486-41997-5

FIFTY CHALLENGING PROBLEMS IN PROBABILITY WITH SOLUTIONS, Frederick Mosteller. Remarkable puzzlers, graded in difficulty, illustrate elementary and advanced aspects of probability. These problems were selected for originality, general interest, or because they demonstrate valuable techniques. Also includes detailed solutions. 88pp. 5 3/8 x 8 1/2. 0-486-65355-2

EXPERIMENTAL STATISTICS, Mary Gibbons Natrella. A handbook for those seeking engineering information and quantitative data for designing, developing, constructing, and testing equipment. Covers the planning of experiments, the analyzing of extreme-value data; and more. 1966 edition. Index. Includes 52 figures and 76 tables. 560pp. 8 3/8 x 11. 0-486-43937-2

STOCHASTIC MODELING: Analysis and Simulation, Barry L. Nelson. Coherent introduction to techniques also offers a guide to the mathematical, numerical, and simulation tools of systems analysis. Includes formulation of models, analysis, and interpretation of results. 1995 edition. 336pp. 6 1/8 x 9 1/4. 0-486-47770-3

INTRODUCTION TO BIOSTATISTICS: Second Edition, Robert R. Sokal and F. James Rohlf. Suitable for undergraduates with a minimal background in mathematics, this introduction ranges from descriptive statistics to fundamental distributions and the testing of hypotheses. Includes numerous worked-out problems and examples. 1987 edition. 384pp. 6 1/8 x 9 1/4. 0-486-46961-1

Browse over 9,000 books at www.doverpublications.com

Astronomy

CHARIOTS FOR APOLLO: The NASA History of Manned Lunar Spacecraft to 1969, Courtney G. Brooks, James M. Grimwood, and Loyd S. Swenson, Jr. This illustrated history by a trio of experts is the definitive reference on the Apollo spacecraft and lunar modules. It traces the vehicles' design, development, and operation in space. More than 100 photographs and illustrations. 576pp. 6 3/4 x 9 1/4. 0-486-46756-2

EXPLORING THE MOON THROUGH BINOCULARS AND SMALL TELESCOPES, Ernest H. Cherrington, Jr. Informative, profusely illustrated guide to locating and identifying craters, rills, seas, mountains, other lunar features. Newly revised and updated with special section of new photos. Over 100 photos and diagrams. 240pp. 8 1/4 x 11. 0-486-24491-1

WHERE NO MAN HAS GONE BEFORE: A History of NASA's Apollo Lunar Expeditions, William David Compton. Introduction by Paul Dickson. This official NASA history traces behind-the-scenes conflicts and cooperation between scientists and engineers. The first half concerns preparations for the Moon landings, and the second half documents the flights that followed Apollo 11. 1989 edition. 432pp. 7 x 10.
0-486-47888-2

APOLLO EXPEDITIONS TO THE MOON: The NASA History, Edited by Edgar M. Cortright. Official NASA publication marks the 40th anniversary of the first lunar landing and features essays by project participants recalling engineering and administrative challenges. Accessible, jargon-free accounts, highlighted by numerous illustrations. 336pp. 8 3/8 x 10 7/8. 0-486-47175-6

ON MARS: Exploration of the Red Planet, 1958-1978--The NASA History, Edward Clinton Ezell and Linda Neuman Ezell. NASA's official history chronicles the start of our explorations of our planetary neighbor. It recounts cooperation among government, industry, and academia, and it features dozens of photos from Viking cameras. 560pp. 6 3/4 x 9 1/4. 0-486-46757-0

ARISTARCHUS OF SAMOS: The Ancient Copernicus, Sir Thomas Heath. Heath's history of astronomy ranges from Homer and Hesiod to Aristarchus and includes quotes from numerous thinkers, compilers, and scholasticists from Thales and Anaximander through Pythagoras, Plato, Aristotle, and Heraclides. 34 figures. 448pp. 5 3/8 x 8 1/2.
0-486-43886-4

AN INTRODUCTION TO CELESTIAL MECHANICS, Forest Ray Moulton. Classic text still unsurpassed in presentation of fundamental principles. Covers rectilinear motion, central forces, problems of two and three bodies, much more. Includes over 200 problems, some with answers. 437pp. 5 3/8 x 8 1/2. 0-486-64687-4

BEYOND THE ATMOSPHERE: Early Years of Space Science, Homer E. Newell. This exciting survey is the work of a top NASA administrator who chronicles technological advances, the relationship of space science to general science, and the space program's social, political, and economic contexts. 528pp. 6 3/4 x 9 1/4.
0-486-47464-X

STAR LORE: Myths, Legends, and Facts, William Tyler Olcott. Captivating retellings of the origins and histories of ancient star groups include Pegasus, Ursa Major, Pleiades, signs of the zodiac, and other constellations. "Classic." – *Sky & Telescope*. 58 illustrations. 544pp. 5 3/8 x 8 1/2. 0-486-43581-4

A COMPLETE MANUAL OF AMATEUR ASTRONOMY: Tools and Techniques for Astronomical Observations, P. Clay Sherrod with Thomas L. Koed. Concise, highly readable book discusses the selection, set-up, and maintenance of a telescope; amateur studies of the sun; lunar topography and occultations; and more. 124 figures. 26 halftones. 37 tables. 335pp. 6 1/2 x 9 1/4. 0-486-42820-6

Browse over 9,000 books at www.doverpublications.com